T0199319

Victorian Scientific Naturalism

Victorian Scientific Naturalism

Community, Identity, Continuity

EDITED BY GOWAN DAWSON
AND BERNARD LIGHTMAN

The University of Chicago Press
Chicago and London

Gowan Dawson is a senior lecturer in Victorian studies at the University of Leicester, UK.

Bernard Lightman is professor of Humanities at York University in Toronto.

The University of Chicago Press, Chicago 60637
The University of Chicago Press, Ltd., London
© 2014 by The University of Chicago
All rights reserved. Published 2014.
Printed in the United States of America

23 22 21 20 19 18 17 16 15 14   1 2 3 4 5

ISBN-13: 978-0-226-10950-3 (cloth)
ISBN-13: 978-0-226-10964-0 (e-book)
DOI: 10.7208/chicago/978-0-226-10964-0.001.0001

Library of Congress Cataloging-in-Publication Data

Victorian scientific naturalism : community, identity, continuity / edited by Gowan Dawson and Bernard Lightman.
        pages ; cm
    Includes bibliographical references and index.
    ISBN 978-0-226-10950-3 (cloth : alkaline paper) — ISBN 978-0-226-10964-0 (e-book)
    1. Naturalism—History—19th century.   2. Naturalism—Religious aspects.
    3. Science—Great Britain—History—19th century.   4. Science—Social aspects—
    Great Britain—19th century.   5. Great Britain—Intellectual life—19th century.
    I. Dawson, Gowan, editor.   II. Lightman, Bernard V., 1950– editor.
    Q127.G4V454 2014
    501—dc23
                                                                    20130336679

♾ This paper meets the requirements of ANSI/NISO Z39.48-1992 (Permanence of Paper).

Dedicated to the late Frank Miller Turner,
whose work on scientific naturalism continues to inspire

Contents

Introduction

GOWAN DAWSON & BERNARD LIGHTMAN

In the prologue to his *Essays upon Some Controverted Questions* (1892), Thomas Henry Huxley offered a retrospective defense of what he called the "principle of the scientific Naturalism of the latter half of the nineteenth century."[1] It has often been assumed that, in so doing, Huxley introduced another new term, akin to his earlier coinage *agnosticism*, into the lexicon of nineteenth-century science. Huxley, as James G. Paradis has proposed, "coined 'scientific naturalism' as an antithetical term to 'supernaturalism.'"[2] He was, after all, acutely aware of the power of language, and never hesitated to invent new words when it furthered his agenda.[3] While critics like Arthur James Balfour, in *The Foundations of Belief* (1895), rigorously kept the two terms separate, asserting that "science preceded naturalism, and will survive it," Huxley, who acknowledged that in Balfour's book the "word 'Naturalism' has a bad sound and unpleasant associations," boldly conflated them, although ceding priority to the capitalized *Naturalism*.[4] It was certainly preferable to the considerably more contentious term *scientific materialism* coined by his close friend John Tyndall twenty years earlier.[5] Huxley's putative neologism, seemingly emerging from the intellectual disputes of the late nineteenth century, was widely adopted in the following century, most notably by Frank M. Turner, as a historiographic category denoting the secular creeds of the generation of intellectuals who, in the wake of Charles Darwin's *On the Origin of Species* (1859), wrested cultural authority from the old Anglican establishment and installed themselves as a new professional scientific elite.[6] More recently, it has also been used as a philosophical designation in contemporary debates on science and religion.[7] In both these modern incarnations, the origins of the term are generally traced back to Huxley.

The ideas encompassed by "scientific Naturalism," according to Huxley, had a much longer lineage, and in his prologue he called it the "principle . . . in which the intellectual movement of the Renascence has culminated, and which was first clearly formulated by Descartes."[8] A year later, in *Evolution and Ethics* (1893), he went back still further, claiming "scientific naturalism took its rise among the Aryans of Ionia."[9] However, the actual term itself had never previously appeared in any of Huxley's voluminous published writings, and, with him beginning to use it only at the very end of his long career, and even then only sparingly, it has all the appearance of a rhetorical afterthought that conveniently summed up his ideas and aspirations as perhaps the most active and visible exponent of a scientific worldview in Victorian Britain.

Since Turner's pioneering work in the 1970s, scientific naturalism has been vitally significant for the understanding of science in the Victorian period and even into the present day. It encapsulates a particular set of values and assumptions among nineteenth-century men of science that began to crystallize in the crisis of authority in the mid-1850s that followed the aristocratic mismanagement of the Crimean War and the trauma of the Indian mutiny, and that led to the centrality of scientific methods and epistemologies in all aspects of the modern technocratic state, from education and medicine to agriculture and the military. The consequences of this campaign to reform and modernize Victorian Britain, including the tensions between science and religion that it ostensibly created, can still be felt in most industrialized societies today. Despite this continuing importance, though, it has never previously been recognized that the term *scientific naturalism* was actually used widely long before Huxley's purported coinage in 1892, while its suffixed practitioner form, *scientific naturalist*, has an even longer and more intricate etymology. Such attention to semantic detail is hardly inconsequential in studying the history of science, where sensitivity to so-called actors' categories has long constituted the prevalent methodology. Recent exercises in "word history" or "historical semantics" have shown that, for other nineteenth-century neologisms like *altruism*, the relation between words and concepts is never simply neutral, and the changing fortunes of a term have significant implications for the construction and communication of the various ideas it might entail.[10] At the same time, appellations such as "Man of Science," which emphasized the nature of the person rather than the activity undertaken, helped fashion a particular sense of identity, imbued with the values of masculine self-renunciation and public-spirited integrity, among scientific practitioners in mid-Victorian Britain.[11] Language, as historians have increasingly come to recognize, is not passively reflective, but is itself an agent of intellectual change.

By giving this volume the title *Victorian Scientific Naturalism*, rather than a host of other potential alternatives, including "evolutionary naturalism,"

"Darwinism," or "Agnosticism," we hope to signal its relation—admiring, but not uncritical—to Turner's seminal contribution to the field, as well as to reclaim the term as an authentic actors' category. This reclamation, moreover, is part of an attempt to go beyond Turner's emphasis on intellectual history and instead employ more recent scholarly agendas, including historical semantics, to reevaluate the place of scientific naturalism in the broader landscape of Victorian Britain. Contingent uses of particular terminologies, after all, help us to better understand what was at stake in the diverse contexts in which nineteenth-century science was conducted. The deployment of *Victorian Scientific Naturalism* in conjunction with the three nouns in our subtitle—*Community, Identity, Continuity*—also acknowledges the significance of specific names and terms in the construction of identity and community, and their continuity across lengthy periods of time, issues that are central concerns of the essays in the collection. Language, of course, is a crucial resource for forging identities and consolidating a sense of community, and, in negotiating the residual complexities and vested interests of Victorian society, the scientific naturalists often had to enter flexible coalitions with other groups of cultural leaders while nevertheless maintaining their own distinct sense of communal identity.

To fully appreciate what *scientific naturalism*, or *scientific naturalist*, signified to Huxley in 1892, as well as to assess their validity as widely invoked historiographic categories, it is necessary to begin this introduction by recovering the prehistory of the terms, examining their earlier usages in both public statements, such as lectures and journalism, as well as private contexts like correspondence and society meetings. Subsequent sections will survey the historiography of Victorian scientific naturalism over the last four decades, and assess how the essays collected in this volume offer new perspectives, focusing on issues of community, identity, and continuity, which, together, move this historical work forward.

Actors' Categories and Word Histories

Two years after first employing the term *scientific Naturalism*, Huxley complained that the designation *scientist* was a vulgar American neologism that "must be about as pleasing a word as 'Electrocution' . . . to any one who respects the English language."[12] He was unaware that, far from being another jarringly modern Americanism, *scientist* had in fact been coined by the Cambridge polymath William Whewell as far back as the 1830s, and had only subsequently been adopted across the Atlantic after failing to catch on in Britain. Ironically, it was the term he had himself apparently invented in 1892 whose origins were actually American. *Scientific naturalism* seems to have first come into usage among American evangelicals at the end of the 1840s, and may have been coined—if anyone can lay claim to it—by Tayler Lewis, professor of Greek at Union College in Schenectady. In July 1849 Lewis lectured on "Naturalism, in its various phases, as opposed to the true Scriptural doctrine of the Divine Imperium." Advancing an argument for the constant interaction of a "Supernatural Power" with the "physical machinery" of the phenomenal world, Lewis insisted that this doctrine was "*rational,* and rational, too, after the highest order of rationality, although it may seem to stand opposed to some of the most confident assumptions of a merely scientific naturalism."[13] The adjective *scientific,* Lewis implied disdainfully, only reduced the stature of this particular strain of naturalism, although he refrained from alluding to any specific instances of what this new term might entail.

Two years later he was more explicit about where some of its overly confident assumptions might be found. In the April 1851 number of *Bibliotheca Sacra,* the principal journal of American Congregational evangelicism, Lewis commended Laurens Perseus Hickok's *Rational Psychology* (1849) for affording a "complete armory of weapons against the scientific naturalism of such books as the Vestiges of Creation."[14] *Scientific naturalism,* in its initial American incarnation, was a pejorative epithet with which evangelicals could label the torrent of problematic secular publications, such as the evolutionary *Vestiges of the Natural History of Creation* (1844), then arriving from the Old World. Back in Britain at this time, the young Huxley was himself no less eager to traduce the "unfortunate scientific *parvenu*" who had anonymously authored the "spurious, glib eloquence" that Lewis designated mere scientific naturalism.[15]

By the 1860s this new term had crossed the Atlantic, appearing sporadically in the British religious press. The scandalous infidelity of the 1840s had now become a mainstay, if in a more respectable form, of the higher journalism, with Huxley, Tyndall, and others parading their scientific skepticism in conspicuously open-minded periodicals like the *Fortnightly Review* and the *Contemporary Review.* The "flat denial of *the efficacy of prayer,* by reason of the immutability of nature's laws," a topic that would soon be taken up by Tyndall in the pages of the latter, was, according to David Brown in the *Sunday Magazine* for December 1866, "one of the melancholy results of this scientific naturalism." For Brown, a leading Scottish Free Church theologian, all of the manifestations of this scientific naturalism, even including "Dr. Darwin's . . . now celebrated work on the 'Origin of Species,'" were imbued with an inescapable sense of melancholy.[16] Similarly, when, in the late 1870s, the *Contemporary* became more doctrinally conservative under new ownership, the prolific journalist and hymn writer William Brighty Rands used its pages to bemoan the "moral decay" and "diminish[ing] political earnestness" of the present day, tracing these "startlingly rapid changes" to the "postulate of scientific naturalism . . . that a science of morality can be constructed without help from theology." For the inadequate moralists who held to this debasing postulate, Rands avowed, "Jesus of Nazareth was no more and no other than the late Professor Clifford held him to be."[17] The *Contemporary's* readers would have required no reminder that the iconoclastic William Kingdon Clifford considered Christ the "son . . . of a celestial despot" who revenged his "death . . . effected by unknowing agents" by a "fixed resolve to keep his victims alive for ever, writhing in horrible tortures, in a place which his divine foreknowledge had prepared beforehand."[18] As in America in earlier decades, *scientific naturalism*, now being ascribed to close associates of Huxley such as Darwin and Clifford, was almost exclusively a negative and pejorative term employed by religious writers when it appeared in the British press from the 1860s to the 1880s.

Scientific naturalism's potential as a more affirmative label for a certain set of shared doctrines was nevertheless soon recognized, especially among the expanding but increasingly diverse community of secularists and freethinkers. In January 1878 the *Secular Review*, published and edited by Charles Watts, printed a letter protesting against the men of science, including "Darwin, Huxley, Tyndall, Proctor, Clifford . . . &c.," who "inferentially dismissed . . . with supreme contempt" all spiritualistic phenomena. The correspondent ended his letter, "To me Spiritualism is a phase of Naturalism, and possesses that amount of vital energy which will never be extinguished by the gratuitous fulminations of those who decry it without fair examination," before signing off, significantly, under the pseudonym "Naturalist."[19] A fortnight later another letter, from a correspondent who signed himself "Draco," after the first legislator of ancient Athens, retorted that "if 'Naturalist' holds on to his plea in the dark, he must drop the cognomen 'Naturalist,' for a true Spiritualist grasps the supermundane principle . . . in short, an immaterial substance, called spirit." "Draco" defended the "men of science [who] have laid the subject well open, and . . . find that the immaterial element is all nonsense," before closing his own missive: "If I mistake not, I think 'Naturalist' is trying to kick over the traces of Spiritualism, and wants to make a couplet of the two—Spiritualism and Naturalism. But they cannot clash together under the idea of Scientific Naturalism."[20] This, importantly, seems to have been the first time that the term was used, at least in public, as an entirely positive designation for the scientific rejection of all nonmaterial phenomena.

"Naturalist" quickly fired back a impassioned rejoinder, stating, "Your correspondent, 'Draco,' is exercising the office of Secular Archon without due veneration for free-inquiry." After defending his attempt to merge spiritualism and naturalism, he insisted:

> "Draco" may answer for himself in these words: "Because they cannot clash together under the idea of Scientific Naturalism." As an old secularist, I say—in sorrow, not in anger—that when correspondents either cannot or will not deduce fair inferences from analogous premises, or examine and discuss facts with logical arguments, and agreeableness to rational moderation, mutual improvement in useful knowledge may be relegated to the advent of Greek calends.

Scientific naturalism was, at least to this "independent Freethinker, who loathes sectarianism in science with feelings of unmitigated disgust and abhorrence," a narrow and divisive term, imposed by draconian authority, that explicitly excluded those secularists for whom an interest in spiritualism was a valid part of their freethinking rejection of establishment theology.[21] Its utility to Huxley and his associates, not just in spite of but actually because of the complaints of men like the spiritualistic "Naturalist," was becoming evident, and this fierce exchange of correspondence may well have been read by Tyndall, who had taken out a subscription to the *Secular Review* at this time.[22] The introduction of the term *scientific naturalism* into the pages of the *Secular Review* by "Draco" nevertheless associated it with a populist strain of secularism that Huxley was not entirely comfortable with. He was indignant when, in 1883, Watts's spin-off venture the *Agnostic Annual* appropriated his own appellation, snorting at how these "free thinkers 'make free'" and asserting, "I have a sort of patent right in Agnosticism—it is my 'trade mark.'"[23] Huxley would later gain a measure of revenge by himself making free with what, a decade afterward, would become his other patented trademark, and it may be that Huxley's adoption of *scientific naturalism* in the early 1890s was directly related to his continuing inability, throughout the 1880s, to control the range of meanings ascribed to *agnosticism*.[24]

Before that could happen, however, it was necessary for *scientific naturalism* to be denuded of the grubby taint of Watts's secular penny press, as well as extricated from its awkward origins as a pejorative label employed by evangelicals on both sides of the Atlantic. In his 1892 prologue, Huxley was careful not only to dwell "upon scientific Naturalism . . . in its critical and destructive aspect," but also to show "that it builds up, as well as pulls down." He enumerated a "common body of established truths," including evolution (though not natural selection) and the material basis of consciousness, which represented the "foundation" on which scientific naturalism was "already raising the superstructure."[25] In particular, this superstructure would afford a "clear apprehension of the natural sanctions . . . which constitute the common foundation of morality and of law." Tellingly, Huxley concluded his prologue with an assurance that, even with the recognition of the naturalistic basis for morality, the "human race is not yet, and possibly may never be, in a position to dispense with . . . the Jewish and Christian Scripture." The Bible, after all, "speaks no trash about the rights of man," and Huxley's own "respect and . . . attention" to its conservative "ethical contents," he reminded readers, was demonstrably long-standing.[26] Whatever else it might be, Huxley's version of scientific naturalism was unimpeachably respectable, scrupulously cleansed of all the deleterious ethical and political connotations it had accrued since first coming into usage in the 1840s. The lengthy intellectual lineage that, as seen earlier, he claimed for scientific naturalism similarly invested it with a tacit respectability and implied that it long predated the more recent outlooks of both its evangelical critics and secularist usurpers. Rather than arising sui generis as many historians have assumed, the "scientific Naturalism" Huxley outlined in his 1892 prologue was crucially shaped by the different contexts in which the term had been used over the previous half century.

Part of the reason that scientific naturalism appealed to Huxley as a retrospective encapsulation of his central ideas and principles, notwithstanding the suspect range of connotations that this conjunction of adjective and noun had accumulated over the years, may have been that its suffixed practitioner form had long had a much more precise and positive set of meanings. To be a scientific naturalist in the first half of the nineteenth century was simply to be taken seriously as an expert and specialist practitioner of the life sciences, with the *Edinburgh Journal of Science* in 1831 dismissing a dilettante museum owner as "rather a collector of curiosities than a scientific naturalist."[27] The term even had a particular, and highly positive, meaning in jurisprudence. As George Cornewall Lewis observed in his *An Essay on the Influence of Authority on Matters of Opinion* (1849), in some legal situations,

> a knowledge of the proper science, and a peculiar training of the senses, are requisite, and therefore . . . a witness who possess these qualifications is far more credible than one who is destitute of them. For example, a scientific naturalist who reports that he has seen an undescribed animal or vegetable in a remote country, is far less likely to be mistaken than a common traveller. . . . A skilled witness of this sort may be considered, in a certain sense, as a *witness of authority*.[28]

The opinions of a scientific naturalist were authoritative and implicitly trustworthy, free from any of the suspicions that clung to the testimony of mere common—and therefore likely of a lower social class—observers. It was an epithet that implied not only expertise, but gentility, good character, and truthfulness.

After the middle of the century, when *scientific naturalism* increasingly came to signify a particular set of principles antithetical to the supernatural or spiritual, *scientific naturalist* too began to assume a more specific meaning. In America, Tayler Lewis used it to describe the misguided exponents of his earlier coinage, scorning, in 1855, the "contradiction which our scientific naturalists are sometimes so fond of using,—*ex nihilo nihil*—nothing can ever come from nothing."[29] Back in Britain, Huxley himself, at least when speaking in public, generally stuck to the traditional usage of *scientific naturalist* as little more than a synonym for an expert practitioner.[30] By the early 1850s, however, he had already begun to use the term in private to also refer more specifically to a new cohort of men of science who found their ideal in practitioners of an earlier generation whose pursuit of institutional preferment had not compromised their intellectual independence. As Huxley wrote to the entomologist William Sharp Macleay in November 1851, "[Edward] Forbes has more influence by his personal weight and example upon the rising generation of scientific naturalists than Owen will have if he write from now till Doomsday."[31] As a famously knowledgeable and highly skilled practitioner, Richard Owen was an exemplar of the older meaning of *scientific naturalist*, but his willing acceptance of the patronage of the conservative Anglican establishment made him anathema, according to Huxley, to a new breed for whom the term also implied intellectual freedom. Huxley, recently returned from HMS *Rattlesnake*'s global voyage and only starting to make a name for himself, was not yet personally acquainted with many of this rising generation whom he labeled *scientific naturalists*. The term nevertheless afforded an early sense of group identity that would be made increasingly real, especially in opposition to Owen, over the coming years.

In the following decades the epithet *scientific naturalist* retained both its specific reference to the Huxleyan model of a specialist practitioner and its older undertow of gentile trustworthiness. The Metaphysical Society was founded in 1869 to allow members of all religious denominations, including those without any faith, to debate the most pressing philosophical issues of the day within an atmosphere of mutual tolerance and civility. This unusual combination of sociability with controversy, as Paul White argues in his contribution to this volume, necessitated the strict maintenance of gentlemanly manners, and it was in this polite atmosphere that Huxley coined *agnosticism* to differentiate his unbelief from more contentious labels like *atheism*. In a typically heated debate at the Metaphysical Society in December 1878, Richard

Holt Hutton, who had first publicly named Huxley as a "great and even severe Agnostic" eight years earlier, was so unpersuaded by George Romanes's materialistic explanation for complex instincts in animals that he asked those of his fellow members who were more likely to concur with Romanes, or who at least had specialist knowledge of the subject, if they could provide any better evidence for his arguments.[32] Hutton observed, "I am anxious . . . to elicit from some of the scientific naturalists and biologists in our Society, their view as to the best criterion, if there be any such criterion, by which we may identify an instinct which is the *débris* of an intelligent and deliberate habit."[33] As well as Huxley, the Metaphysical Society numbered Tyndall, Clifford, and Leslie Stephen among its diverse membership, and in endeavoring to group them together under a suitable soubriquet that would not violate the gentlemanly, though nonetheless vehement, tone of the debate, Hutton settled on *scientific naturalists*. Notably, Hutton invoked biologists as distinct from the scientific naturalists, presumably because St. George Mivart, the Catholic anatomist who had been ostracized by Huxley since the early 1870s, was also a member of the Metaphysical Society (and it seems unlikely that he intended the adjective *scientific* to qualify both nouns, as "scientific biologists" would be a tautology). Even if *scientific naturalism* still remained predominantly a pejorative term with troublesome connotations of amorality and infidelity at this stage, Huxley, with Hutton speaking directly before him, would have recognized that its suffixed practitioner form continued to have a very different, and potentially much more advantageous, range of meanings.

Both scientific naturalism and scientific naturalist were, then, actors' categories for much of the nineteenth century, and their shifting and overdetermined original meanings need to be recognized by scholars who, following Turner, employ them simply as retrospective historiographic categories. In recent years *evolutionary naturalism* has been adopted by many scholars as an alternative that, while retaining Turner's naturalistic emphasis, elides any hint of hard-edged scientism and instead foregrounds the contribution of Darwin. With the continuing controversies over evolution in certain parts of the world, *evolutionary naturalism* perhaps also has a more vigorously secular and less dispassionate ring in modern ears. *Evolutionary naturalism*, though, is very much a twentieth-century historians' term, even if Michael Ruse's claim to have coined it as recently as the mid-1990s is easily invalidated by its appearance in book titles from the 1960s.[34]

The acceptance of *scientific naturalism* as an actors' category might strengthen its validity as a historiographic term, showing that the kind of group identity among its proponents assumed by Turner and others was actively forged, by linguistic means as well as others, by the historical actors

themselves. At the same time, it also presents particular problems. As a polemical construct employed by both evangelicals and secularists even before
it was taken up by the archpolemicist Huxley, scientific naturalism, as an actors' category, tends to deliberately polarize differences and mask those continuities and coalitions between ostensibly opposed groups that are brought to
light in the chapters in the "Broader Alliances" section of this volume. To employ it unreflexively as a historiographic category therefore risks allowing the
partisan, self-serving perspective of participants to reify how historians understand the complex dynamics of nineteenth-century intellectual debate.[35]
To insist on *scientific naturalism* as the sole label for a large and diverse group
of writers and men of science in the second half of the nineteenth century
also foregrounds, and perhaps exaggerates, the significance of Huxley—what
Ruth Barton has termed "the Huxley problem"—at the expense of, say, the
hugely influential but less rhetorically incisive Herbert Spencer.[36]

As a shorthand for a particular outlook among men of science and associated intellectuals in Victorian Britain, though, *scientific naturalism* has many
pragmatic virtues, not least its general recognizability, in the wake of both
Huxley's usage and Turner's *Between Science and Religion* (1974), to scholars outside the specific field, students, and readers beyond academic institutions. And, as Ralph O'Connor has recently suggested regarding the cognate
category of *popular science*, such labels ought not to be regarded as analytical tools in themselves and instead are merely necessarily equivocal ways of
grouping phenomena and practices that, if used appropriately, can facilitate
more nuanced historical understandings.[37] Attention to its status as an actors' category, utilizing what Jim Secord has called the "electronic harvest"
of wholesale textual digitization (although it is important to acknowledge
that there is much that is still undigitized), nevertheless reminds us, as do
the other new perspectives presented in this volume, that Victorian scientific
naturalism remains a complex and contested issue that continues to require
careful and innovative scholarly analysis.[38]

Changing Historiographies and New Historical Actors

Scientific naturalism's fate as a historiographic term in the twentieth and
twenty-first centuries is no less intricate and unstable than its fortunes as a
nineteenth-century actors' category. Turner was not the only historian responsible for establishing it as a key component of the terminology of modern scholarship, with Robert M. Young also making frequent allusions to the
scientific naturalists in the essays that were later collected together in *Darwin's Metaphor* (1985). Turner and Young were both part of a historiographic

turn to contextualism in the history of science that began in the 1970s. The scientific naturalists, with their aspirations for social and cultural authority, were attractive subjects for Victorian scholars working in this mode. In a series of essays originally published in the late 1960s and early 1970s, Young, a neo-Marxist historian then based in Cambridge, used the term to describe a wider movement of which Darwinism was but a part. According to Young, "The river of nineteenth-century naturalism was fed by many streams" embracing a number of naturalistic approaches to the earth, life, and humanity. Tributaries feeding the mainstream of scientific naturalism included political, psychological, and philosophical movements such as utilitarianism, associationism, phrenology, mesmerism, Owenite socialism, positivism, and scientific historical criticism of the Bible. There were also tributaries flowing from scientific research in geology, paleontology, natural history, and the study of breeding.[39]

Young additionally implied that tributaries originating from the mainstream of Christianity also fed the scientific naturalism of the second half of the nineteenth century. He criticized scholars for concentrating too much on the conflict between Christians and scientific naturalists, pointing out that while natural theology was built on an explicitly theological theodicy, scientific naturalism similarly rested on a secular theodicy based on biological conceptions and the assumption of the uniformity of nature. Natural theology was not naturalistic, but like scientific naturalism, it was "designed" to rationalize the same set of assumptions about the existing order.[40] The fight between the two groups, in Young's view, was merely over the best way to preserve the status quo. The appeal to God or revelation was not as persuasive as the appeal to unchanging laws of nature, according to the scientific naturalists. If evolutionary science confirmed that social change must be gradual—that radicalism was contrary to the natural order—then it was rational to be reconciled to the gradual elimination of evil.

Turner reinforced the use of *scientific naturalism* as a valid scholarly term to describe a group of intellectuals with a common agenda in chapter 2 of his *Between Science and Religion*. He asserted that scientific naturalism represented a distinctly Victorian movement, as it "arose and flourished in circumstances unique to the period 1850 to 1900." The scientific naturalists, Turner argued, attempted to redefine science so as to eliminate the influence of natural theology. They were naturalistic in that they permitted no recourse to causes not present in empirically observed nature, and they were scientific because nature was interpreted according to three major midcentury scientific theories: the atomic theory of matter, the conservation of energy, and evolution. Turner pointed out that, having redefined science, the scientific naturalists pushed for new interpretations of humanity, nature, and society derived primarily from the theories, methods, and categories of the new science they envisioned.[41]

Unlike Young, Turner was not interested in the continuities between scientific naturalism and Christian modes of thought. Instead, in a now classic article published four years after his book, "The Victorian Conflict between Science and Religion" (1978), he stressed the tensions between science and religion. Here he pointed out that historians had not appreciated how much the fight involved more than disagreement over the epistemological status of science. The debates were also about "the character of the Victorian scientific community, its function in society, and the values by which it judged the work of its members." Issues related to the professionalization of science, according to Turner, were actually the driving force behind the heated controversies over evolution. The scientific naturalists worked to organize a more professionally oriented scientific community, and not just a redefinition of science. Within the scientific community they attempted to raise standards of competence, build a common bond of purpose, and subject practitioners to the judgment of peers rather than external authorities. Outside they intended to establish the independence of a professionalized science, including its right of self-definition. These aims brought them into conflict with two groups. The first, supporters of organized religion, wanted to maintain control over education and to retain religion as the source of moral and social values. The second, the religiously minded sector of the preprofessional scientific community, fought to retain a theological framework for scientific thought.[42] Turner developed this approach further in a series of essays later collected together in *Contesting Cultural Authority* (1993), where he discussed the conflicts over a range of cultural issues between the Anglican clergy and the scientific naturalists.[43]

Yet at the end of the 1980s Peter Allan Dale noted both the vital importance of Turner's work, but also a scholarly neglect of scientific naturalism. He asserted that the rise "of 'scientific naturalism,' as F. M. Turner has testified, and too few students of the period have taken sufficiently to heart, was the single most important intellectual phenomenon of the postromantic nineteenth century."[44] Dale was principally addressing literary scholars in his *In Pursuit of a Scientific Culture* (1989), and for many historians, and historians of science in particular, Young's and Turner's treatments of scientific naturalism had proved influential. By the time of Turner's *Contesting Cultural Authority*, the concept of a distinct group of Victorian intellectuals committed to scientific naturalism was increasingly accepted. Bernard Lightman's *The Origins of Agnosticism*, the first full monograph on the subject since Turner's

Between Science and Religion, was published in 1987, and Ruth Barton's influential article on the internal politics of the X Club, the mutually supportive London dining club established in 1864, followed three years later.[45] Darwin and Huxley, meanwhile, were the subjects of best-selling biographies in the early 1990s.[46]

Although by the end of the twentieth century the scientific naturalists were widely recognized as forming a distinct and important group of intellectuals, the trajectory of more recent scholarship has tended to draw attention away from them. In the last twenty years, historians of Victorian science have produced a picture of the period that has moved the scientific naturalists away from the center of things. There has been an eclipse of scientific naturalism—to modify the title of Peter J. Bowler's classic study of the initial fate of Darwin's theory of natural selection.[47] The canonical figures, including Darwin, were put aside, at least for the time being. This historiographical shift was most evident in 2009 when the celebrations of the two hundredth anniversary of the birth of Darwin and the one hundred fiftieth anniversary of the publication of the *Origin of Species* placed scholars in an awkward position. Called upon to participate in the celebrations, scholars in the field threatened to ruin the party by telling stories about science in Darwin's age that seemed to push him to the sidelines.[48]

The move away from the canonical figures began with the publication of Adrian Desmond's *The Politics of Evolution* (1989), although he subsequently coauthored one of the finest biographies ever written on Darwin and wrote a superb biography of Huxley.[49] Desmond proposed a history of biology "from below" as an antidote to the habitual fascination with the scientific elite.[50] His gripping account of radical anatomists who used Lamarckian evolution as a means of undermining Anglican privilege brought into view a little-known corner of the British scientific world in the second quarter of the nineteenth century. After that, scholars were drawn, more and more, to the study of new groups of historical actors. Anne Secord's groundbreaking article on artisan botanists, "Science in the Pub" (1994), pointed to the existence of working-class science in a location where few had thought to look for it.[51] Jim Secord's *Victorian Sensation* (2000) demonstrated that more attention had to be given to popularizers of science, like Robert Chambers, who were not scientific practitioners but who had the ear of the Victorian reading public.[52] Notably, Secord's book focused on Chambers's anonymous best seller *Vestiges of the Natural History of Creation*, which, as seen earlier, was impugned by Tayler Lewis as a manifestation of the dangerous creeds of scientific naturalism as early as the 1850s. Both *Victorian Sensation* and Desmond's *The Politics of Evolution* also helped to push back the establishment of a secular naturalistic tendency in British science into the 1830s and 1840s, an argument developed further by John van Wyhe in *Phrenology and the Origins of Victorian Scientific Naturalism* (2004).[53]

In the same vein as Desmond and the Secords, Lightman's *Victorian Popularizers of Science* (2007) greatly expanded the cast of self-consciously popular writers who, in the second half of the nineteenth century, combined instruction with entertainment to make science an enticing commodity that was eagerly purchased by an unprecedentedly large and diverse set of consumers.[54] Religious themes and notions of design continued as an important component of much popular science until the very end of the century, and even when popularizers seemed to accept Darwinian evolution, they often imbued it, as did Arabella Buckley, with moral sympathy and religious meaning. Similarly, the work of Evelleen Richards, Suzanne Le-May Sheffield, Ann Shteir, and Barbara Gates led to an increased awareness of the role of women in science, whether as writers, illustrators, painters, or social activists.[55] Even new work on the Victorian scientific elite, such as Crosbie Smith's *The Science of Energy* (1998), tended to put the scientific naturalists in the background.[56] Smith dealt with a faction of physicists and engineers from Cambridge and Glasgow, which he labeled the North British group, who effectively opposed the metropolitan proponents of scientific naturalism through their construction of the science of energy physics. Finding the perceived materialism of the scientific naturalists unacceptable, they worked to undermine the authority of Huxley and his allies and promoted a natural philosophy in harmony with Christian belief.

Significantly, in examining these new groups of historical actors, scholars discovered that many of them did not accept the authority of the scientific naturalists. Turner argued that the intellectual conflicts of the second half of the nineteenth century could be understood chiefly as a by-product of the contest for cultural authority between the Anglican clergy and the scientific naturalists. But it has become evident that the contest for cultural authority was actually far more complex than Turner had imagined. Competition was not limited merely to two groups within the intellectual elite. There were a number of groups vying for cultural authority, and many of them considered science to be a key component in their strategy. That made the scientific naturalists, who claimed to speak on behalf of science, prime targets not only for popularizers and North British physicists, but also for the feminist antivivisectionists, led by Frances Power Cobbe; for a new generation of scientifically informed, aristocratic Anglicans with an interest in spiritualism; for socialist intellectuals who dismissed Darwin and Huxley as procapitalists; and for idealists like Thomas Henry Green, who equated Darwinism with

materialism.[57] Even some of those who accepted most aspects of Darwinian evolution, including the codiscoverer of natural selection Alfred Russel Wallace, contested other elements of the scientific naturalist agenda, as
Wallace did on the supernatural origins of human intelligence and ethics.
Like Mivart, Wallace was brutally excommunicated by his erstwhile friends,
with Huxley, at Darwin's behest, pairing them as "Mr. Darwin's Critics" in
a ferocious rejoinder for the *Contemporary Review*.[58] But far from wielding
the unconstrained power and potency that the famously bellicose rhetoric of
Huxley, Tyndall, and Clifford might imply, the scientific naturalists were in
fact confronted by critics on every side, and what cultural authority they did
have was heavily contested at all stages. The aggressiveness of leading scientific naturalists in the public sphere might therefore actually be an indication
of weakness rather than of strength.

Although attention had shifted to nonelite groups and to those who contested the authority of the scientific naturalists, there was still interest in Huxley and his allies. Substantial studies of important scientific naturalists were
published in the first decade of the twenty-first century.[59] Many of them complicated our picture of the scientific naturalists by looking at them in relation
to new issues. Paul White, for example, demonstrated that Huxley's creation
of a new identity for the scientific practitioner was drawn, in part, from his
understandings of domesticity, literature, and religion.[60] Gowan Dawson's
examination of how Darwin's enemies linked his work to the immorality of
avant-garde art and literature, in particular aestheticism, has shown that the
scientific naturalists were compelled to construct their model of professional
scientific authority in line with their opponents' standards of respectability.[61]
Dawson's *Darwin, Literature and Victorian Respectability* (2007) shows why
it is so important to take account of the critics of scientific naturalism, as it
was their attacks that often shaped the strategies and modes of representation of Huxley, Clifford, and others. Jim Endersby's analysis of Joseph Dalton
Hooker's career in *Imperial Nature* (2008) has illustrated how some scientific
naturalists attempted to distance themselves from those who embraced the
notion of "professional" as a self-serving interest group with a commercial
stake in advancement. Endersby's book also emphasized Hooker's botanical
collection and classification, pushing scholars to explore in more detail the
scientific practice of the other scientific naturalists.[62]

Since the historiography of nineteenth-century science has changed so
much over the past two decades, and has raised so many questions about the
nature and power of scientific naturalism, the time is right to return to these
canonical figures, in the light of the new scholarly agendas, and reevaluate
their status as icons of the Victorian scientific scene. That is the purpose of

this volume, which brings together new essays by some of the most important scholars, historians of science as well as literary critics, working on scientific naturalism over the last three decades.

New Perspectives and Expanded Maps

Victorian Scientific Naturalism is divided into four sections, which examine different aspects of the interrelated issues of community, identity, and continuity. In the first, "Forging Friendships," the chapters discuss the formation of scientific naturalism in the mid-1850s and the shared interests and concerns that helped establish a cogent sense of community, and often also of close and genuine friendship, over subsequent decades. The scientific naturalists are generally assumed to be a relatively cohesive group, but hitherto there has been little attention given to how their group identity was actually forged—what brought them together and allowed them to continue to operate as an identifiable group? These chapters illustrate how their mutual opponents, communal hobbies, and shared philosophical commitments helped form them into a recognizable cadre of intellectuals. Gowan Dawson examines the paleontological dispute from 1856 to 1857 between Huxley, Hugh Falconer, and Richard Owen, and its role in forging a sense of solidarity among future defenders of Darwin. In this dispute Huxley criticized Georges Cuvier's method of paleontological reconstruction—it relied on the law of necessary correlation, to Huxley a quasi-theological doctrine. Hooker, Darwin, and Spencer (as well as, seemingly, Charles Lyell) backed Huxley, and even Falconer eventually changed his mind and vehemently endorsed his erstwhile antagonist. The Cuvierian controversy begun by Huxley brought together the leading lights of the emergent scientific young guard more than three years before the publication of the *Origin of Species*, providing them with a key secular principle, as well as a mutual bête noire in Owen, which would both be important later during the debates over Darwin's naturalistic mode of species transmutation.

Dawson's chapter is complemented by Michael S. Reidy's study of how a shared love of mountains established a bond between the scientific naturalists. He examines the camaraderie between those who accompanied Tyndall on his ascents of Mount Blanc in the late 1850s, as well as the mountaineering adventures of Hooker in Nepal and William Spottiswoode in eastern Russia, Croatia, and Hungary. Focusing on Tyndall and Leslie Stephen as two of the foremost Victorian mountaineers, Reidy explains what drew them to Alpine peaks: the need for a place where they could experience the sublime in nature and where they could think about topics others associated with religion. In the section's third chapter, George Levine switches from mutual enemies and communal hobbies to explore the underlying philosophical and aesthetic assumptions that, in their reciprocal inconsistency, also helped to forge a sense of group identity. Focusing primarily on Spencer, Tyndall, and Huxley, he considers the paradoxical nature of the scientific naturalists' vision. They were antimetaphysicians who depended on metaphysically unprovable assumptions; they were empiricists who, as with Karl Pearson, also espoused idealism; and they were moralists who, like Huxley, argued that the ethical process was in antagonism with the cosmic process of which it was a part. To Levine, dwelling on paradox was a shared artistic strategy pursued by the leading lights of scientific naturalism to engage their readers in imagining alternative ways of seeing the world.

The second section, "Institutional Politics," examines how this nascent sense of group identity established among the members of the X Club, the metropolitan dining society that offers a nineteenth-century equivalent of Harry Collins's "core set" at the heart of the development of scientific naturalism, was integral to the complicated institutional maneuverings through which Huxley and his key allies sought to reshape science in the 1870s.[63] Bernard Lightman and James Elwick deal with Huxley as educator and statesman of science. Lightman's analysis of Huxley's work on the Devonshire Commission demonstrates the complexity of his strategies for increasing state funding for science. Huxley's priorities were enlarging the research grant to the Royal Society, ensuring the creation of a general Science School, defending Hooker from Owen's attempt to gain control of Kew Gardens, and establishing a national system of education that began in elementary school. But, notably, Huxley was opposed to increasing state funding to science if it compromised scientific autonomy. Elwick analyzes how Huxley's educational career was driven by the examination system. Like other Victorian educators caught up in the exam mania of the third quarter of the nineteenth century, Huxley believed that exams were far more important than actual teaching. Elwick shows how the exam system worked in practice, and how Huxley was able to use it to strengthen the cultural authority of scientific naturalism.

In the section's third chapter, Jim Endersby complicates the conventional picture of the X Club and its ideological commitments through his investigation of Hooker and his idiosyncratic sense of identity. Endersby argues that Hooker was a gentleman first and a scientific naturalist only second, as is vividly illustrated by the political complexities of the so-called Ayrton affair, in which Hooker took the opposite approach to Huxley in resisting the imposition of governmentally mandated meritocratic standards at Kew (although, as Lightman's chapter shows, Huxley, demonstrating the strength of his communal loyalty, nevertheless supported Hooker, and had his own qualms about the implications of state funding for science). Hooker sought neither to be a professional nor to be a vocal proponent for a secular society. For Hooker, as Endersby asserts, scientific naturalism was merely a form of politeness, a way of avoiding unnecessary controversy—particularly over religion—in order to maintain the sociability of a gentleman. Endersby's treatment of Hooker raises important questions about seeing naturalism, secularism, and professionalism as aspects of a deliberate program to modernize and reform Victorian Britain.

In the third section, "Broader Alliances," the chapters by Ruth Barton, Paul White, and Matthew Stanley all illustrate the fluid nature of the identity constructed by scientific naturalists between the 1850s and the 1880s. Expanding the internal focus of the opening section, they examine the continuities and coalitions that existed between the scientific naturalists and ostensibly antagonistic groups. Barton discusses the creation of Sunday lecture societies as vehicles to oppose Sabbatarian legislation. They were but one of the many campaigns conducted in the Victorian period against legislation that privileged particular religious beliefs and practices, especially Anglican ones. Her analysis of the patrons, presidents, vice presidents, and speakers involved with these societies, among which were several prominent scientific naturalists, identifies several constituencies that shared the secularizing objectives of Huxley and his colleagues. Barton's point is that the scientific naturalists were but one group within a broad coalition that included Unitarians and other Nonconformists, secularists, and liberal Anglicans who worked together to secularize the state. Barton emphasizes how they were willing to suspend their own identity as a distinct group in situations where responding to a common foe dictated a coordinated strategy.

White's chapter draws attention to how scientific naturalism was submerged into a larger community of Victorian intellectuals in the meetings of the Metaphysical Society during the 1870s. In the Metaphysical Society, where community was built on the basis of one's conduct toward belief rather than on shared beliefs, the boundaries between scientific naturalism and other, supposedly opposed, schools of thought seemed to dissolve in the course of critical engagement. White uses his study of the Metaphysical Society to raise important questions about how we construct our notion of the scientific naturalists as a distinct group of elite intellectuals. As with Endersby's chapter in the previous section, White uses Turner's particular conception of scientific naturalism as a foil against which to develop a new understanding of the concerns and allegiances of scientific practitioners like Hooker and Huxley. Stanley points to the seeming contradiction that both Christian theists and scientific naturalists saw the concept of the uniformity of nature as integral to their worldviews. Whereas the scientific naturalists believed that the principle of the uniformity of nature left no room for divine action in science, Christian theists, such as Lord Kelvin, James Clerk Maxwell, and William Benjamin Carpenter, insisted that the consistency of natural law pointed toward the existence of a divine being. Stanley argues that adherence by both groups to this axiom created a common space for them to work together despite disagreement on the larger meaning of scientific theories.

The fourth, and final, section of the volume is entitled "New Generations" and examines later incarnations of scientific naturalism that demonstrate the continuity of the sense of group identity that was first forged in the 1850s, even while important aspects of the approach to the natural world taken by Huxley and his allies were transfigured or discarded. Theodore M. Porter argues that the role of science in general culture was more important to the scientific naturalists than the cultivation of professional autonomy and specialized disciplines. Especially by the late 1870s, the scientific naturalists, struggling to deal with increased working-class radicalism, wanted to restore the same unity of elite culture that had existed in the early Victorian period. But with the ascent of technical ideals in science, the impulse to make science the shared idiom of a public culture gradually faded, which led to the eventual decline of scientific naturalism. Melinda Baldwin investigates the younger generation of scientific naturalists in her chapter. She asserts that figures such as E. Ray Lankester, George John Romanes, William Turner Thiselton-Dyer, and Raphael Meldola adopted the journal *Nature* as a central organ of scientific communication, unlike the members of the X Club who had been their mentors. Due to its publication speed, the second generation preferred the weekly *Nature* to monthly general periodicals like the *Nineteenth Century* as the forum for announcing their discoveries. Peter J. Bowler examines the evolution of agnosticism into rationalism in the early twentieth century. Whereas Huxley insisted on keeping his distance from rationalists, as he thought they lacked respectability, the second and third generation of scientific naturalists had no such reservations. Bowler discusses how Lankester, Arthur Keith, Julian Huxley, and J. B. S. Haldane became an important part of the rationalist movement and how they functioned within it.

Taken together, the chapters in this volume offer a series of new perspectives on Victorian scientific naturalism that, drawing on the recent historiographic shifts and recovery of previously neglected groups of historical actors that were surveyed earlier, produce a radically different understanding of the movement centering on the issues of community, identity, and continuity. There remain, inevitably, links to the groundbreaking scholarship of Young

and Turner in the early 1970s (not least the retention of their original no-
menclature), but there are also significant departures from their respective
conceptions of an identifiable and relatively unified cadre of powerful, and
ultimately successful, reformist modernizers. While the labels *scientific naturalist*—now recognized as an actors' category—and *evolutionary naturalist*
are employed extensively throughout this volume, the actual category of "scientific naturalism" is shown to be much more fluid and mutable than historians have hitherto acknowledged. In certain chapters, such as White's and, to a
lesser extent, Barton's, it even risks being discarded altogether as a capricious
and prohibitively narrow designation that obscures other no less significant
social formations and alliances among late nineteenth-century learned elites,
while also tending to homogenize the divergent opponents of so-called scientific naturalism as merely religious, unscientific, or amateur.

Reflecting the nature of scholarship in the field, this volume attempts to
accommodate a range of competing perspectives that contribute to an ongoing debate over what has hitherto been termed scientific naturalism. Indeed, other chapters keep with the same terminology as Young and Turner,
although endeavoring to expand it sufficiently to include, for instance, the
gentlemanly protocols and distinctly nonprofessional aspirations of Hooker,
as in Endersby's contribution, or the metaphysical and deist attitudes discussed by Levine. Such divergent views often had to be suppressed in order to
present a united front in the face of hostile criticism, but they were nonetheless real and significant. At the same time, scientific naturalism, even when
expanded to encompass such discrepancies, still effectively encapsulates the
values and assumptions of many mid- and late Victorian men of science and
related intellectuals. They are just not necessarily the same ones that Turner
propounded when he originally defined the category in *Between Science and
Religion.*

The close attention to local contexts and specific circumstances exemplified in many of the chapters also unveils a much more expansive, but also a
more variable, uneven, and complex map of Victorian scientific naturalism
than scholars have hitherto worked with. Similarly, the shift from a focus
on disembodied ideas or abstract theorizing to an emphasis on practice, the
physical activities involving tangible objects that take place in specific times
and places and within the broader context of everyday life, whether risk-
ing life and limb climbing in the Alps, maintaining aristocratic protocols of
conduct, or marking interminable piles of examination papers, affords new
insights into the various ways that scientific naturalism operated in the scientific landscape of the second half of the nineteenth century.

But while Turner insisted that scientific naturalism "arose and flourished

in circumstances unique to the period 1850 to 1900," perceiving its abrupt
demise as seemingly coincidental with the century's end, the revised understanding of the movement presented in this volume does not conform to
such convenient—as well as entirely arbitrary—historiographic boundaries.[64] Rather, as with comparable nineteenth-century cultural categories
such as Romanticism, scientific naturalism also had a second generation who,
like Shelley, Byron, and Keats, both venerated and simultaneously critiqued
their celebrated forebears, maintaining a continuity with certain aspects of
their agenda into the early twentieth century, while transmuting others into
markedly different outlooks such as rationalism. The sustained onslaught
against positivist modes of thought at the close of the nineteenth century
charted by Alex Owen in *The Place of Enchantment* (2004) might have weak-
ened scientific naturalism, but, as several of the chapters in this volume show,
it nevertheless survived, if in new forms and under revised headings, long
into the twentieth century.[65]

In fact, scientific naturalism is alive and well in the twenty-first century, as
creationists such as Philip E. Johnson loudly bemoan in print and across the
Internet.[66] And, as was evident in the 2009 bicentenary celebrations, Darwin
is still its internationally renowned poster boy. It was Darwin's most loyal
supporters, men such as Huxley, Tyndall, and Hooker, who provided the
philosophical and professional underpinning for the modern incarnation of
scientific naturalism. This is why historians can never afford to lose sight of
Victorian scientific naturalism for too long.

Notes

1. Thomas H. Huxley, prologue to *Essays upon Some Controverted Questions* (London: Macmillan, 1892), 35.

2. James G. Paradis, *T. H. Huxley: Man's Place in Nature* (Lincoln: University of Nebraska
Press, 1978), 180.

3. See James E. Strick, *Sparks of Life: Darwinism and the Victorian Debates over Spontaneous
Generation* (Cambridge, MA: Harvard University Press, 2000), 90–91.

4. Arthur James Balfour, *The Foundations of Belief* (London: Longmans, Green, 1895), 134;
and Leonard Huxley, *Life and Letters of Thomas Henry Huxley*, 2 vols. (London: Macmillan,
1900), 2:396.

5. John Tyndall, "Scope and Limit of Scientific Materialism," in *Fragments of Science for
Unscientific People* (London: Longmans, Green, 1871), 107–24.

6. See Frank Miller Turner, *Between Science and Religion: The Reaction to Scientific Naturalism in Late Victorian England* (New Haven, CT: Yale University Press, 1974).

7. See, for instance, David Ray Griffin, *Religion and Scientific Naturalism: Overcoming the
Conflicts* (Albany: State University of New York Press, 2000).

8. T. H. Huxley, prologue, 35.

9. Thomas H. Huxley, *Evolution and Ethics* (London: Macmillan, 1893), 49.

10. See Thomas Dixon, *The Invention of Altruism: Making Moral Meanings in Victorian Britain* (Oxford: Oxford University Press, 2008).

11. See Ruth Barton, "'Men of Science': Language, Identity and Professionalization in the Mid-Victorian Scientific Community," *History of Science* 41 (2003): 73–119; and Paul White,*Thomas Huxley: Making the "Man of Science"* (Cambridge: Cambridge University Press, 2003).

12. "The Word 'Scientist,'" *Science-Gossip* 1 (1894–95): 242.

13. "Professor Lewis's Naturalism," *Literary World* 6 (1850): 560. Lewis's lecture was later published as *Nature, Progress, Ideas* (Schenectady, NY: G. Y. Van Debogert, 1850).

14. Tayler Lewis, "Hickok's Rational Psychology," *Bibliotheca Sacra* 8 (1851): 377.

15. [T. H. Huxley], "The Vestiges of Creation," *British and Foreign Medico-Chirurgical Review* 13 (1854): 433, 438.

16. David Brown, "On Some Recent Utterances of Scientific Men," *Sunday Magazine* 4 (1867), 174, 170.

17. M. A. Doudney [W. B. Rands], "Ideas of the Day," *Contemporary Review* 37 (1880): 838, 844, 839, 843.

18. W. K. Clifford, "The Unseen Universe," *Fortnightly Review*, n.s., 17 (1875): 780.

19. "Pleas of a Convert," *Secular Review* 2 (1878): 45.

20. "Spiritualism," *Secular Review* 2 (1878): 77–78.

21. "Not 'Spiritualism,' but Freethought," *Secular Review* 2 (1878): 110.

22. See A. Gowans Whyte, *The Story of the R.P.A., 1899–1949* (London: Watts, 1949), 22.

23. Quoted in Adrian Desmond, *Huxley: Evolution's High Priest* (London: Michael Joseph, 1997), 145.

24. See Bernard Lightman, "Huxley and Scientific Agnosticism: The Strange History of a Failed Rhetorical Strategy," *British Journal for the History of Science* 35 (2002): 271–89.

25. T. H. Huxley, prologue, 37, 40.

26. Ibid., 49, 52, 53.

27. "Mr. Johnston's Account of the Meeting of Naturalists at Hamburgh," *Edinburgh Journal of Science*, n.s., 4 (1831): 212.

28. George Cornewall Lewis, *An Essay on the Influence of Authority on Matters of Opinion* (London: John W. Parker, 1849), 25.

29. Tayler Lewis, *The Six Days of Creation* (Schenectady, NY: G. Y. Van Debogert, 1855), 201–2.

30. See Thomas Henry Huxley, "On the Educational Value of the Natural History Sciences" [1854], in *Lay Sermons, Addresses and Reviews* (London: Macmillan, 1870), 91.

31. L. Huxley, *Life and Letters of Thomas Henry Huxley*, 1:94.

32. [R. H. Hutton], "Pope Huxley," *Spectator* 43 (1870): 135.

33. R. H. Hutton, "Is 'Lapsed Intelligence' a Probable Origin for Complex Animal Instincts?" [17 December 1878; 2], in *Papers Read at the Meetings of the Metaphysical Society*, 3 vols. (London: privately printed, 1869–80), 3:2657e.1025, Bodleian Library, Oxford.

34. Michael Ruse, introduction to *Evolutionary Naturalism: Selected Essays* (London: Routledge, 1995), 1. See, for example, Paul F. Boiler Jr., *American Thought in Transition: The Impact of Evolutionary Naturalism, 1865–1900* (Chicago: Rand McNally, 1969).

35. On the problems of using only actors' categories, see Harry Collins, "Actors' and Analysts' Categories in the Social Analysis of Science," in *Clashes of Knowledge: Orthodoxies and*

Heterodoxies in Science and Religion, ed. Peter Meusburger, Michael Welker, and Edgar Wunder (Heidelberg, Germany: Springer, 2008), 101–10.

36. Barton, "Men of Science," 75.

37. Ralph O'Connor, "Reflections on Popular Science in Britain: Genres, Categories, and Historians," *Isis* 100 (2009): 333–45.

38. James Secord, "The Electronic Harvest," *British Journal for the History of Science* 38 (2005): 463–67.

39. Robert M. Young, *Darwin's Metaphor: Nature's Place in Victorian Culture* (Cambridge: Cambridge University Press, 1985), 1–2, 4, 68, 79.

40. Ibid., 191, 240.

41. Turner, *Between Science and Religion*, 9–12, 24–30.

42. Frank M. Turner, "The Victorian Conflict between Science and Religion: A Professional Dimension," *Isis* 69 (1978): 358, 360, 364–65.

43. Frank M. Turner, *Contesting Cultural Authority: Essays in Victorian Intellectual Life* (Cambridge: Cambridge University Press, 1993).

44. Peter Allan Dale, *In Pursuit of a Scientific Culture: Science, Arts, and Society in the Victorian Age* (Madison: University of Wisconsin Press, 1989), 6.

45. Bernard Lightman, *The Origins of Agnosticism: Victorian Unbelief and the Limits of Knowledge* (Baltimore: Johns Hopkins University Press, 1987); and Ruth Barton, "'An Influential Set of Chaps': The X-Club and Royal Society Politics, 1864–85," *British Journal for the History of Science* 23 (1990): 53–81.

46. Adrian Desmond and James Moore, *Darwin* (London: Michael Joseph, 1991); Janet Browne, *Charles Darwin: Voyaging* (London: Jonathan Cape, 1995); and Desmond, *Huxley*.

47. Peter J. Bowler, *The Eclipse of Darwinism: Anti-Darwinian Evolution Theories in the Decades around 1900* (Baltimore: Johns Hopkins University Press, 1983).

48. See Gowan Dawson, "First among Equals," *Times Literary Supplement*, 9 January 2009, 7–8.

49. Desmond and Moore, *Darwin*; and Desmond, *Huxley*.

50. Adrian Desmond, *The Politics of Evolution: Morphology, Medicine, and Reform in Radical London* (Chicago: University of Chicago Press, 1989), 20.

51. Anne Secord, "Science in the Pub: Artisan Botanists in Early Nineteenth-Century Lancashire," *History of Science* 32 (1994): 269–315.

52. James A. Secord, *Victorian Sensation: The Extraordinary Publication, Reception, and Secret Authorship of "Vestiges of the Natural History of Creation"* (Chicago: University of Chicago Press, 2000).

53. John van Wyhe, *Phrenology and the Origins of Victorian Scientific Naturalism* (Aldershot, UK: Ashgate, 2004).

54. Bernard Lightman, *Victorian Popularizers of Science: Designing Nature for New Audiences* (Chicago: University of Chicago Press, 2007).

55. Evelleen Richards, "Redrawing the Boundaries: Darwinian Science and Victorian Women Intellectuals," *Victorian Science in Context*, ed. Bernard Lightman (Chicago: University of Chicago Press, 1997), 119–42; Suzanne Le-May Sheffield, *Revealing New Worlds: Three Victorian Women Naturalists* (London: Routledge, 2001); Ann B. Shteir, *Cultivating Women, Cultivating Science: Flora's Daughters and Botany in England, 1760 to 1860* (Baltimore: Johns Hopkins University Press, 1996); and Barbara Gates, *Kindred Nature: Victorian and Edwardian Women Embrace the Living World* (Chicago: University of Chicago Press, 1998).

56. Crosbie Smith, *The Science of Energy: A Cultural History of Energy Physics in Victorian Britain* (Chicago: University of Chicago Press, 1998).

57. Richards, "Redrawing the Boundaries," 128–35; Donald Luke Opitz, "Aristocrats and Professionals: Country-House Science in Late-Victorian Britain" (PhD diss., University of Minnesota, 2004), 33, 101, 109, 256–57; Erin McLaughlin-Jenkins, "Common Knowledge: Science and the Late Victorian Working-Class Press," *History of Science* 39 (2001): 445–65; and Sandra Den Otter, *British Idealism and Social Explanation: A Study in Late Victorian Thought* (Oxford: Clarendon Press, 1996).

58. See Robert J. Richards, *Darwin and the Emergence of Evolutionary Theories of Mind and Behavior* (Chicago: University of Chicago Press, 1987), 176–84, 226–27.

59. See Theodore M. Porter, *Karl Pearson: The Scientific Life in a Statistical Age* (Princeton, NJ: Princeton University Press, 2004); Peter Morton, *"The Busiest Man in England": Grant Allen and the Writing Trade, 1875–1900* (Basingstoke, UK: Palgrave Macmillan, 2005); Mark Francis, *Herbert Spencer and the Invention of Modern Life* (Ithaca, NY: Cornell University Press, 2007); Michael Taylor, *The Philosophy of Herbert Spencer* (London: Continuum, 2007); and Mark Patton, *Science, Politics and Business in the Work of Sir John Lubbock: A Man of Universal Mind* (Aldershot, UK: Ashgate, 2007).

60. White, *Thomas Huxley*.

61. Gowan Dawson, *Darwin, Literature and Victorian Respectability* (Cambridge: Cambridge University Press, 2007).

62. Jim Endersby, *Imperial Nature: Joseph Hooker and the Practices of Victorian Science* (Chicago: University of Chicago Press, 2008).

63. See Harry M. Collins, "The Place of the 'Core-Set' in Modern Science: Social Contingency with Methodological Propriety in Science," *History of Science* 19 (1981): 6–19.

64. Turner, *Between Science and Religion*, 11.

65. Alex Owen, *The Place of Enchantment: British Occultism and the Culture of the Modern* (Chicago: University of Chicago Press, 2004).

66. See, for example, Philip E. Johnson, *Reason in the Balance: The Case against Naturalism in Science, Law and Education* (Downers Grove, IL: InterVarsity Press, 1995); and Larry Vardiman, "Scientific Naturalism as Science," Institute of Creation Research, accessed 3 July 2012, http://www.icr.org/article/422/.

Forging Friendships

1

"The Great O. versus the Jermyn St. Pet": Huxley, Falconer, and Owen on Paleontological Method

GOWAN DAWSON

His health irreparably damaged by long exposure to the tropical climate of northern India, Hugh Falconer retired from his post as superintendent of the Royal Botanic Gardens at Calcutta in the spring of 1855 and journeyed back to Britain by the overland route. During the long voyage he passed through the Crimean peninsula just as the yearlong siege of Sebastopol reached its critical juncture, with British, French, and Turkish troops launching their last and most sustained bombardment of the Russian fortifications.[1] A year earlier the *Spectator* had observed that the rationale of the initial attack was that "if we would overthrow the power [of the Russian empire], we must sweep away its foundations," while in October, a month after the siege's bloody conclusion, the *Edinburgh Review* proclaimed that "to attack Sebastopol was . . . to assail the stronghold of Russia in the East."[2] Back in London at the end of 1855, where newspaper coverage of the war was still transfixing public attention, Falconer soon encountered a no less "remarkable" verbal onslaught in which, like the besieged ramparts of Sebastopol, the "very foundations of palæontology, as they have hitherto been understood, are assailed." This ferocious "attack on Cuvier and his followers" was launched in February 1856, the very month when peace was being negotiated at the Congress of Paris, by a "man of science, of recognized standing, [who] assails generally admitted principles and established reputations" and whose bellicose manner was, Falconer considered, such "as to require some notice."[3] Although regularly bedridden and with a "constitutional aversion from the hispid walks of controversy," the veteran of Sebastopol felt compelled to single-handedly defend the besieged foundations of Cuvierian paleontology.[4]

Despite its timing, the onslaught that prompted Falconer's concern in fact only rarely resorted to the pugnacious militaristic rhetoric that was still

FIGURE 1.1. Hugh Falconer in the early 1860s. *Palæontological Memoirs and Notes of the Late Hugh Falconer*, ed. Charles Murchison, 2 vols. (London: Robert Hardwicke, 1868), vol. 1., frontispiece. © The British Library Board, 7204.df.20.

dominating the press, though leavened by criticism of the bungling aristo-cratic high command, in the early months of 1856.[5] The triumphant asser-tion that "Cuvier himself . . . surrenders his own principle" of "physiological correlation" was the most conspicuous exception to its generally unmartial tone. The infallibility of the celebrated axiom in which each element of an animal is presumed to correspond mutually with all the others and thus any part, even just a single disarticulated bone, necessarily indicates the configu-ration of the integrated whole had been given up, it was implied, in the same way as Russia, as the *Times* reported, was just then enduring the "surrender of . . . part of her territory" in Paris.[6] The author of this otherwise carefully worded attack, the printed form of which, as Falconer noted, was signed and "authenticated with his initials," had nevertheless displayed a zealous enthu-siasm for military matters only months earlier when writing for the anony-mous *Westminster Review*.[7] There he had expressed his "taste for ordnance of all kinds, sea fights, Minié rifles, and Crimea expeditions," as well as revealing a detailed knowledge of the tactics by which the "allied armies [in] their siege operations [at] the south side of Sebastopol" could have secured "a *coup-de-main* . . . without serious loss to the assailants."[8] This aficionado of siege warfare even identified himself, again in the pages of the *Westminster*, with the "youthful vigour" of the ongoing Islamic insurgency against the imperialistic "Greco-Russian Czar-worship, misnamed Christianity," in the Caucasus.[9]

The unsigned author was Thomas Henry Huxley, who, as an assistant surgeon in the Royal Navy at the start of the war, had only narrowly avoided active service in the Crimea, all the while insisting that "nobody can accuse me of an objection to facing the Rooshians."[10] The image of naturalistic science as engaged in perpetual warfare against a corrupt orthodox theology that Huxley continued to hone throughout his long career—the famous "military metaphor"—was one that had its origins in the real battlefields of the mid-1850s. The metaphorical conjunction of science and militarism, however, was certainly not exclusive to Huxley, and was, in any case, a decidedly reciprocal interchange. The very paleontological principles on which Huxley turned the ersatz guns of his military metaphor had themselves been invoked metaphorically by the Crimean conflict's new breed of war correspondents reporting from the front line, with William Howard Russell of the *Times* proposing after the end of the Sebastopol siege:

> If CUVIER . . . could reconstitute the whole structure of some antediluvian animal from the mere glance at some joint or fragment of bone, it is sufficient for us to examine the emaciated body and empty havresac of any one of the wretched Russian soldiers. . . . From this we can infer fairly enough the condition to which the empire has been reduced.[11]

Back on the home front, moreover, it was Falconer, with the artillery fire of Sebastopol still ringing in his ears, who interpreted Huxley's lecture at the Royal Institution in explicitly martial terms, even noting the speaker's quasi-militaristic title as "an officer on the palæontological staff of the Museum of Economic Geology."[12] It is evident that Huxley's opponents were equally willing to apply the language of warfare to scientific controversies during this period, and, unlike the self-proclaimed "Prophet-Warrior," to do it under their own names.[13]

This, of course, was a more defensive version of the metaphorical juxtaposition of science and war. However, while discussions of the so-called military metaphor in nineteenth-century science have tended to emphasize its relation only to hostile offensive tactics (Adrian Desmond, for instance, depicts Huxley "shouldering his .45 to shoot over the ranks of obstructive

Anglicans"), when it first began to be employed in the mid-1850s, there was considerable interest in, as well as great admiration for, the defensive strategies adopted by both sides in the Crimea.[14] As James J. Reid has observed, the "Crimean War was mostly a defensive war in which aggressive forward movements had limited aims. This defensive-minded strategy . . . originated in the post-Napoleonic perception of balance of powers that sought to avoid conquests and far-ranging military campaigns."[15] In September 1855 the Crystal Palace in Sydenham opened a Crimean Court with a scale model of Sebastopol showing the impregnable earthen ramparts built by the Russians according to the defensive principles outlined in *A Proposed New System of Fortification* (1849) by James Fergusson. "Mr. Fergusson's model," as the official guidebook explained, demonstrated the "great principle of his system" in "arming the ramparts and . . . flanking defences," and was something that at "the present moment the visitors will be particularly interested with."[16] The extensive grounds of the Crystal Palace also featured gigantic brick-and-mortar models of extinct creatures modeled by Benjamin Waterhouse Hawkins under the guidance of Richard Owen, whose astonishing "ability to reproduce accurate models of the entire structure and correct proportions of extinct animals from the discovery of a single bone" had, as Hawkins effusively acknowledged, made the tableau of prehistoric life possible.[17] The upholders of the paleontological principles celebrated in these three-dimensional models on the same site as Fergusson's Crimean Court now felt the need—utilizing a more defensive version of the military metaphor—to publicly render the foundations of their Cuvierian methods of reconstruction no less unassailable from concerted attacks.

Huxley's lecture, delivered on 15 February 1856 only hours after dispatches reporting the final destruction of the Sebastopol docks were received by the War Department, was one of the Royal Institution's weekly Friday Evening Discourses, which attracted large, fashionable audiences made up of the institution's members and their guests.[18] The subject seemed calculated to interest men of science while at the same time not perturbing the fashionable ladies among Huxley's auditors. He was simply to "set forth . . . an estimate of the science of Natural History" as a means of instilling "*knowledge*," "*power*," and "*discipline*." The opening discussion of natural history as a form of knowledge, however, soon afforded Huxley a reason to adopt a less conciliatory tone, especially when he turned to the "works of Paley and the natural theologians," who had discerned evidence of a "utilitarian adaptation to benevolent purpose" throughout the natural world. In his *Moral and Political Philosophy* (1785) William Paley had reconciled Christian ethics with those of utilitarianism, and Huxley now attributed a similar emphasis on pragmatic expediency and end-directed teleology to his argument from design in *Natural Theology* (1802).

Despite the evident teachings of recent investigations of nature, the natural theological "principle of adaptation of means to ends" continued to be trumpeted, even "in the writings of men of deservedly high authority," as the "great instrument of research in natural history."[19] Huxley traced this overhyped "doctrine to its fountain head," finding that it was "primarily put forth by Cuvier." The only viable means of explaining this, he went on, was to conclude that this "prince of modern naturalists . . . did not himself understand the methods by which he arrived at his great results," and that his "master-mind misconceived its own processes." This might, as the thirty-year-old who was still yet to publish any paleontological papers of his own acknowledged to the scientific grandees of the Royal Institution, seem "not a little presumptuous." But if all the "arguments be justly reasoned out" without regard to celebrity or patronage, then it would be recognized, Huxley insisted, that "it is correct." Huxley risked the justifiable charge of impudence in front of such an august audience because it was on the basis of the doctrine of adaptation to purpose that it had been "handed down from book to book, that all Cuvier's restorations of extinct animals were effected by means of the principle of the physiological correlation of organs."[20] This, as Huxley knew well, was also the method deployed in the famous paleontological reconstructions from just fragmentary remains—celebrated at the Crystal Palace—of the "*British* Cuvier," Owen, whose once-supportive relationship with Huxley had grown increasingly rancorous over the previous three years (and who, Huxley quipped, "stands in exactly the same relation to the French [Cuvier] as British brandy to cognac").[21]

Huxley impugned the inductive abilities of Cuvier and Owen, asserting that the ostensibly infallible law of physiological correlation, the unerring veracity of which afforded crucial evidence of the harmonious design of organic structures, was in reality based on prosaic empirical observations of customary correspondences and had nothing to do with genius or any other purportedly preternatural capabilities. Assumptions about the absolute necessity of such correspondences were made on the basis of authority, both scientific and religious, and not reason or logic. Although Huxley did not mention it directly, his audience would have been aware that the same unquestioning obedience to rank and authority was, at that very moment, being widely blamed for the dreadful administrative failures of the patrician and nepotistic civil service during the Crimean conflict, most notably by Charles Dickens in his satirical serial novel *Little Dorrit* (1855–57).

This particular paleontological dispute, which raged throughout 1856 and into the following year, is not unknown to historians of Victorian science, but it has never before received detailed consideration and thus its crucial importance in the emergence of scientific naturalism, during the crisis of authority in the mid-1850s precipitated by the aristocratic mismanagement of the Crimean War, has been overlooked.[22] In the dispute, Huxley received invaluable support from other young naturalists similarly seeking to make a remunerative career in science, including Joseph Dalton Hooker and Herbert Spencer, as well as from a reclusive contemporary of Owen's who had long harbored a clandestine theory of species change that was no less naturalistic and contrary to conventional conceptions of divine design than Huxley's attitude to paleontological method. The spring of 1856 was precisely the moment at which Charles Darwin began revealing elements of his theory of natural selection to a carefully chosen group of potentially sympathetic naturalists, and, as this chapter will argue, Huxley's ferocious and exactly contemporaneous paleontological dispute, initially with Falconer and then with Owen, helped forge a crucial sense of solidarity among those who would soon emerge as the principal advocates of *On the Origin of Species* (1859) and whom historians have grouped together as the spokesmen of the nascent doctrines of scientific naturalism. While the scientific naturalists have generally been assumed to constitute a relatively cohesive group, little attention has hitherto been paid to the factors—mutual enemies as much as shared intellectual commitments, as well as informal ties like friendship, domestic visits, humor, and gossip—that helped them establish a cogent sense of community in the middle of the nineteenth century.

Falconer v. Huxley

Huxley's vehement attack on Cuvier immediately posed a delicate problem for Darwin and Hooker, who, in May 1856, abandoned an attempt to get their headstrong new friend elected to the Athenæum because of concerns that Owen would sway the club's committee against him. Darwin fearfully imagined Owen with "a red face, dreadful smile & slow & gentle voice" asking "what Mr Huxley has done, deserving this honour; I only know that he differs from, & disputes the authority of Cuvier . . . as of no weight at all."[23] But in reality it was instead Falconer whom Darwin "found . . . very indignant at the manner in which Huxley treated Cuvier in his R. Inn Lecture."[24] Significantly, for Darwin and Hooker, Falconer had been "a mutual friend so dear to us both" for almost a decade before either had even met Huxley, with Falconer having stayed at Darwin's home at Downe in the mid-1840s during an earlier period of leave from India.[25] At this time Darwin regarded him as one of the "most rising naturalists in England" and continually found it "wonderful how much heterogeneous information he has about all sorts of things." Falconer's particular area of expertise was prehistoric proboscideans, and his return to India in 1848 was, Darwin lamented, a "grievous loss to palæontology."[26] Hooker, who himself journeyed to India a year later, confirmed Darwin's worst fears with the news that a fever contracted during the "hot season" had left Falconer incapable of "doing any thing in Science." Although he later regained his "health & spirits," even growing "fat & look[ing] far better than he did in England," paleontology generally had to take second place to his paid "pursuits as Botanist horticulturalist & Landscape Gardener" at Calcutta's Royal Botanic Gardens.[27]

Darwin was evidently anxious to renew the acquaintance when the once more ailing Falconer came back from Calcutta—via Sebastopol—for the last time in late 1855, telling Hooker in April of the following year, "Has Falconer appeared in [the] world yet; if so & you know his address, I wish you would let me have it."[28] Hooker had already been in touch with his fellow veteran of the subcontinent's stifling climate, and passed on his new address in Piccadilly with the proviso that Falconer had "vow[ed]" he would "not go anywhere." Refusing to be put off, Darwin pleaded with Falconer that "I do so want to see you," and implored him to "be a good man & cast your resolutions to the dogs" by coming to stay once more at Downe along with "Hooker and one or two others."[29] Ironically, one of those other guests at the proposed weekend party was to be Huxley, who had been invited to Downe for the very first time. Falconer, even without knowing the identity of the rest of the company, held to his resolution, and when Darwin finally did meet his erstwhile friend again a month later in May, he must have soon realized that the planned introduction of Falconer to Huxley would have been unexpectedly incendiary.

Falconer's final task before leaving India in early 1855 had been to catalog the paleontological collections in the museum of the Asiatic Society of Bengal. Working "under the pressure of an approaching departure for Europe," he had employed Cuvier's customary methods to quickly bring taxonomic order to the "appalling confusion, disorder and dilapidation" of the miscellaneous fossils that were "huddled together in heaps . . . without a label or mark of any kind whatsoever to indicate whence they came!" even if the lack of time meant that some identifications had to remain "simply conjectural."[30] A year earlier he had written to his niece from Calcutta, explaining the procedures employed in his paleontological work:

A tooth or the end of a joint . . . is as conclusive evidence of the former existence of an animal as if all the structure—skin, flesh and blood, and living
limbs—were before us. . . . The evidence is fragmentary and inductive, but . . .
clear and conclusive. . . . For the Almighty has so ordained it that reason can
safely reproduce all that has been lost, and restore to the tooth all that was
correlative to it in life.[31]

Now back in London, Falconer was loath to have precisely these same methods repudiated by a brash novice whom he had probably never heard of before reading his lecture in the *Proceedings of the Royal Institution,* Huxley
having been in Australia as a lowly assistant surgeon on HMS *Rattlesnake*
when Falconer had returned to India in 1848.

Falconer's response appeared in the June 1856 number of the *Annals and
Magazine of Natural History,* and, alongside the military language that was
noted earlier, his tone was unmistakably terse, impatient, and condescending.
Huxley's charges against Cuvier were "remarkable" and "startling," and the
necessity of "making practical refutations" of them led Falconer to exclaim
testily, "All this is familiar knowledge; the only marvel is, that one should
have to adduce the facts at the present day in such an argument."[32] The Cuvierian axiom of "necessary correlation" that Huxley had found so untenable
was simply "*necessary* in the sense of being demonstrable in such a way that
the contrary involves an absurdity and is inconceivable," so that "Mr. Huxley,
with the skeleton of a hawk before him, might as well say that, for any physiological necessity to the contrary, that creature might have its jaws with teeth,
and its internal organs arranged, like those of a tiger."[33] When applied to
actual paleontological reconstructions, Huxley's unwillingness to acknowledge that certain associations in anatomical structures were necessary might
permit the creation of absurd, incongruous monstrosities.

The end "result" of all this belligerence, Falconer concluded, was "that
after the encounter the law of correlation stands exactly as Cuvier found and
left it . . . wholly uninjured by its latest assailant."[34] Like Fergusson's famously
indestructible defensive earthworks, which, as the *Methodist Quarterly Review*
reported, had enabled the "Russians in Sebastopol, behind their impregnable
ramparts, [to] laugh at Minnie rifles and Lancaster guns," Falconer's defense
of the besieged foundations of Cuvierian paleontology had decisively repelled
the great savant's brash assailant, leaving the axiom of necessary correlation
entirely unimpaired.[35] Not knowing his adversary's nascent tenacity, Falconer
probably assumed that the matter was now settled. Huxley would have to
accept a humiliating peace like that agreed between the victorious European
powers in the Treaty of Paris. Falconer even felt able to retire from the fray to
the very city where the Crimean conflict had been concluded three months

earlier. In a letter from Paris "he alluded to the subject as if he had eaten Huxley without salt & left no bones at all," as his old friends Hooker and Darwin noted between themselves with considerably less assurance.[36]

Ensconced at Downe, Darwin was "most heartily sorry at the whole dispute," which he worried would "prevent two very good men from being friends." Reluctant to "give up the time to form a very certain judgement to my own satisfaction, in Falconer v. Huxley," Darwin found himself genuinely torn between old and new friends. In a letter to Hooker written over two days in June 1856, he kept veering abruptly—sometimes even mid-sentence—between their rival claims, stating first that Falconer's "article struck me as very clever," then noting, "I rather lean to the Huxley side" and "I think Huxley[']s argument best," before adding, "But to deny all reasoning from adaptation & so called final causes [as Huxley had done], seems to me preposterous," and finally concluding, "I deprecate the contemptuous tone of Huxley." Darwin's own tone of anxious equivocation was at last resolved only in a short postscript, apparently added later, when he told Hooker, "I have just reread your note & it seems to me that there is great justness in your remarks on Huxley & the general question, being discussed as it has been discussed."[37] Although this particular note seems no longer to exist, Hooker's attitude toward the dispute, which apparently decided Darwin's own, can be inferred from another brief missive he sent to Huxley earlier in the same month.

Leafing through "old quarterlies," Hooker had come across a "passage that," as he gleefully informed Huxley, "will amuse you and rile Falconer."[38] The passage appeared in the yellowing pages of a twenty-seven-year-old copy of the *Quarterly Review* and described Cuvier's confidence in his methods as a "delusion."[39] Hooker pointedly underlined this word in his own précis of the passage, emphatically urging Huxley that "it is worth your reading."[40] While Hooker later conceded, "I do not think that Huxley's original lecture [at the Royal Institution] was particularly good at all," he nevertheless came out decisively on his side in the dispute with Falconer, whom, having fallen out with him over his uncouth conduct at the Asiatic Society while the two were in Calcutta, he was happy to see riled. Hooker's clear stance helped clarify Darwin's own position on the matter, and he soon explained retrospectively, "I thought from the first that he [i.e., Huxley] was right, but was not able to put it clearly to myself."[41] Across the Channel, meanwhile, Falconer was blithely unaware that these two old friends of more than ten years' standing had opted to support, and even furnish with strategically advantageous passages from periodicals, the headstrong young naturalist they had only recently got to know.

Falconer would have been still more surprised because, until now, Darwin had given every impression of being a staunch advocate of Cuvierian methods in paleontology. After all, in his chapter on "Geology" for John Herschel's *Manual of Scientific Enquiry* (1849), Darwin had advised Royal Navy officers that "bones . . . from any formation are sure to be valuable; even a single tooth, in the hands of a Cuvier or Owen, will unfold a whole history. . . . Every fragment should be brought home."[42] The fragmentary bones that Darwin himself brought back from South America had been passed on to Owen, whose "Fossil Mammalia" volume of the *Zoology of the Voyage of HMS Beagle* (1839) was a veritable textbook of orthodox functional correlation, replete with allusions to Cuvier's "beautiful and justly celebrated reasoning" on a single phalanx bone.[43] Professing his "entire ignorance of comparative Anatomy," Darwin never challenged the Cuvierian basis of Owen's reconstructions, even while his own, more idiosyncratic approach to reading fossils emphasized environmental rather than functional considerations.[44]

Furthermore, although Huxley represented physiological correlation as ineluctably a quasi-theological doctrine, Darwin remained interested in the nonadaptational correlative changes in an organism's organization that might be effected by natural selection modifying any one part (contra Cuvier's insistence that such complex and holistically coordinated alterations were impossible).[45] In the late 1830s he had jotted in his secret notebook on transmutation, "Thinking of effects of my theory, laws probably will be discovered. of co relation of parts, from the laws of variation of one part affecting another," a principle that, twenty years later, he designated "correlation of growth" in the *Origin*.[46] As Timothy Shanahan has proposed, "Darwin was genuinely pluralistic in his approach to explaining biological phenomena . . . [and] much impressed by Cuvier's principle of the 'Correlation of Parts', which treated organisms as integrated systems."[47] Darwin's closest confidant in his clandestine evolutionary speculations, Charles Lyell, also discerned a potential support for his friend's provocative theory in Cuvierian correlation, noting in his own private journal on the species question "Cuvier's observation that each organ of an individual bears a relation to the whole. This supports view of indefinite variation as a tree-climbing creature will acquire all the necessary changes."[48] It is important to recognize that, in the mid-1850s, when even Lyell was beginning to entertain the so-called development hypothesis, Huxley still held to a stridently nonprogressive line on species, and, at this time, there was no connection whatsoever between his fierce antagonism toward Cuvier and Darwin's as yet undisclosed, and Cuvierian-inflected, evolutionism. Falconer had therefore presumably confided in Darwin, in May 1856, about his indignation at the manner in which Huxley's

Royal Institution lecture treated Cuvier in the firm belief that his old friend would feel the same way.

In the absence of any clear scientific impetus, part of the reason for Darwin's sudden apostasy from Falconer and Cuvierian correlation in June 1856—personal as much as intellectual—likely went back to the weekend party in April that Falconer had declined to attend. Darwin had used the occasion to tentatively reveal aspects of his thinking on species mutation to a carefully chosen group of naturalists who could assist him with pertinent questions and objections, and whom he suspected might be susceptible to his circumspect promptings. Had he made the railway journey from central London to rural Kent, Falconer could have provided some much-needed "rudimentary knowledge" of the "Pigeons-skeletons" that littered Darwin's home, and, although "included . . . in the category of those who have vehemently maintained the persistence of specific characters," Darwin was "fully convinced" that, with a little logical persuasion, he would become, as he told him, "less fixed in your belief in the immutability of species."[49] While Falconer later avowed that he had "long enjoyed the privilege of intimate intercourse with Charles Darwin," and had "been for many years familiar with the gradual development of his views on the Origin of Species," it was instead Huxley, arriving late because of a heavy workload, who was taken into Darwin's confidence that spring at Downe.[50]

Initially no less opposed to transmutation than Falconer, Huxley was increasingly swayed by the evidence and arguments presented by Darwin, who rejoiced at the "change in . . . Huxley's opinions on species" that was tangible over the course of the weekend.[51] Darwin had been gently insisting on the possibility of organic forms *generally undergoing* further development" in letters to Huxley for the previous three years, and only a month after his first stay at Downe the erstwhile scourge of those who were "unorthodox about species" was forecasting in the *Medical Times and Gazette* that "the 'Theory of Progressive Development' will present by far the most satisfactory solution" to the thorny question of the emergence of new species.[52] Thus, by the early summer of 1856 it was evident that Huxley, with his youthful vigor and rhetorical self-assurance, represented a more effectual ally for Darwin, and his heterodox evolutionism, than the frail old friend who was still socially and intellectually isolated after eight long years in India.

The "Law of Necessary Correlation" Is—Nowhere

Falconer in fact became the source of "splendid joke[s]" among Darwin's inner circle, and his rancorous dispute with Huxley, which coincided with a crucial moment in Darwin's strategic disclosure of his long-harbored speculations, helped forge a sense of solidarity between those who, three years later, would emerge as the principal supporters of the controversial theories presented in the *Origin*.[53] Recent accounts of both Huxley and Hooker, for instance, have emphasized their considerable differences on issues like gentlemanly conduct, religious authority, and professionalization, which were often overlooked by earlier historians who conflated them into a homogenous young guard of self-conscious anticlerical professionalizers.[54] The Cuvierian contretemps, however, established a strong point of connection between the two men that, as has been seen, had them conspiratorially swapping suggestions for riling the unfortunate Falconer. They likely continued to plot when "Huxley and Mrs H staid with us," as Hooker told Darwin, at the very height of the controversy, with Huxley's unrepentant riposte to Falconer—to be discussed further shortly—avowing that he was "supported by every botanist with whom I have spoken" on the matter of whether morphological rather than physiological "laws of correlation guide the botanical palæontologist."[55] These unnamed botanists would certainly have included Hooker, the initial part of whose *Genera Plantarum* (1862), written in Latin with George Bentham, raised "many philosophical questions bearing upon physiological and morphological correlation," according to the *Natural History Review*.[56]

Although Hooker confessed that he had "only half understood the question before" having read the various contributions to the dispute, he subsequently maintained a virulent and long-standing antagonism to the very Cuvierian axioms that Huxley had alerted him to, writing to Darwin in 1869 exulting at his "tremendous upset to Owen's doctrines . . . the 'law of necessary correlation' is—nowhere."[57] Even as late as 1895, Hooker maintained that he had "only one criticism" of Walter Lawry Buller's *Illustrations of Darwinism*, "which is as to the necessary correlation" its author invoked to explain the association of the structure of the "huge forms of Palæozoic life with a tropical vegetation."[58] The acrimonious dispute over paleontological method in the opening months of 1856 helped clarify the strategic alliances that would be vital to the defense of Darwinism in the still more bitter controversies to come in the following years and decades. In fact, even though the main participants in what Darwin squeamishly designated "Falconer v. Huxley" were, at the time, equally opposed to species transmutation, and Darwin's own understanding of growth and variation could, in any case, be readily accommodated with aspects of Cuvierian method, an opposition to necessary correlation became enshrined as a key component of the broader agenda of scientific naturalism.

This was made particularly evident when, in October 1857, the most prominent public advocate of the development hypothesis, the philosopher Herbert Spencer, weighed into the "controversy now going on among zoologists . . . respecting the alleged necessary correlation subsisting among the several parts of any organism." With even less experience of actual paleontology than Huxley, Spencer loftily identified the "flaw[s] in Cuvier's principle" by "recourse to a mechanical analogy" more suited to his own theoretical methods, before concluding, "We agree with Professor Huxley. . . . Palæontology must depend upon the empirical method. Necessary correlation cannot be substantiated."[59] Huxley told Spencer that these statements in the anonymous *National Review* had "been ascribed to Huxley himself; 'and that by no less a person than by Dr. Hooker,'" prompting Spencer to boast to his father, "I have heard Huxley say that there are but four philosophical naturalists in England—Darwin, Busk, Hooker, and himself. Thus the article has been ascribed *by* one of the four *to* one of the four."[60] Two years ahead of the *Origin*'s publication, opposition to Cuvierian correlation afforded a vital rallying call that brought together the different constituencies of the scientific young guard then emerging in London.

These tactical alliances were only hardened by Huxley's prompt rejoinder to Falconer in the very next number, for July 1856, of the *Annals and Magazine of Natural History*. Far from backing down, Huxley simply reiterated his contention that there was "nothing more erroneous than the popular notion . . . that [the Cuvierian] method essentially consisted in reasoning from supposed physiological necessities."[61] Cuvier and his heirs had actually fallen into a logical trap in which "when the result to which a combination tends is obvious, we commonly imagine we can see the reason for that combination."[62] All suggestions that a "correlation is in any case to be called *necessary*" were thus mistaken flights of fancy, given free rein by an unfortunate overemphasis on function rather than structure. At times Huxley did appear to adopt a more conciliatory tone, acknowledging that the "brusque attack from Dr. Falconer" in the previous month's number "caused me at first, I must confess, no slight alarm . . . coming as it did from the pen of a palæontologist of high repute." This was only a tactical retreat, though, and he immediately resumed his assault, insisting that the

> perusal of Dr. Falconer's essay . . . soon relieved me from any real source of uneasiness, by demonstrating very clearly that Dr. Falconer had been far too much in a hurry either to master the real question in dispute, to read what I had written with attention, or to quote me with common accuracy and fairness.

Haste and prejudice had resulted in an "entire misconception of the point at issue," which meant that Huxley "left untouched many points in Dr. Falconer's essay, not because they cannot be answered, but because I conceive they will answer themselves."[63] It was not just Cuvier who was on the receiving end of Huxley's impudence in 1856. Falconer, the discoverer of the world's richest deposits of tertiary mammals in the Siwalik hills in India and a "palæontologist of high repute," was treated with the same laconic disdain by the young tyro still without any paleontological publications of his own.

The aspects of Falconer's article that Huxley did deign to comment on often related more to appropriate conduct than to paleontological method, and Huxley exploited his adversary's long isolation, both on account of his lengthy sojourn in India and because of his frail health and prickly temperament, from the wider scientific community. As Claudine Cohen has observed, Falconer was "a somewhat marginal figure in relation to the scientific establishment of his day," a point that Huxley, perhaps having been apprised of this by Hooker, emphasized with brutal relish.[64] The "singular bad taste" of many passages would, Huxley counseled, "cause Dr. Falconer, in his cooler moments, far more annoyance than they have occasioned to anyone else, except his friends."[65] One of those alleged friends may already have tipped Huxley off about "how ill Falconer behaved" during scientific disputes back in India, where he regularly "gave great offence at the Asiatic Society," for he now insisted, rather disingenuously given Darwin's and Hooker's travails over his own election to the Athenæum, that Falconer had not behaved in the manner expected of participants in metropolitan scientific society.[66]

Oddly, Lyell, who, like Darwin and Hooker, was friends with both Falconer and Huxley, congratulated the latter for "rendering subordinate the personal & controversial part" of their dispute. Lyell, who when he had initially "read Falconer . . . thought he had the best of the argument," now considered that the "tables are turned." He nevertheless signed off, "I shall ask Falconer what he can say to it," suggesting that he might still be won back to the Cuvierian cause.[67] Others were less tentative in their conversion, with Hooker exclaiming, "I have read Huxley's response to Falconer with eminent gusto—how admirably neatly and clearly he puts the whole question. . . . He has put forth his strength here & will I think startle old Falconer." Darwin concurred: "I am delighted at what you say about Huxley's answer & I agree most entirely; it is excellent & most clear."[68] Notwithstanding Huxley's staunch and sometimes scornful tone, there was now not even a hint of equivocation in Darwin's approbation for his new friend, whose adversarial manner was beginning to seem potentially very useful.

Not everyone agreed, of course, and James McCosh and George Dickie, a philosopher and a botanist who worked together at Queen's College in Belfast, afford an instructive case in point. In April 1856 they had sent Huxley a copy of their natural theological treatise *Typical Forms and Special Ends in Creation* with an accompanying letter expressing the hope that "it may be a means of bringing into more general notice certain important and admirable views researched[?] by yourself."[69] Despite their enthusiasm for Huxley's early studies of marine invertebrates, which, for them, revealed the underlying plan of the divine creation, in the book's second edition a year later McCosh and Dickie added a lengthy footnote observing that the famous "controversy" of 1830 between Cuvier and Étienne Geoffroy Saint-Hilaire "has not yet died out. Since the first edition of this work was published, we have this very discussion (in a somewhat confused form) in Huxley's Lecture at [the] Royal Institution[,] . . . Falconer's Examination of this lecture[,] . . . and Huxley's reply." In neither of his contributions, they insisted, had Huxley "succeeded in proving that palæontology, in restoring extinct forms, would be valid, even 'if we knew nothing of final causes or adaptations to purpose.'"[70] Darwin had initially had similar reservations, remarking, as was seen earlier, that "to deny all reasoning from adaptation & so called final causes, seems to me preposterous," though these had apparently been resolved by Huxley's reiteration of his argument in the *Annals and Magazine of Natural History*. That McCosh and Dickie were not similarly won over shows how the dispute over paleontological method in 1856 helped instantiate a clear breach between the nascent scientific naturalists and those, like the authors of *Typical Forms and Special Ends in Creation*, who still endeavored to reconcile the latest findings of modern science, including Huxley's own researches, with natural theology.

More Cuvierian Than Cuvier

Huxley had baited Falconer with the suggestion that "if the master's words be studied carefully, it will be discovered that his followers are more Cuvierian than Cuvier," but in reality the principal defender of necessary correlation in the summer of 1856 was himself not a particularly steadfast advocate of the French naturalist's putative methods.[71] Falconer's approach to paleontological reconstruction, honed in isolation in the foothills of the Himalayas, was as much self-taught as derived from the pages of Cuvier's writings. As Lyell had explained to the Geological Society when news of the Siwalik fossils first reached London in the late 1830s,

> When Captain [Proby] Cautley and Dr. Falconer first discovered these remarkable remains their curiosity was awakened . . . but they were not versed

in fossil osteology, and being stationed on the remote confines of our Indian possessions, they were far distant from any living authorities or books
on comparative anatomy to which they could refer. . . . From time to time
they earnestly requested that Cuvier's works on osteology might be sent out
to them, and expressed their disappointment when, from various accidents,
these volumes failed to arrive. The delay perhaps was fortunate, for being
thrown entirely upon their own resources, they soon found a museum of
comparative anatomy in the surrounding plains, hills, and jungles, where they
slew the wild tigers, buffalos, antelopes, and other Indian quadrupeds. . . .
They were compelled to see and think for themselves while comparing and
discriminating the different recent and fossil bones, and reasoning on the laws
of comparative osteology, till at length they were fully prepared to appreciate
the lessons which they were taught by the works of Cuvier.[72]

When this autodidactic Cuvierianism improvised in the field by Falconer and
Cautley was finally compared with Cuvier's long-awaited volumes, it actually
showed that the conclusions regarding fossil elephants that the metropolitan-
based master had developed in his Parisian museum were frequently erroneous. As they wrote in their *Fauna Antiqua Sivalensis* (1846), "Notwithstanding his array of authority, we cannot help thinking that Cuvier was premature
in his conclusion [that all fossil elephants comprised only a single species],
and that the identity of forms has rather been assumed against the evidence,
than proved by it." There were "not sufficient materials," Falconer and Cautley protested, "for making such a comparison when Cuvier wrote," and it
was merely the "great weight of Cuvier's authority [that] has given an undue
influence to his statement . . . which has biased the observations of some
later writers."[73] This was strikingly close to Huxley's own complaint, a decade
later in the *Annals and Magazine of Natural History*, that the supporters of
"Cuvier's law of correlation" were "guided more by authority than by right
reason."[74] Huxley presumably included Falconer among these slavish adherents to authority, but the latter's more maverick approach to paleontological
practice instead suggested that there were actually many similarities in their
opinions.

Even in his fierce rebuke to Huxley, Falconer had acknowledged that the
"principle of correlation . . . must not, in practice, be pushed too far in palæontology," conceding that there were "numerous instances on record, in which,
in attempting to determine extinct forms from a single bone or tooth . . .
very erroneous conclusions have been arrived at; among others, even by Cuvier himself."[75] Nor was Falconer's defense of Cuvier at all inflected with any
natural theological agenda; the laws of correlation were necessary only in the
sense that all other alternatives were inconceivable. The casual invocation of

the "Almighty" quoted earlier in Falconer's 1854 letter to his niece was likely only meant to suit the feelings of a pious young woman, and, in any case, elsewhere in the same letter he maintained that most modern scientific developments had initially been "denounced as a heresy opposed to the Bible" and thus "when . . . in a good cause, the imputation of *infidelity* is raised, one need not be ashamed of it."[76] In fact, while Huxley only identified with the Muslim "Prophet-Warrior" Schamyl in print, Falconer had adopted Islamic dress during a period in Afghanistan and "was taken for a pilgrim going to Mecca to worship at the shrine of Mahomet."[77]

The truth was that Falconer and Huxley had much more in common than either, in the heat of battle, had been willing to recognize. The parallels between them were certainly apparent to mutual friends, who were swiftly smoothing the ground for a rapprochement. In December 1856 Lyell alerted Huxley to some potentially elephantine teeth he had noticed at the Museum of Economic Geology and told him, "I should much like you without delay to send them (or still better to take them) to Dr. Falconer 31 Sackville St. which is quite in your neighbourhood," explaining that the visit could not be reciprocated because the "doctor will not allow him to go out of his room not even as far as Jermyn St."[78] It is not known how Huxley responded to Lyell's gentle peacemaking, although within a year or two he and the still frail Falconer had begun tentatively corresponding. By this time Falconer had come to regret the whole war of words, telling his niece, "I must keep my temper. . . . What a grumpy old uncle you have got. Like an infuriated Toro . . . having an encounter with Huxley."[79] More significantly, the self-appointed defender of Cuvierian correlation was himself, in a remarkable volte-face, increasingly converted to his adversary's position.

By the early 1860s the "infuriated Toro" had moved on from Huxley and was now "goring Owen, and his jackall [*sic*]."[80] This was a rueful reference to the disputed identification of a mammalian jawbone from the Purbeck beds in southern England, which Falconer assigned to an order of herbivorous rodent marsupialia and Owen to a predatory, carnivorous marsupial. Owen's ostensibly authoritative classification of the so-called *Plagiaulax*, Falconer contended, was merely "an opinion, professing to be founded on the high ground of a connected series of physiological correlations." Both men acknowledged that "comparative anatomy supplies for our guidance fundamental principles," but there was a clear discrepancy in how they applied them. This "conflict of opinion," as Falconer observed, arose "from different methods having been followed by the observers in dealing with the evidence." Unlike Owen's stringently Cuvierian approach, Falconer examined the creature's curious dentition thus:

Why there should be this plurality of incisors above, and only two invariably occupying the same position below, is wholly unknown to us; but the constancy of the structure makes it certain that there must be a sufficient cause for it in nature; and we employ the generalization, empirically arrived at, with as much confidence as we do the law of necessary correlation. In many critical cases, where the evidence is limited or defective, the empirical is even a safer guide than the rational law, since it is freer from the risk of errors of interpretation.[81]

This was a complete reversal of what he had argued in the *Annals and Magazine of Natural History* in June 1856, where it had been claimed that it was the "constant effort of every philosophical mind . . . to extinguish the empirical character of the phænomena, and bring them within the range of a rational explanation." Falconer had even mocked Huxley's argument that empirical procedures were more reliable, exclaiming that it was "a rare spectacle to see empiricism chosen by preference" to the rational axioms adumbrated by Cuvier.[82] Now in 1862, it would seem, Falconer had come round to his erstwhile opponent's opinion that simple empirical observations that certain structural phenomena invariably occur together were just as trustworthy as the doctrinaire law of necessary correlation, and, in certain difficult cases, were perhaps even more dependable.

Owen himself certainly considered that this "polemical paper of Dr. Falconer" marked a departure from the "principle which Cuvier laid down as our guide in such dark routes in Palæontology," insisting that "loyalty to our common science compels me to say that he fell into his mistake as to *Plagiaulax* by neglecting its fundamental principle . . . Cuvier's . . . rule . . . of correlation."[83] Falconer was now regarded, by friends and foes alike, as categorically on the side of the anti-Cuvierian cause, with Darwin telling him, "I have not been for a long time more interested with a paper than with yours. It gives me a demoniacal chuckle to think of Owen's pleasant countenance when he reads it."[84] When, two years afterward, Huxley sent Falconer a copy of his *Lectures on the Elements of Comparative Anatomy* (1864), he only had time to "cast a hurried glance over the Classification chapter." He was nevertheless effusive in his praise of the characteristic rhetorical force exhibited in this opening section, telling Huxley, "You write with a sledgehammer and tick the points with a mace."[85] That Huxley had, in the chapter "On the Classification of Animals," opined that deductions based on the "law of correlation" were "well calculated to impress the vulgar imagination" and then derided "Cuvier, the more servile of whose imitators are fond of citing his mistaken doctrines as to the methods of palæontology against the conclusions of logic and common sense," did not appear to trouble Falconer.[86]

Eight years after their acrimonious dispute in the summer of 1856, Huxley's continuing onslaught against Cuvierian correlation now elicited Falconer's admiration rather than his indignation.

Falconer's dramatic conversion to Huxley's empiricism in the early 1860s occurred alongside his increasing acceptance of Darwin's evolutionism. As Hooker joked to Darwin, "He seems to me to have just awakened to the fact that there is something in you."[87] The article in which Falconer finally endorsed Darwinian evolution, albeit with some reservations over the adequacy of natural selection, appeared in the January 1863 number of the *Natural History Review*, whose editor in chief was his former antagonist Huxley. That Huxley advised Darwin to look in the "last N.H. Review" to see "a grand paper by . . . good old Falconer" suggests just how far the latter had become accepted as part of the circle of scientific naturalists.[88] This could only happen once Falconer was simultaneously denuded of his doubts over the mutability of species and, no less important, his lingering allegiances to Cuvierian correlation. The closer "good old Falconer" became to Huxley, Darwin, and their inner circle—with even Hooker growing to once more respect him "not only personally, but as a scientific man of unflinching & uncompromising integrity"—the more he renounced his erstwhile Cuvierianism.[89] Once again, and even long after the rancor of the initial skirmishes, the controversy over paleontological method was a crucial barometer of support for a range of broader issues centered around Darwin's evolutionism. Although the extent of his apostasy was particularly notable, Falconer was by no means the only naturalist to forsake an earlier adherence to Cuvier's correlative procedures in the wake of Huxley's attack. Before the end of 1856, however, Huxley's criticisms of Cuvier had attracted the attention of a far more formidable adversary.

Our British Cuvier in His True Place

Although he was well aware of Owen's fierce loyalty to Cuvierian correlation, Huxley's original Royal Institution lecture had nevertheless invoked "Professor Owen's determination of the nature of the famous Stonesfield mammal" as a "striking illustration" of how the "whole process of palæontological restoration depends" on nothing more than a knowledge of the "invariable coincidence of certain organic peculiarities."[90] Falconer saw it as another of the lecture's many egregious errors that "Mr. Huxley holds him up . . . as furnishing a bright example . . . of empirical deduction," adding a slyly flattering footnote observing, "Mr. Owen flies his hawk at a much more ambitious quarry in original research; but is it too much to expect that he may on some

occasion record his protest against Mammalian Palæontology being asserted
to rest merely on empirical correlation, in a pithy foot-note."[91] Falconer continued to curry favor with Owen for the rest of 1856, offering to furnish him
with a "specimen of the *Mastodon longirostris*" from Germany, and in turn
being invited by his fellow champion of the Cuvierian true faith to "come
and see the skeleton of the *Din[ornis] elephantopus*, which I have just set up"
at the British Museum.[92] A curt aside, however, was not the form in which
Owen chose to make his inevitable response to Huxley's provocations, and
instead he bided his time through the long summer recess.

It was only once the metropolitan scientific societies began to reconvene
in the autumn that Owen prepared for his long-awaited intervention in the
controversy, with the news that he would soon enter the fray setting the scientific community abuzz with gossip. At the beginning of November, Huxley
excitedly informed Frederic Dyster, a Welsh doctor with whom he had spent
the summer, "There is going to be a set-to at the Geological [Society] on
Wednesday—the great O. versus the Jermyn St. Pet on the Methods of Palaeontology" (the latter soubriquet an ironic reference to Huxley's ensconced
position at the Government School of Mines on Jermyn Street in Piccadilly),
and he promised him, "You shall know the results."[93] But Dyster received no
more of his friend's customary mordant gossip on the matter, and instead it
was others who swapped hearsay about Huxley's own derisory performance in
the much-anticipated "set-to." The news had certainly reached Hooker, who
told Darwin, "Owen I hear committed a cutting telling & flaying alive assault
on Huxley[']s adaptation views at the Geolog. Soc. & read it with the cool
deliberation & emphasis & pointed tone & look of an implacable foe.—H. I
fear did not defend himself well (though with temper)." Hooker noted that
these continuous "embroglios are very bad indeed & must insensibly have a
bad effect upon Huxley."[94] His pugnacious bravado notwithstanding, Huxley
had in fact been no less unwell than Falconer for much of 1856, and only two
days before the Geological Society's ordinary meeting, he reported to Dyster
that he had "contrived to sit in a draught or something or other so that I have
an attack of rheumatism at this moment."[95] Huxley's indisposition allowed
Owen's Cuvierian rear guard to snatch back the initiative.

On 5 November, Owen brandished before the fellows of the Geological
Society a small mammalian jawbone recently discovered in the oolitic slate of
Oxfordshire. As central London engaged in the traditionally exuberant and
incendiary celebrations of Guy Fawkes night, Owen ignited his own paleontological pyrotechnics. Defying the convention that "papers read to the Ordinary Meetings of the Society were . . . primarily descriptive and fact-based,
and it was the discussion that followed that set them within whatever theo-

retical debate or dispute was current at the time," he made recent methodological controversy the very focus of his talk.[96] The fragmentary fossil, Owen
observed, "excites . . . interest" because it offered an important "test . . . of the
actual value of a single tooth in the determination of the rest of the organization of an animal," as well as affording an opportunity, particularly needful
"in the present state of Palæontology," to "analyse the mental processes by
which one aims at the restoration of an unknown Mammal from a fragment
of jaw."[97] Having expertly "conjectured the nearer affinities of the *Stereognathus*," Owen then asked his audience,

> Can this example . . . be justly cited as showing that there is no physiologi-
> cal, comprehensible, or rational law, as a guide in the determination of fossil
> remains: but that all such determinations rest upon the application of ob-
> served coincidences of structure, for which coincidences no reason can be
> rendered?

This, as Owen well knew, was precisely how Huxley had interpreted the
very same example at the Royal Institution nine months earlier, and, with
the exhausted tyro now sitting across from him on the Geological Society's
parliamentary-style benches, he finally responded with emphatic plainness,
"I do not believe this to be the case." If the method of determining the structure and habits of the *Stereognathus* had simply been an "empirical one,"
as Huxley proposed, then the "narrowness of its support from observation"
could never "leave the mind free from a sense of the possibility of its being
liable to be proved an erroneous conclusion."[98] The "higher law" of necessary correlation, on the other hand, afforded an assurance and certainty that
Huxley's indefinite empiricism could never aspire to.[99]

Still worse, the "small and unfruitful minority" of naturalists who persisted
in maintaining the "inapplicability of the law of correlation" were, Owen
claimed, closet adherents of the earthly, materialist "tenets of the Democratic
and Lucretian schools" formulated in ancient Greece and Rome. Their "insinuation and masked advocacy of the doctrine subversive of a recognition
of the Higher Mind" was the result of "some, perhaps congenital, defect of
mind, allied or analogous to 'colour-blindness,'" that rendered them devoid
of humanity's usual "instinctive, irresistible impression of a design or purpose" in nature.[100] Huxley was not named directly in this hyperbolic diatribe,
or indeed in the paper as a whole. Owen's auditors, though, would have been
fully aware that it was his criticisms of Cuvier that were now depicted as not
only erroneous, but actually immoral and unhealthy (an accusation likely
made more vexing by Huxley's current poor state of health).

Spencer defended his friend in the *National Review*, remarking on the

"questionable propriety" of the way in which "Professor Owen avails himself
of the *odium theologicum* . . . in his defence of the Cuvierian doctrine."[101]
What would have been still more questionable for Spencer and his fellow
scientific naturalists was that, in direct contrast to Falconer four months earlier, Owen specifically endorsed, in a zealously homiletic idiom, the natural
theological implications of necessary correlation. While Cuvier himself had
rarely mentioned the role of God in nature, across the Channel his paleontological method was rapidly amalgamated with a rationalist Paleyite argument concerning the evidence of design in the natural world, especially by
William Buckland and his supporters at Oxford. It has long been recognized
that Owen's own anatomical researches had, since at least the 1840s, diverged
sharply from those of Cuvier in prioritizing form over function and tracing the osteological homologies that revealed an underlying skeletal arche-
type.[102] More recently, Nicolaas A. Rupke has shown how Owen nevertheless
simultaneously maintained his erstwhile Cuvierian functionalism in order to
retain the support of powerful institutional patrons like Buckland.[103] It was
certainly Buckland's distinctively anglicized version of Cuvierian correlation
that Owen was upholding at the Geological Society in November 1856, and
it may actually have been Huxley's strategic intention to provoke Owen into
once more publicly asserting the Cuvierian credentials that, privately, he was
increasingly rescinding.

Despite the success of this putative strategy, however, Huxley appeared
discomfited and unable to muster his usual acerbic wit in the ensuing discussion of Owen's paper. Soon after, he even expressed an uncharacteristic desire
to "set an example of abstinence from petty personal controversies."[104] By the
end of 1856 the campaign against Cuvierian correlation appeared to have run
into the doldrums. Owen pressed home his advantage at the beginning of the
following year, agreeing to give a course of guest lectures on prehistoric mammals at the Government School of Mines, where Huxley had been employed
for the last two years. Ahead of the initial lecture, Owen announced his position in John Churchill's annual *Medical Directory* as "Professor of Comparative Anatomy and Palaeontology, Government School of Mines, Jermyn-St."
This was precisely the title that Huxley, the self-proclaimed "Jermyn St. Pet,"
considered his own, and he complained to the directory's publisher that his
rival's spurious claim was "calculated to do me injury."[105]

This infamous spat between Owen and Huxley has been frequently examined by historians, who have generally considered it to mark the formal ter-
mination of any lingering vestige of polite relations between the two men.[106]
The quarrel was therefore a hugely significant moment in the enduring personal dispute that, with Owen and Huxley emerging after 1859 as the *Origin's*

FIGURE 1.2. Richard Owen lecturing on fossil mammals at the Museum of Practical Geology in spring 1857. "Professor Owen's Lectures," *Illustrated Times* 4 (1857): 252. Author's collection.

most high-profile critic and defender, would soon have such a determining impact on the reception of Darwinism. What has never previously been recognized, though, is that right from the beginning of this course of guest lectures, Owen also continued his defense of Cuvier's axiom of correlation, and this time in Huxley's own institutional stronghold. And just as with their anticlimactic "set-to" at the Geological Society, Huxley again failed to live up to his bellicose imprecations to Dyster, this time proclaiming that Owen had "reckoned without his host—and will have to eat a large leek," a threat that drew self-consciously on the martial rhetoric of Shakespeare's *Henry V*.[107]

In the initial lecture, on 26 February 1857, Owen evidently spent some of the time considering the recent criticisms of correlation, for on the following day Roderick Murchison told a correspondent:

> I never heard so thoroughly eloquent a lecture as that of yesterday. . . . It is the first time I have had the pleasure of seeing our British Cuvier in his true place, and not the less delighted to listen to his fervid and convincing defence of the principle laid down by his great precursor. Every one was charmed, and he will have done more (as I felt convinced) to render our institution favourably known than by any other possible event.[108]

This introduction to the wide-ranging, twelve-lecture course must have contained a great deal besides, but for Murchison it was Owen's fervid defense of his esteemed precursor's "principle" that was its conspicuous highlight. There would, of course, have been at least one noticeable exception to Murchison's blithe assumption that everyone was "charmed" by Owen's "defence," and Huxley would have been still less happy had he realized that the director of his own institution was directly linking the growth of its public renown to his archantagonist's ardent vindication of Cuvierian correlation.

The course of twelve lectures, in which Owen continued to defend Cuvier, concluded at the beginning of April with a "grand party" organized by Murchison and attended by many of the older generation of gentlemen of science such as the astronomer Edward Sabine. As he accepted the congratulations of the illustrious guests at Murchison's glittering end-of-term soiree, Owen would certainly have considered that he had decisively repulsed the previous year's attacks on the central axiom of Cuvierian paleontology, even taking the fight into the very lecture theater and museum that the brash antagonist of correlation considered his own territory. The self-styled "Jermyn St. Pet" had been neutered in his own backyard.

Owen, however, had merely won the battle, but decisively lost the war. The very arguments in favor of empirical deduction rather than necessary correlation that he presumed had now been expunged had already gained the support, albeit expressed only in private letters, of Hooker, Darwin, and perhaps Lyell, and, as has been seen, would soon convince even Falconer. Spencer too would tie his colors to the mast of empiricism on "this vexed question in physiology" in an article for the *National Review* in October 1857.[109] The Cuvierian controversy initiated by Huxley at the beginning of the previous year had brought together the leading lights of what, by 1858, he was proprietorially calling "my scientific young England," providing an empirical and secular approach to organic structure to which they could consent alongside the equally naturalistic mode of species transmutation that Darwin began privately revealing to them at precisely this same moment.[110] In the principal upholder of correlative paleontological methods, moreover, they found an opponent who, at least in Huxley's carefully constructed version of this "servile follower," embodied the corrupt old world of scientific patronage and gentlemanly deference, and whom they would go on to challenge on several other fronts, from simian brain anatomy to museum politics.[111]

Notes

1. See "Biographical Sketch," in *Palæontological Memoirs and Notes of the Late Hugh Falconer*, ed. Charles Murchison, 2 vols. (London: Robert Hardwicke, 1868), 1:xli.

2. [James Dixon], "The War with Russia," *London Quarterly Review* 4 (1855): 240; and [Henry Reeve], "The Results of the Campaign," *Edinburgh Review* 102 (1855): 575.

3. Hugh Falconer, "On Prof. Huxley's Attempted Refutation of Cuvier's Laws of Correlation," *Annals and Magazine of Natural History*, n.s., 17 (1856): 477–78.

4. Ibid., 476; and Hugh Falconer, "Note on a Correction of Published Statements Respecting Fossil Quadrumana," in *Palæontological Memoirs*, 1:314.

5. See Stefanie Markovits, *The Crimean War in the British Imagination* (Cambridge: Cambridge University Press, 2009), 12–62.

6. Thomas Henry Huxley, "On Natural History, as Knowledge, Discipline, and Power," *Notices of the Proceedings of the Royal Institution* 2 (1854–58): 192; and *Times*, 23 February 1856, 10.

7. Falconer, "Huxley's Attempted Refutation," 476.

8. [T. H. Huxley], "Contemporary Literature: Science," *Westminster Review*, n.s., 7 (1855): 562–63.

9. [T. H. Huxley], "Schamyl, the Prophet-Warrior of the Caucasus," *Westminster Review*, n.s., 5 (1854): 517, 491–92.

10. Quoted in Adrian Desmond, *Huxley: The Devil's Disciple* (London: Allen Lane, 1994), 196.

11. [William Howard Russell], "The Fall of Sebastopol," *Times*, 11 September 1855, 6.

12. Falconer, "Huxley's Attempted Refutation," 477.

13. [T. H. Huxley], "Schamyl," 511.

14. Adrian Desmond, *Huxley: Evolution's High Priest* (London: Allen Lane, 1997), 252.

15. James J. Reid, *Crisis of the Ottoman Empire: Prelude to Collapse, 1839–1878* (Stuttgart: Steiner, 2000), 255.

16. Samuel Phillips, *Guide to the Crystal Palace and Park*, 5th ed. (London: Bradbury and Evans, 1855), 45.

17. "Twenty-Second Ordinary Meeting," *Journal of the Society of Arts* 2 (1854): 447.

18. On the dispatches to the War Department, see *Times*, 16 February 1856, 10.

19. T. H. Huxley, "On Natural History," 189–90.

20. Ibid., 190–91.

21. Leonard Huxley, *Life and Letters of Thomas Henry Huxley*, 2 vols. (London: Macmillan, 1900), 1:161.

22. Adrian Desmond gives a brief but effective account of the dispute's social and ideological contexts in *Archetypes and Ancestors: Palaeontology in Victorian London, 1850–1875* (Chicago: University of Chicago Press, 1982), 57–58.

23. Frederick H. Burkhardt et al., eds., *The Correspondence of Charles Darwin*, 20 vols. (Cambridge: Cambridge University Press, 1985–), 6:106.

24. Ibid., 6:112.

25. Ibid., 4:172. Darwin and Hooker first met Huxley, respectively, only in 1853 and 1851.

26. Ibid., 3:272, 3:300, 3:302.

27. Ibid., 4:171, 4:329.

28. Ibid., 6:72.

29. Ibid., 7:496.

30. Hugh Falconer, *Descriptive Catalogue of the Fossil Remains of Vertebrata . . . in the Museum of the Asiatic Society of Bengal* (Calcutta: Baptist Mission Press, 1859), 7, 2, 4.

31. Grace Prestwich, *Essays: Descriptive and Biographical* (Edinburgh: William Blackwood, 1901), 6.

32. Falconer, "Huxley's Attempted Refutation," 476, 485, 483, 486.

33. Ibid., 488, 483, 487.

34. Ibid., 492.

35. Charles Collins, "Spiritual Despotism," *Methodist Quarterly Review* 39 (1857): 38.

36. Burkhardt et al., *Correspondence*, 6:176.

37. Ibid., 6:147.

38. Leonard Huxley, *Life and Letters of Sir Joseph Dalton Hooker*, 2 vols. (London: John Murray, 1918), 1:427.

39. "On Systems and Methods in Natural History," *Quarterly Review* 41 (1829): 313.

40. L. Huxley, *Hooker*, 1:427.

41. Burkhardt et al., *Correspondence*, 6:176, 6:178.

42. Charles Darwin, "Geology," in *A Manual of Scientific Enquiry*, ed. John Herschel (London: John Murray, 1849), 168.

43. Richard Owen, "Fossil Mammalia," in *Zoology of the Voyage of HMS Beagle*, ed. Charles Darwin, 5 vols. (London: Smith, Elder, 1839–43), 1:64.

44. Burkhardt et al., *Correspondence*, 1:368. See Stan P. Rachootin, "Owen and Darwin Reading a Fossil: *Macrauchenia* in a Boney Light," in *The Darwinian Heritage*, ed. David Kohn, 155–83 (Princeton, NJ: Princeton University Press, 1985).

45. See Stephen Jay Gould, *The Structure of Evolutionary Theory* (Cambridge, MA: Belknap Press, 2002), 333–39.

46. Paul H. Barrett et al., eds., *Charles Darwin's Notebooks, 1836–1844* (Ithaca, NY: Cornell University Press, 1987), 410; and Charles Darwin, *On the Origin of Species* (London: John Murray, 1859), 5.

47. Timothy Shanahan, *The Evolution of Darwinism: Selection, Adaptation, and Progress in Evolutionary Biology* (Cambridge: Cambridge University Press, 2004), 111.

48. Leonard G. Wilson, ed., *Sir Charles Lyell's Scientific Journals on the Species Question* (New Haven, CT: Yale University Press, 1970), 226.

49. Burkhardt et al., *Correspondence*, 7:161, 7:368; and Hugh Falconer, "On the American Fossil Elephant," *Natural History Review*, n.s., 3 (1863): 77.

50. Falconer, "American Fossil Elephant," 77.

51. Burkhardt et al., *Correspondence*, 6:100.

52. Ibid., 5:133, 5:213; and Thomas H. Huxley, "Lectures on General Natural History," *Medical Times and Gazette* 12 (1856): 483.

53. Burkhardt et al., *Correspondence*, 7:246. It was John Lubbock who made the "splendid joke about Falconer."

54. See Paul White, *Thomas Huxley: Making the "Man of Science"* (Cambridge: Cambridge University Press, 2003), 109–10; and Jim Endersby, *Imperial Nature: Joseph Hooker and the Practices of Victorian Science* (Chicago: University of Chicago Press, 2008), 332. See also Endersby's chapter in this volume.

55. Burkhardt et al., *Correspondence*, 6:176; and T. H. Huxley, "On the Method of Palæontology," *Annals and Magazine of Natural History*, n.s., 18 (1856): 51.

56. "Bentham and Hooker's *Genera Plantarum*," *Natural History Review*, n.s., 3 (1863): 34.

57. Burkhardt et al., *Correspondence*, 6:176, 17:26.

58. L. Huxley, *Hooker*, 2:322.

59. [Herbert Spencer], "The Ultimate Laws of Physiology," *National Review* 10 (1857): 345, 351, 350, 348.

60. David Duncan, *The Life and Letters of Herbert Spencer* (London: Williams & Norgate, 1911), 83.

61. T. H. Huxley, "Method of Palæontology," 50.

62. Ibid., 43.

63. Ibid., 50, 54.

64. Claudine Cohen, *The Fate of the Mammoth: Fossils, Myth, and History*, trans. William Rodarmor (Chicago: University of Chicago Press, 2002), 132.

65. T. H. Huxley, "Method of Palæontology," 50, 54.

66. Burkhardt et al., *Correspondence*, 4:172, 4:267.

67. C. Lyell to T. H. Huxley, 18 July 1856, Huxley Papers, 6.11, Imperial College of Science, Technology, and Medicine Archives, London. This and subsequent letters are quoted by permission of the Archives, Imperial College London.

68. Burkhardt et al., *Correspondence*, 6:175, 6:176, 6:178.

69. G. Dickie to T. H. Huxley, 16 April 1856, Huxley Papers, 13.144.

70. James McCosh and George Dickie, *Typical Forms and Special Ends in Creation*, 2nd ed. (Edinburgh: Thomas Constable, 1857), 445n.

71. T. H. Huxley, "Method of Palæontology," 48.

72. Charles Lyell, "Anniversary Address Delivered by the President" [1837], *Proceedings of the Geological Society* 2 (1833–38): 508–9.

73. Hugh Falconer and Proby T. Cautley, *Fauna Antiqua Sivalensis* (London: Smith, Elder, 1846), 13, 20.

74. T. H. Huxley, "Method of Palæontology," 44.

75. Falconer, "Huxley's Attempted Refutation," 493.

76. Prestwich, *Essays*, 6.

77. See ibid., 2.

78. C. Lyell to T. H. Huxley, 13 December 1856, Huxley Papers, 6.12.

79. Patrick J. Boylan, *The Falconer Papers, Forres* (Leicester: Leicestershire Records Service, 1977), 13.

80. Ibid.

81. Hugh Falconer, "On the Disputed Affinity of the Mammalian Genus *Plagiaulax*," *Quarterly Journal of the Geological Society* 18 (1862): 350, 351, 352.

82. Falconer, "Huxley's Attempted Refutation," 492.

83. Richard Owen, *Monograph on the Fossil Mammalia of the Mesozoic Formations* (London: Palæontographical Society, 1871), 104, 111, 108, 96.

84. Burkhardt et al., *Correspondence*, 10:524.

85. H. Falconer to T. H. Huxley, 8 April [1864], Huxley Papers, 16.1.

86. Thomas Henry Huxley, *Lectures on the Elements of Comparative Anatomy* (London: John Churchill, 1864), 4, 5.

87. Burkhardt et al., *Correspondence*, 11:14.

88. Ibid., 11:29.

89. Ibid., 13:48.

90. T. H. Huxley, "On Natural History," 194.

91. Falconer, "Huxley's Attempted Refutation," 488n.

92. Richard S. Owen, *The Life of Richard Owen*, 2 vols. (London: John Murray, 1894), 2:21.

93. T. H. Huxley to F. Dyster, 3 November 1856, Huxley Papers, 15.78.

94. Burkhardt et al., *Correspondence*, 6:260.

95. T. H. Huxley to Dyster, 3 November 1856, Huxley Papers, 15.78.

96. John C. Thackray, *To See the Fellows Fight: Eyewitness Accounts of the Meetings of the Geological Society of London and Its Club, 1822–1868* (Stanford in the Vale: British Society for the History of Science, 1999), vii.

97. Richard Owen, "On the Affinities of the *Stereognathus Ooliticus* (Charlesworth), a Mammal from the Oolitic Slate of Stonesfield," *Quarterly Journal of the Geological Society of London* 13 (1857): 4–5.

98. Ibid., 6.

99. Ibid., 5.

100. Ibid., 8, 9.

101. [Spencer], "Ultimate Laws," 347.

102. See Evelleen Richards, "A Question of Property Rights: Richard Owen's Evolutionism Reassessed," *British Journal for the History of Science* 20 (1987): 129–71; and Adrian Desmond, *The Politics of Evolution: Morphology, Medicine, and Reform in Radical London* (Chicago: University of Chicago Press, 1989), 341–45.

103. Nicolaas A. Rupke, *Richard Owen: Victorian Naturalist* (New Haven, CT: Yale University Press, 1994), 117–23.

104. L. Huxley, *Life and Letters of Thomas Henry Huxley*, 1:151.

105. T. H. Huxley to J. Churchill, 22 January 1857, Huxley Papers, 12.194.

106. See Desmond, *Huxley: Devil's Disciple*, 230; and White, *Huxley*, 52.

107. T. H. Huxley to F. Dyster, December 1856, Huxley Papers, 15.80.

108. Owen, *Life of Richard Owen*, 2:61.

109. [Spencer], "Ultimate Laws," 352.

110. L. Huxley, *Life and Letters of Thomas Henry Huxley*, 1:161.

111. Quoted in White, *Huxley*, 52.

2

Evolutionary Naturalism on High: The Victorians Sequester the Alps

MICHAEL S. REIDY

> The world is not all radiant and harmonious; it is often savage
> and chaotic. . . . The nature-worshipers are blind and deaf to the
> waste and the shrieks which meet the seekers after truth.
> —"COSMIC EMOTION," New York Times, 28 August 1881

Introduction

Mont Blanc, 12 August 1857. A professional Swiss guide was leading three British mountaineers amid massive boulders under the shadow of the Aiguilles du Midi. As they slowly made their way through a confusion of crags, they ascended to the Glacier Pelrins, where the famous "Humboldt of the mountains," Horace-Bénédic de Saussure, had nearly lost his life in 1787. Two more hours and they stood together on the Glacier du Bossons, where they strapped woolen leggings up to their knees and tied ropes around their waists. One particularly tricky snow bridge, the midsection of which had melted away, required a precarious leap over an immense chasm. John Tyndall made the jump first, followed by his lifelong friend Thomas Archer Hirst, but Thomas Henry Huxley stood frozen at the edge, overwhelmed with fear. The year before he had made a false step on the Grimsel Pass, which according to Tyndall had nearly cost Huxley his life. His foot had slipped and he had fallen some way before coming to an abrupt stop with his legs overhanging a deep abyss.[1] That memory came foremost to Huxley's mind, now mixed with fatigue and uncertainty, as he tentatively made the jump. Once safely across, he emptied his flask of brandy in one large gulp. "Tyndall," he exclaimed, "I am quite exhausted. . . . I have determined not to attempt the ascent. I thought as I sat beside that crevasse that it was hardly fair to those at home to incur such peril."[2]

With steep steps cut into the ice by their guide, all three made it safely to the Grand Mulets, and then farther up to a primitive "cabane" made of boards stretched between upright posts. Huxley lay in the corner, motionless, while the guide discussed with Tyndall the next day's summit attempt.

FIGURE 2.1. An exposed snow bridge on the approaches to Mont Blanc, near the Grand Mulets. From George Baxter, *Leaving the Grand Mulets* (1855), N. 2, The Ascent of Mont Blanc, Lewis 336a, Harvard Art Museums / Fogg Museum, Gift of Charles H. Taylor, M4486.

All three couldn't make it, the guide whispered, especially Huxley. "What does that fellow say?" Huxley demanded, an absent expression on his face. "Does he mean that I am not to go to the top? I will go or else throw myself into a crevasse."[3] Tyndall realized immediately that Huxley was asleep, and he would have no recollection whatsoever of his rather bizarre outburst.

Beginning early the next morning, Hirst and Tyndall completed the ascent, the first time either had stood on the summit of Europe's highest peak. It was an arduous climb, across crevassed glaciers, rocky precipices, and seemingly endless snowfield. It took them much longer than they had planned, and they did not return to the makeshift hut until well after dark. Hirst was so exhausted that he could barely move, while Tyndall looked "altogether shabby," with both arms of his coat torn and one toe kicked out of his boots. But it was Huxley who seemed worse for all his worrying. He had spent over seventeen hours "a prisoner" in the hut, and his relief upon hearing their return was palpable. "To the end of my life, Tyndall," he said, exasperated, "I can never forget the sound of your batons when you reached the rock."[4]

I begin with this vignette because, in and of itself, it is quite striking: three future members of the X Club clambering together on the approaches to Mont Blanc—three young researchers who would all become recognized authorities in their respective fields. Imagine the conversations they must have had during their days trekking through the mountains. Even more striking is that this camaraderie in the high mountains was far from an unusual occurrence. Most members of the X Club were mountaineers.[5]

Though overlooked today, Hirst was well connected in Victorian scientific circles.[6] He first met Tyndall in June 1846 while both worked as railway surveyors. He then followed Tyndall first to Marburg, then to Queenwood, replacing Edward Frankland in 1853 as a teacher of geometry and surveying until 1856. After a few years of wandering through Europe after the tragic death of his wife, he returned to London in 1860, where he embarked on a successful career as an educator, administrator, and acute mathematician. He accepted the chair of mathematical physics at University College, London, in 1865. He was Tyndall's first climbing partner, a "fellowship of the rope" that matured into a lifelong friendship. He always viewed Tyndall as his mentor, "my angel of mercy, my guiding star," he wrote.[7] After accompanying Tyndall to the top of Europe's highest peak in August 1857, he returned to the Alps several times in the early 1860s to help his mentor take measurements of glacier motion. Six feet two inches tall, he had a perfect climber's build, able to match, step by step, Tyndall's alpinist gait. They shared a different type of fellowship as well. "I belonged to the pagan persuasion," he admitted to Francis Galton later in life, a "bias toward freedom of thought in religious matters."[8]

Joseph Dalton Hooker had made two long treks into Sikkim and Nepal in 1849 and 1850.[9] On his second journey, a six-month hike into western Nepal, he climbed above the Donkai Pass to well beyond nineteen thousand feet before slipping into the forbidden land of Tibet. Upon his return to Sikkim, his climbing partner Archibald Campbell was arrested and beaten, and both were detained for almost two months. Whenever and wherever he climbed, he gathered plant specimens at different elevations, particularly species of rhododendrons, to document their variations in changing environments. It is not often stressed that Charles Darwin dedicated an entire chapter of *On the Origin of Species* (1859) exclusively to mountain biogeography, material that he had gathered largely from his confidant Hooker in the Himalayas.

Edward Frankland, one of the most highly respected chemists in the Victorian era, spent twenty-two hours on the top of Mont Blanc in August 1859, accompanying Tyndall on his third ascent of the mountain. They were the first two mountaineers to spend the night atop Europe's highest peak. They performed experiments of all kinds as they ascended, including the burning of candles at different elevations. Over the next decade, Frankland continued this research in his laboratory, demonstrating that varying atmospheric pressures affect luminosity.[10] In collaboration with Norman Lockyer, he applied his mountaineering observations to advance our understanding of the gases composing the sun. He also kept climbing, using other mountain ascents to study the relation of bodily energy to food consumption. As his biographer has noted, this research "marked the first recognition of the 'calorific value' of food."[11]

The mathematician William Spottiswoode traveled widely throughout Europe, including to the mountainous regions of eastern Russia, Croatia, and Hungary. He used his vast experiences to publish an important paper entitled "On Typical Mountain Ranges, an Application of the Calculus of Probabilities to Physical Geography," an attempt to find the causes and orientation of the formation of mountain ranges. Francis Galton, another admirer of the Swiss Alps, cited this paper as his first inspiration in applying statistics to the social sciences.[12]

Herbert Spencer went on his first Swiss tour in the summer of 1853. It included trips up the Rhone and Gorner Glaciers, over the Grimsel Pass, and to the top of several snow-clad peaks. He penciled a letter to his brother from the top of the Faulhorn, describing with ecstasy his position "surrounded by a vast panorama of mountains . . . hourly sending down avalanches which sent a peal of thunder across the valley."[13] He also rambled throughout the Scottish Highlands. In 1871, in fellowship with both Tyndall and Hirst, he climbed Glen Nevis "up to its top, where it becomes Swiss like in character."[14] He returned to the Swiss Alps for a second time the following summer in the company of fellow X Clubber George Busk. Together, they explored the glaciers at the base of the Jungfrau.

The point is this: a disproportionately large number of X Clubbers went to the mountains. If you include the most prominent evolutionary naturalists, the list grows longer, though the high proportions stay the same.[15] All were not lifelong mountaineers. Some were just ramblers, what we would call hikers, and what Leslie Stephen disparagingly referred to as tourists. Two of them, however, towered over the sport in the mid-nineteenth century. They were simultaneously the most vocal of the evolutionary naturalists and the two most accomplished climbers of their age—John Tyndall and Leslie Stephen. We can catch a glimpse of the relationship between evolutionary naturalism and high Alpine terrain in the life and works (both intellectual and physical) of these two intrepid mountaineers. Both strove to be the first on the summit of Alpine peaks, and in their published accounts, they focused almost exclusively on the awe-inspiring beauty and exhilarating adventure they experienced in the high mountains. Their journals and personal correspondence, however, tell a different story. They were searching for something far more ephemeral than first summits. Something drew them to the mountains. The point of this chapter is to figure out what.

Tyndall, Stephen, and the Allure of the Alps

Tyndall first went to the Alps in 1849 with Hirst on a vacation from his studies in Marburg, but he did not begin climbing until the mid-1850s. He spent his first year on a glacier in 1856 accompanied by both Huxley and Hooker. He took quickly to the sport, becoming one of the pioneering climbers during the "golden age of mountaineering." He narrowly missed being the first to summit the Matterhorn, reaching the penultimate peak of the mountain, now known as Pic Tyndall, in 1862. In 1868 he became the first person to climb the more difficult traverse up from Breuil and down to Zermatt, turning the once-inaccessible mountain into a pass. He also made numerous first ascents of other more formidable peaks throughout the Alps, including the majestic Weisshorn on 19 August 1861, by far the most difficult route to have been accomplished. Unlike other mountaineers, Tyndall preferred to keep to the rocks rather than the ice and snow, and he often climbed without guides. He thus contributed significantly to the advance of both rock climbing and guideless climbing. He spent over twenty-five summers in the Alps, building a second home with his wife, Louisa, above the Bel Alp Hotel, a favorite haunt of British mountaineers.[16]

If there was a greater mountaineer at that time, it could only be Leslie Stephen. His *Playground of Europe* (1871) became an instant classic and helped turn the high Alps into a playground for the British, particularly the British agnostic. His crowning achievement was the first ascent of the Schreckhorn on 16 August 1861. His list of other first ascents, too long to mention here, was well known to all mountaineers in the mid-nineteenth century. Tall and lanky, with striking blue eyes and later an equally striking red beard, on more than one occasion he has been described as walking "from peak to peak like a pair of compasses."[17] Over a forty-year period beginning in 1855, he went to the Alps twenty-five times, his final year in 1894. As his first biographer noted, "His love of the Swiss mountains expired with his last breath."[18]

Bernard Lightman and Ruth Barton have both analyzed the relationship between mountaineering and naturalism during this period. Lightman has argued that the wonders of nature replaced Tyndall's and Stephen's need for religion, that their agnosticism was "synonymous with the natural order" that they found in the Alps.[19] Tyndall, in particular, was searching for a replacement for Christianity during the 1850s; according to Lightman, he "yearned for a sense of direction that could only be supplied by a definite creed."[20] As

FIGURE 2.2. John Tyndall on an ascent of the Lauwinen Thor in August 1860, scaling the rock face. He often climbed on the rocks while his guides and porters felt safer on the snow couloir. Note the barometer carried by the last porter. From John Tyndall, *Hours of Exercise in the Alps* (London, 1871), facing p. 11.

Ruth Barton has noted, Tyndall found in the wonders of the Alps the direct engagement with nature that he needed to formulate a "complete philosophy of life."[21] It was in the Alps while mountaineering that he learned to connect the two sides of human nature: feeling and intelligence. This chapter is in the spirit of refining these previous stances rather than overturning them. What did Tyndall and Stephen find so appealing about the mountains? How did mountaineering lead to a "complete philosophy of life"? Fortunately, these were the exact questions that both Tyndall and Stephen contemplated in their journals and attempted to answer in print. Stephen, in particular, was an adamant defender of the faith, in this instance the faith in the complete and utter goodness of mountaineering.

So, why did Tyndall, Stephen, and so many other evolutionary naturalists go to the mountains? An initial, crude answer can be gleaned from the opening vignette about Huxley on the Grand Mulets: the terror he experienced at the edge of the bottomless crevasse. In the Alpine environment, perhaps more than in any other, all climbers eventually feel real terror for their lives. There comes a point in all of Tyndall's and Stephen's writing where something has gone wrong and the result is the real possibility of death. Tyndall often spoke of the calmness that came over him at those moments.[22] It was slow paced, an extremely focused experience. Once that happened, it stayed with them forever, at least in a philosophical sense. It enabled them to critically analyze death—and life—in a manner that others could not. As Stephen wrote, "Your mind is far better adapted to receive impressions of sublimity when you are alone, in a silent region, with a black sky above and granite cliffs all round, with a sense still in your mind, if not of actual danger, still of danger that would become real with the slightest relaxation of caution."[23]

Both Tyndall and Stephen wrote extensively of their experiences with danger and death in the Alps. In 1865, the year of the Matterhorn tragedy when four climbers fell over three thousand feet to their deaths, Tyndall spent several days searching for their mutilated bodies. Later that same summer, he scaled the cliffs of the Riffelhorn, which had taken the life of Knyvett Wilson, a Rugby schoolmaster, only a week earlier. Tyndall also climbed the Schilthorn that summer, where Alice Arrbuthnot, a twenty-one-year-old on her honeymoon, had the previous month been struck and killed by lightning.[24] In his publications, Tyndall downplayed the dangers he experienced, but his journals and letters, especially to Huxley and Hooker, are filled with derring-do of all kinds, as he recounts the desperate chances he often took falling into crevasses, sliding down avalanches, and tiptoeing across cliff bands.[25] Stephen, likewise, often acted as if anyone could summit a peak in the Alps, though he did acknowledge the irresponsible positions he often put himself into.[26]

Bruce Hevly quite correctly argued that the Victorians used these dangerous experiences to add reliability to their claims. "Heroism," Hevly noted, "with its elements of direct action, lonely commitment, and manly risk, helped to shape arguments over glacier physics." Authority, that is, was gained primarily through the "physical discomfort, if not immediate danger" that each scientist experienced as he climbed.[27] Yet, though we certainly need to acknowledge the gendered notions at work, the immediacy of the danger was also what led directly to a sophisticated inward analysis of one's own psyche. It was one of the reasons Tyndall and Stephen went to the Alps.

In their journals and publications, Tyndall and Stephen closely associated the danger that they experienced in the mountain environment with their own insignificance and place within the natural order. Tyndall, while climbing the Aiguilles du Midi, wrote in his journal of 4 August 1857:

> It was a fine, and in some degree a fearful sight; for amid these gigantic chasms and ridges, plates and pinnacles, we lacked the assurance that acquaintance even with dangerous places brings. Our guide was never here before, no guides come here, and the twisted[,] pitted and riven ice is most treacherous looking. . . . Thus there is a dash of awe connected with our sciences—stirring up the feelings, which mix themselves with the intellect and invade its cold tranquility. What is man physically speaking amid the agents which nature here exhibits[?] An ice pebble would crush him to atoms, an ice furrow if he fell into it, would be his grave.[28]

It was through such experiences that Tyndall experienced otherworldliness. After a day among the fissures of the Aletsch Glacier, he wrote, "Beside such might, man feels his physical helplessness, and obtains the conception of a power superior to his own. His emotions are stirred. His fear, his terror, his admiration[;] he ends his survey breathing into the rushing cataract a living soul."[29] The rugged terrain of the Alps forced Tyndall to grapple with the mystery beyond life; it stirred his emotions, his fear, terror, and admiration.

In his own writings, Stephen echoed a similar spirit. On the sides of the Eiger-Joch, he opined, "The mountains represent the indomitable force of nature to which we are forced to adapt ourselves; they speak to man of his littleness and his ephemeral existence; they rouse us from the placid content in which we may be lapped when contemplating the fat fields which we have conquered and the rivers which we have forced to run according to our notions of convenience."[30] Experiencing the mountain by climbing it, for Stephen, was a way in which to learn about one's place in nature. It focused his impressions of the sublime.

Other reasons for going to the mountains are more straightforward, more tactile. As Tyndall always believed, "The aspects of nature are more varied and impressive in Alpine regions than elsewhere."[31] The mountains enabled him to experience nature firsthand, to see its laws in action, in situ. In the mountains, more than anywhere else (excepting perhaps the littoral environment), one could experience the laws of nature at work in all its aspects: in deep time in the formation of mountains and the carving out of valleys, in shallow time in the movement of glaciers, and in a single day by traveling through Humboldt's vertical zones. After his first day on the Mer de Glace, Tyndall wrote, "It is difficult, in words, to convey the force of the evidence which this glacier presents to the observer who *sees* it; it seems in fact like a grand laboratory experiment made by Nature."[32] Stephen also focused on this aspect of his experience: "The Alpine fabric includes an inexhaustible wealth of natural wonders. No other mountain chain of Europe includes, like the Alps, the flora of three zones. The arctic and the temperate join hands with the tropical, and we find representatives of the vegetation of more than thirty degrees of latitude in a short space."[33] In the mountains, a personal laboratory opened up for those who could reach it.[34]

For both Tyndall and Stephen, the mountains provoked an instinct to think of the deep past, and thus to learn firsthand about the geology of the earth. "It is in the mountains," Stephen opined, "that we instinctively ask what force can have carved out the Matterhorn, and placed the Wetterhorn on its gigantic pedestal."[35] As Tyndall noted on the sides of the Schleckhorn on 4 August 1861, "The mind involuntarily reverts to the past, and from a few scattered facts we restore a state of things which existed and had ceased before the advent of man upon the earth. Whence this power?"[36] For Tyndall and other agnostics, the questions posed by mountains were especially important.

Tyndall approached the mountains as a practicing scientist, intent on formulating long-term research projects. As he pointed out, he was not equipped as some men were with the kind of brain that has "those quick single flashes which illuminate detached problems . . . and thus make quick work of the question."[37] While grappling with the glacier questions, for instance, as with all other scientific questions, he admitted that his "thoughts ripen slowly, and it is only by degrees that I see the importance of certain observations and measurements." Using a mountain metaphor to explain the scientific process, he noted that, "as a general rule, the dominant result does not stand alone, but forms the culminating point of a vast and varied mass of inquiry."[38] Tyndall had to be immersed in nature to fully contemplate it, to experience it slowly day after day, climb after climb, year after year, similar to the slow movement of the glaciers he was studying.

The mountains enabled Tyndall and Stephen to contemplate nature's intricacies, its beauty and desolation, its order and chaos, its past and present. Tyndall, as one of the premier definers of the new specialty coalescing at that time, used his experience in the mountains to help define physics.[39] It included those subjects he experienced on a personal level within the Alpine environment, those "which lie nearest to human perception:—light and heat, colour, sound, motion, the loadstone, electrical attraction and repulsions, thunder and lightning, rain, snow, dew, and so on."[40] In his public lectures at the Royal Institution and in his more popular publications, Tyndall freely mixed his two pursuits, helping to advance both the definition of physics and the acceptance of the new sport of mountaineering.

It would be incorrect, however, even for Tyndall, to view his experiences as primarily scientific. For Tyndall, and certainly for Stephen, something more austere was at work than merely nature's laws. There was a mystery behind nature that the mountaineer was in a propitious position to uncover. That mystery, moreover, was deeply personal. When finally reaching the summit of the Weisshorn, Tyndall wrote in his journal, "We formed the centre of an Alpine circle of unparalleled grandeur, Switzerland, Savoy, Italy, all spread their mountain treasures before us. I opened my note book to write a few words concerning the scene, but I was absolutely unable to do so. I delivered myself up to the silent contemplation of it. Completely overpowered and subdued by its unspeakable magnificence."[41] Often, the beauty and chaos of it all could not be studied or written about; it could only be breathed in. For what Tyndall was searching for, feelings and emotions always trumped science and the intellect.

While Tyndall viewed the Alps through the lens of science, Stephen approached them more as a poet. Like other evolutionary naturalists, he had come from a family steeped in the Anglican evangelical tradition. He was a weak, frail, and sickly child.[42] His father had attended Trinity Hall in Cambridge and had become the regius professor of modern history at Trinity College. The young Leslie followed his father to Trinity Hall, where he took his first major steps toward athleticism, becoming an avid rower. He also kept to his studies. Coached by Isaac Todhunter, an accomplished mathematician and the first biographer of William Whewell, Stephen scored well enough on the Mathematical Tripos Exams to attain a fellowship at Trinity Hall, an accomplishment that required him to take orders.[43] He was made deacon in December 1855 and ordained a priest the next year.[44]

These were also the years Stephen first went to the Alps, in 1855 to the Bavarian Highlands, and again in 1857, his first real climbing season on glaciers, including a remarkable ascent with Francis Galton to the top of the Col de Geant.[45] He returned in 1859 to make the first ascent of the Bietschhorn. He also ascended the Dom that year, the third highest peak in the Alps, and he attempted the mighty Weisshorn, though without success. He returned to the Weisshorn in 1862 to make the second successful summit of the mountain.

The timing here is suggestive. Stephen was ordained a priest in 1856, and his major achievements in climbing came in 1859, 1860, and 1861. He was thirty years old and the most accomplished mountaineer in Europe. He also

experienced a crisis of faith and could no longer take part in religious services.[46] He resigned his tutorship in 1861. "I now believe in nothing, to put
it shortly," he admitted, "but I do not the less believe in morality, etc. etc., I
mean to live and die like a gentleman if possible."[47] That same year, his first
book publication appeared in print. It was not his own work, but rather a
translation of Baron Hermann von Berlepsch's *Die Alpen*, translated as *The
Alps; or, Sketches of Life and Nature in the Mountains* (1861).

Once we know of Stephen's later publications, this seems an odd beginning to his publishing career. It was a book on mountaineering, plain and
simple. Baron Berlepsch's approach to the mountains was unquestionably
secular in nature. It was not God that one finds in the mountains, but the un-
varying nature of nature's laws. The story is told in the vertical, as the baron
(and Stephen) take the reader from the civilized world of towns and beautiful ladies, up through the larch and stone pine forests, past rock and ice, to
a climax on a hypothetical summit. The section on mountain summits ends
with the question "What is the use of going up there?"[48] The baron's answer
is also Stephen's answer:

> It is the feeling of spiritual power that glows in him, and drives him to over-
> come the dead horrors of nature; it is the charm of measuring the power pecu-
> liar to man, the infinite capacity of an intelligent will, against the rough oppo-
> sition of dust; it is the holy impulse to seek out, in the service of the everlasting
> science of the earth's life and framework, for the mysterious connection of all
> creation; it is perhaps the longing of the lord of the earth to place the seal on
> his consciousness of a relationship to the infinite, by a bold free deed on the
> last conquered height, looking round on the world lying at his feet.[49]

Though couched in religious terminology, the "longing" is a personal quest
to comprehend the beauty and chaos of nature. The experience is emotional
and spiritual but not religious. The "lord of the earth" seeking the "mysterious connection of all creation" is Stephen himself.

In his own *The Playground of Europe*, published ten years later, Stephen
used Berlepsch's account of mountains and their meaning as his model. The
text begins as a history of climbing and the transition from the "old school"
to the "new school" of climbers. Surprisingly, the key transitional figure is not
de Saussure or some other notable pioneering mountaineer, but Jean-Jacques
Rousseau. Note the religious imagery used to formalize the complete secularization of the mountains: "Rousseau was the arch-heretic who instituted
a regular and avowed worship of the Alps. . . . [He] set up mountains as objects of human worship. . . . Rousseau, though partly anticipated, and though
his revelation had to be completed by various supplementary prophets, may
be called, without too much straining of the language, the Columbus of the

Alps, or the Luther of the new creed of mountain worship."[50] Implicit in this dubious history is the complete ousting of natural theology from the realm of mountaineering, a parallel to the work of Tyndall and others in the sciences.[51] "The love of mountains," Stephen averred, "came in with the rights of man and victory of the philosophers."[52] By dispelling God and godliness from his narrative, Stephen is free to focus exclusively on what is so meaningful about Alpine excursions: "Its charm," according to Stephen, "lies in its vigorous originality" and connection with "all that is noblest in human nature."[53] The point of the book is to show that this nobility comes at a price: "those love [the mountains] best who have wandered longest in their recesses, and have most endangered their own lives and those of their guides in the attempt to pen out routes amongst them."[54] Danger brought excitement, but also knowledge and meaning. It cleared the head. As he put it, "Sluggish imaginations require strong stimulants."[55] Yet, for that to occur, those experiences had to be extremely tangible, dangerous, and perhaps even terrifying, "food highly spiced enough for such robust digestion."[56]

The text then recounts his most famous mountaineering accomplishments, including the first ascent of the Schreckhorn. The mountain had been attempted before, by the Swiss naturalist Joseph Hugi in 1828, for instance, and by the Swiss geologist Pierre Desor in 1842. Stephen climbed it by ascending the upper Schreck couloir to the ridge and then following the southeast ridge to the top. His successful first ascent gave Stephen an "acute attack of the climbing fever."[57] A climb up the Rothhorn was the perfect antidote, as the final summit consists of a sharp ridge that climbers are forced to sit astride, with a leg hanging over each side of the precipice. "I found myself fumbling vaguely with my fingers at imaginary excrescences, my feet resting upon rotten projections of crumbling stone, whilst a large pointed slab of rock pressed against my stomach."[58] This sounds curiously like his hypothetical near-death experience, published as "A Bad Five Minutes in the Alps," discussed below. Obviously proud of his own accomplishment, he allowed himself to boast of his climbing prowess. "I will not tell at length how I was sometimes half suspended like a bundle of goods by the rope; how I was sometimes curled up into a ball, and sometimes stretched over eight or nine feet of rock. . . . I conceived myself to be resting entirely on the point of one toe upon a stone coated with ice and fixed very loosely in the face of a tremendous cliff."[59] It was only by "dallying with danger" that one could learn to "appreciate the real majesty of an Alpine cliff."[60] And only through such an appreciation could true knowledge be attained.

Stephen ended his text, just as the baron had, with a hypothetical climb to a distant peak, in an attempt to express the reasons why he climbed, why

FIGURE 2.3. Leslie Stephen climbing the approaches to the Rothhorn. From Leslie Stephen, *The Playground of Europe* (London: Longmans, Green, 1871), frontispiece.

everyone should climb. "Now the first merit of mountaineering is that it enables one to have what theologians would call an experimental faith in the size of mountains—to substitute a real living belief for a dead intellectual assent."[61] If the main influence of mountains consisted in their enormous size, their steepness, their terror, then the mountaineer had the advantage of actually experiencing these aspects firsthand. It enabled one "to measure that magnitude in terms of muscular exertion instead of bare mathematical units."[62] One could only do this by experiencing the mountain personally, on its highest crags, not from the confines of a hut or low-lying valley.

This intimacy with nature was also exactly what Stephen experienced on nonhypothetical summits, such as the Wetterhorn. "Now any one standing at the foot of the Wetterhorn," Stephen wrote, "may admire their stupendous massiveness and steepness; but, to feel their influence enter in the very marrow of one's bones, it is necessary to stand at the summit, and to fancy the one little slide down the short ice slope, to be followed apparently by a bound into clear air and fall down to the houses, from heights where only the eagle ventures to soar."[63] The experience is always mediated through danger, which is what so excited the imagination. On a mountain slope, avowed Stephen, the powers of nature "are impressed upon the mountaineer with tenfold force and intensity . . . open[ing] up new avenues of access between the scenery and his mind."[64] Stephen realized that the word *beautiful* was misleading when attempting to describe mountains; rather, "they have a marvelously stimulating effect upon the imagination."[65] In short, mountains made you think.

Think, yes, but not of God. Stephen's mountaineering and his crisis of faith seemed to be inextricably mixed. Thus, it is fitting that Stephen chose mountaineering as the subject to first describe his agnosticism.[66] Published in *Freethinking and Plainspeaking* in 1873, a long article entitled "A Bad Five Minutes in the Alps" was his chance to write about facing death. According to his first biographer, it was "perhaps the most readable bit of theological controversy in existence."[67] The tale begins in a remote mountain hut, with the smoke of tobacco pipes inside mimicking the mist outside. Stephen had been reading an old journal announcing that "an energetic controversy was raging as to the efficacy of prayer," which some "bold man" had argued was "an obsolete superstition."[68] The stupidity of the arguments on both sides led him to quit the hut, brave the weather, and take a stroll. After several hours tramping through the rain, and turning into a human sponge in the process, he decided to take a shortcut, which, like most shortcuts in the mountains, caused him to lose his way. The tall tale continues with a short crossing of a cliff face for just a few yards, with only one hard climbing move, that was it, enabling him to regain the trail. He slipped in his attempt. One small handhold kept him from the torrent some two hundred feet below and the prospect of turning his body into a "heap of mangled flesh and bones."[69] He tried to get his feet secure on the rock, but the entire surface had been polished by an ancient glacier. "A geologist would have been delighted with this admirable specimen of the planing power of nature," Stephen acknowledged. "I felt, I must confess, rather inclined to curse geology and glaciers,"[70] a feeling he extended to all of science, particularly "to gravitation and the laws of motion."[71]

As he hung suspended, he tried to calm himself by thinking of past sermons that he had heard, but he could not get hateful thoughts out of his mind. What was he doing on the side of a cliff in the first place? He hated the cliff, but he also had "an unpleasant sense that my hatred could do it no harm."[72] It dawned on him that the "whole doctrine preached by the modern worshippers of sublime scenery seemed inexpressibly absurd and out of place." This led him to ask, "What is this universe in which we live, and what is, therefore, the part we should play in it?"[73] He duly ran though the answers given by "Protestants, Catholics, Epicureans, Positivists, Broad Churchmen, Pantheists, and a vast variety of sects."[74] Yet, each sect seemed to contain "gratuitous falsehoods." His main objection to Christianity was the belief that all humans were accomplished sinners—vile, bad people. It was a good world as far as worlds went, he thought, and the people in it were good too, if a bit lazy in their thinking, "mechanically repeating fragments of an old melody from which all sense has departed."[75] Luckily, another doctrine was at hand: Christianity, but without all the damnation and vile humanity bit. The corruption of mankind, in this view, was just a "biblical way of stating Mr. Darwin's doctrines."[76] Yet, this seemed far too Panglossian for his taste. If one got rid of hell and God fire, could one really still believe in heaven and God's benevolence, directed at each individual soul? He doubted it. He decided, therefore, to stick to the brute facts: he was made of flesh and bone, "a machine, with food for fuel, grinding out so much thought, motion, and producing sundry chemical and mechanical changes in surrounding objects."[77] He would simply transform after death: "What was me will be part of the glacier stream, increasing the deposits on the flanks of the mountains."[78] Yet even this answer fell short, for he cared nothing for his body parts once they were disconnected from his actual body. This sort of brutal materialism was, in his mind, unfruitful.

Truth is, Stephen realized, nothing had meaning except visceral experiences, and the desire to keep holding on even though all other options were played out "and one theory of the universe seemed to be about as uninteresting as another." The only answer to make sense was survival and the experience of life. And survive he did. He fell, but merely to a mossy ledge some ten feet below. His last thought in those five minutes was whether he could make it back to the hut for dinner.[79] Eating a hearty meal among friends in a mountain hut was the type of fellowship he required. For Stephen, at least, that was his garden, and he intended to cultivate it.

Noel Annan suggested that "Stephen went to the Alps to climb and for no other reason."[80] But that is simply not true. He went to the Alps to climb, certainly, but there were many other reasons. Or, to put it another way, if he went to the Alps only to climb, he did so because climbing gave him so many other experiences: danger, for one; feelings of the sublime, another; his own

insignificance, yet another. It was more than plain fun. It was outwardly satisfying, but equally inwardly appealing. It was his time for introspection.

Conclusion

There is assuredly morality in the oxygen of the mountains.
— JOHN TYNDALL, *Hours of Exercise in the Alps* (1871)

The two great defenders of agnosticism were also the two greatest climbers in
the golden age of British mountaineering. The height of their climbing came
in the early 1860s, the same years in which they formulated their own evolutionary naturalism. The two great agnostics, the two great mountaineers, the
two defiant defenders of both of those faiths, hopscotched over each other on
the sides of the Swiss Alps.

The paths of these two mountaineers often crossed in the Alps. Neither
was a founding member of the Alpine Club, but both were elected the first
year, in 1857, the only two. In 1858 Stephen climbed the Monta Rosa, the second highest peak in the Alps. The next year, Tyndall climbed the peak alone,
without a guide, the first successful solo attempt on the peak. Tyndall was the
first to climb the Weisshorn, in 1861; the second person to climb it was Stephen, in 1862. They were fiercely competitive, chasing each other up peaks.
As Stephen humorously put it, "Racing in the Alps is an utter abomination,
and I have never been guilty of such a crime; except, indeed, once in an ascent
of Mont Blanc, and again, I fear, in a dash up the Aeggishhorn, and yet once
or twice more on some of the Oberland peaks, and perhaps on a few other
occasions."[81] Sometimes their competitive natures produced tensions. In his
account to the Alpine Club of his ascent of the Rothhorn, Stephen did not
hide his disdain for those who climbed for science: "And what philosophical
observations did you make? will be the inquiry of one of those fanatics who,
by a reasoning process to me utterly inscrutable, have somehow irrevocably
associated alpine traveling with science. To them I answer, that the tempera-
ture was approximately (I had no thermometer) 212 Fahrenheit below freez-
ing point. As for ozone, if any existed in the atmosphere, it was a greater fool
than I take it for."[82] These and similar statements rubbed Tyndall—who was
well known for his study of the ozone and who had rarely left a hut without a
thermometer—and other scientists within the Alpine Club the wrong way.[83]

In the end, however, the two prominent intellectuals shared much more
than either would perhaps have admitted. Both found in the Alps a consolation for their loss of faith, an outlet for their agnosticism, and a perfect

backdrop for their search for meaning. Both brought the same, fundamental question to the mountains: could one be a moral person without believing in a Christian God? One a literary critic, the other a scientist, they both went climbing to find some basis to create a justifiable ethic. This is why there is so much religious language in the mountaineering narratives of both Tyndall and Stephen. Their new secular approach necessarily contained aspects of the language it was created to overturn, especially in its use of religious imagery.[84] Bernard Lightman cites Frederic Harrison, who observed that "the Alps were to Stephen the elixir of life, a revelation, a religion," and he rightly notes that Stephen often spoke of the Alps as a "sacred place."[85] The famous mountaineer Douglas Freshfield once said, "The Alps were for Stephen a playground but they were also a cathedral."[86] As both Barton and Lightman have found, Tyndall's descriptions of the Alps sometimes read like religious experiences.[87] Lightman quotes Huxley as referring to Tyndall as the "mightiest evangel" of that "sect of muscular philosophers whose best known church is the Alpine Club."[88]

Yet, we must be careful here: It is not the mountains that they are revering; it is their visceral experience of the mountains that forms the basis of their faith. The mountains were not holy or sacred; they were not places of worship. Rather, they were perfectly secular spaces where the imagination was allowed to ramble, just as the body was allowed to scramble. Drawing on Humboldt, de Saussure, and Rousseau, they were following decidedly secular approaches to nature in their mountaineering narratives.

The Alps were not a substitute for religion, then, in the simplest sense. Tyndall and Stephen did not worship them as one would worship a deity. Rather, they went to the mountains as the perfect place to think about topics that others associated with religion, like morality and the human condition. Both were searching for feeling and emotion, to hear what the mountains had to say. The mountains spoke to them, often in the harshest of terms, at other times in a soothing, familiar voice. As Stephen wrote, the mountains spoke "in tones at once more tender and more awe-inspiring than that of any mortal teacher. The loftiest and sweetest strains of Milton and Wordsworth may be more articulate, but do not lay so forcibly a grasp upon my imagination."[89] Amid the seracs and glaciers of Mont Blanc, Tyndall wrote in his journal, "We fill our brains with thought and occupy our understanding with problems. Our capacity for such things is our justification for occupying ourselves with them, but amid these sublime summits, looking at these stars, listening to that wailing wind, the mind also proves its capacity for thought more solemn than those which occupy the understanding."[90] Their relationship to nature had to do with understanding the natural world, experiencing it personally and viscerally. Nature was more a companion than a savior, more a guide than a god.

Companions, however, can often be fickle. What united Tyndall and Stephen was their willingness to listen to this aspect of the mountains as well, to include both nature's regularity and nature's chaos in their life ethic. That is, watching a sunset or overlooking a panoramic view does not lead, in and of itself, to a moral vision. Both Tyndall and Stephen were looking for more than that, for more than awe or beauty. In this sense, their approach to the mountains, as Lightman has suggested, was akin to what fellow evolutionary naturalist William K. Clifford referred to as "cosmic emotion." Lightman notes that this entailed "a sense of awe in regard to the order manifested throughout the universe."[91] Yet, it also entailed a call to action, a guide for moral decision making. As Clifford wrote, "We are taught therein to look upon Nature as a divine Order or Cosmos, acting uniformly in all of its diverse parts; which order, by means of its uniformity, is continually educating us and teaching us to act rightly."[92] Above all else, it was this teaching toward a justifiable ethic, ending in the call to act rightly, that created the significance of the cosmic order, what Tyndall and Stephen found through a cosmic emotion.

Cosmic emotion did not arise from simply the order in nature. It could not be found by studying nature's laws. Nature was not just "radiant and harmonious"; it was also "savage and chaotic."[93] Responding to Clifford's ideas about cosmic emotion, Frederic Harrison noted the important distinction between nature worshippers and truth seekers. The former sought beautiful vistas and panoramic views; the latter searched for deeper meaning and purpose. Tyndall and Stephen were seekers of truth.

The way that Tyndall, Stephen, and other scientific naturalists viewed the mountains has had considerable influence on how we view mountaineering today. High Alpine environments are not so much a place of escape as they are a place to explore one's inner self—not so much a place to get away, as a venue to get inside. Today's mountaineers seek answers, but they also relish the questions. It explains why they place themselves in danger, why they search out harrowing summits, and why they have such difficulty explaining their actions to nonclimbers. It also helps explain why climbers view their sport in such deeply personal terms, a moral quest as much as a physical one. Mountaineers speak reverently of what they call a "fellowship of the rope." That fellowship does not exist solely between roped partners, though that is how the phrase originated. It speaks to a sense of identity among all mountaineers that arises from a shared experience of finding meaning in the mountains.

The fellowship that binds today's mountaineers had its origins in the mid-nineteenth century, in the writings of Tyndall, Stephen, and other pioneering alpinists. They were all nature worshippers, but most also aspired to be seekers of truth. Mountaineering as a sport was formed in the turbulent wake of evolutionary theory and the larger questions that it raised. For those Victorians who had experienced a crisis of faith, the dangerous Alpine environment offered a safe sanctuary. A surprising facet of Tyndall's *Hours of Exercise in the Alps* and Stephen's *The Playground of Europe*—both published in 1871, the same year as Darwin's *Descent of Man*—is the complete absence of anything that could be called religious. They are both perfectly and graciously secular. An identity thereby coalesced around both the sport and its practitioners. Alpinists could share in the experience of Godliness without having to write about God. On the sides of mountains, in the midst of all God's wonders, it was safe to be an agnostic.

It should no longer be surprising, then, that most of the X Clubbers and prominent evolutionary naturalists found value in the mountains. Their shared experience in the mountains helped influence their common project of formulating an ethic based on nature rather than God. Naturalism for these intrepid mountain pioneers was infused with the Romantic notions of the sublime, influenced heavily by Humboldt's secular yet socially focused vision of how to act morally. These notions of the sublime lingered well into the nineteenth century through the writings of traveling naturalists in the Humboldtian tradition, including Hooker, Darwin, and others. They certainly were ever present in the mountaineering narratives of the Victorian era. By engaging with nature through sport, Tyndall and Stephen experienced the awe-inspiring beauty and unremitting chaos of the natural world, stripped of its natural theological connotations. Though they shared a desire to be the first on untrammeled Alpine peaks, they went to the mountains in search of something far more important. As clichéd as it may sound, their aim was always to find purpose and meaning. Thousands of pages of Tyndall's journals, volumes of Stephen's letters, and that is what you get.

Tyndall attempted to climb the Matterhorn a handful of times. He compared his third failed attempt, tellingly, as akin to "the breaking down of a religious faith."[94] But, finally, in August 1868, he stood on its very pinnacle. He was struck by the "inexorable decay" of its highest cliffs, and a queer sadness unexpectedly came over him. While on the summit, he pondered, "When I look at the heavens and the earth, at my own body, and my strength and weakness of mind, even at these ponderings, I ask myself, Is there no being or thing in the universe that knows more about these matters than I do? What is my answer?"[95] Stephen had similar experiences. After a climb in 1873 with the painter Gabriel Loppé, Stephen penned perhaps his most famous passage, asking, "Does not science teach us more and more emphatically that nothing which is natural can be alien to us who are part of nature? Where does Mont Blanc end, and where do I begin?"[96] This is the essence of the cosmic emotion. For John Tyndall and Leslie Stephen, the questions that mountains force them to ask, whether through beauty or desolation, order or chaos, was the most important part of the mountaineering experience. They found morality mixed with the oxygen as they breathed in the answers. It was the main reason they went to the Alps.

Acknowledgments

Research for this chapter was funded, in part, by a National Science Foundation Collaborative Research Grant (NSF 0924426). In addition to the volume's editors and anonymous reviewers, I would like to thank Katie Ives (editor, *Alpinist Magazine*) and Dennis Dueñas (my climbing partner) for stimulating discussions concerning the transcendental motivations for mountaineering.

Notes

1. John Tyndall, Journals, 24 August 1856, Royal Institution of Great Britain, London. Hereafter cited as "Tyndall's Journals."

2. Ibid., 12 August 1857.

3. Ibid.

4. Ibid.

5. The members of the X Club included Tyndall, Huxley, Hirst, J. D. Hooker, Edward Frankland, William Spottiswoode, Herbert Spencer, John Lubbock, and George Busk. See Ruth Barton, "John Tyndall, Pantheist: A Rereading of the Belfast Address," *Osiris*, 2nd ser., 3 (1987): 111–34.

6. Secondary material on Hirst is disparagingly sparse. His extensive journals housed in the Royal Institution of Great Britain, however, represent a potential gold mine for researchers interested in Victorian thought and culture. See William H. Brock and Roy M. MacLeod, eds., *Natural Knowledge in a Social Context: The Journals of Thomas Archer Hirst FRS* (London: Mansell, 1980); and Ruth Barton, "Hirst, Thomas Archer (1830–91)," *Dictionary of Nineteenth-Century British Scientists*, ed. Bernard Lightman, 4 vols. (Bristol, UK: Thoemmes Press, 2004), 2:973–77.

7. As quoted in Barton, "Hirst," 973.

8. As quoted in Brock and MacLeod, *Natural Knowledge*, 19, 15.

9. Michael S. Reidy, "From the Oceans to the Mountains: Spatial Science in an Age of Empire," in *Knowing Global Environments: New Perspectives on the Field Sciences*, ed. Jeremy Vetter,

21–48 (New Brunswick, NJ: Rutgers University Press, 2010); Michael S. Reidy, "The Rucksack of Joseph Dalton Hooker," *Alpinist Magazine* (2010–11): 83–86; and Richard Bellon, "Joseph Hooker Takes a 'Fixed Post': Transmutation and the 'Present Unsatisfactory State of Systematic Botany,' 1844–1860," *Journal of the History of Biology* 39 (2006): 1–39.

10. Colin A. Russell, *Edward Frankland: Chemistry, Controversy and Conspiracy in Victorian England* (Cambridge: Cambridge University Press, 1996), esp. 411–33; and Colin A. Russell, "Frankland, Edward (1825–1899)," in *Dictionary of Nineteenth-Century British Scientists*, 2:727–31. Tyndall and Frankland attended the University of Marburg together, where they both studied under Robert Bunsen. In 1862, Frankland accepted the professorship of chemistry at the Royal Institution, again working alongside Tyndall.

11. Russell, "Frankland," 729.

12. William Spottiswoode, "On Typical Mountain Ranges: An Application of the Calculus of Probabilities to Physical Geography," *Journal of the Royal Geographical Society*, 31 (1861):149–54; Nicholas Wright Gillham, *A Life of Sir Francis Galton: From African Exploration to the Birth of Eugenics* (Oxford: Oxford University Press, 2001), 157–58; and Ruth Barton, "Spottiswoode, William (1825–1883)," in *Dictionary of Nineteenth-Century British Scientists*, 4:1889–91. Spottiswoode also joined Tyndall and Frankland as a member of the Royal Institution in 1864, becoming treasurer from 1865 to 1873 and honorary secretary from 1871 to 1878.

13. David Duncan, ed., *The Life and Letters of Herbert Spencer*, 2 vols. (London: Methuen, 1908), 1:497–99.

14. Ibid., 2:231.

15. Following Bernard Lightman, the leading evolutionary naturalists of the period were Stephen, Tyndall, Huxley, Spencer, William K. Clifford, and William Pollock. All were either mountaineers or had spent some time in the Alps. See Bernard Lightman, "Robert Elsmere and the Agnostic Crisis of Faith," in *Victorian Faith in Crisis: Essays on Continuity and Change in Nineteenth-Century Religious Belief*, ed. Richard J. Helmstadter and Bernard Lightman (London: Macmillan, 1990), 292.

16. Joe D. Burchfield, "Tyndall, John (1820–93)," *Dictionary of Nineteenth-Century British Scientists*, 4:2053–58; Roy MacLeod, "Tyndall, John," in *Dictionary of Scientific Biography*, ed. Charles Coulston Gillispie, 16 vols. (New York: Charles Scribner's Sons, 1970–80), 13:521–24; Lord Schuster, "Tyndall as a Mountaineer," in *Life and Work of John Tyndall*, ed. A. S. Eve and C. H. Creasey, 340–92 (London: Macmillan, 1945); Michael S. Reidy, "John Tyndall's Vertical Physics: From Rock Quarries to Icy Peaks," *Physics in Perspective* 12 (2010): 122–45; and John Tyndall, "Life in the Alps: A Sketch by Professor Tyndall," Add.MS.53715/81, British Library, London.

17. Frederic William Maitland, *The Life and Letters of Leslie Stephen* (London: Duckworth, 1906), 143; and Alan Bell, "Stephen, Sir Leslie (1832–1904)," in *Oxford Dictionary of National Biography*, ed. H. C. G. Matthew and Brian Harrison, 60 vols. (Oxford: Oxford University Press, 2004), 52:448.

18. Maitland, *Life and Letters*, 88.

19. Bernard Lightman, *The Origins of Agnosticism: Victorian Unbelief and the Limits of Knowledge* (Baltimore: Johns Hopkins University Press, 1987), 153.

20. See Lightman, "Robert Elsmere and the Agnostic Crisis of Faith," 296.

21. Barton, "John Tyndall, Pantheist," 121.

22. See, for example, Tyndall's Journals, 25 July 1857.

23. Leslie Stephen, *The Playground of Europe* (London: Longmans, Green, 1871), 296. Many writers in the nineteenth century, including W. K. Clifford, suggested that such an embracement of danger related directly to the struggle that agnostics confronted with end of life. See "The Un-

seen Universe," in *Lectures and Essays, by the Late William Kingdon Clifford*, ed. Leslie Stephen and Frederick Pollock, 161–79 (London: Macmillan, 1886).

24. Eve and Creasey, *Life and Work*, 113.

25. As George Levine has argued, Victorian evolutionary naturalists believed that the epistemology of science required intellectual sacrifice; they gave up orthodox beliefs to get at the truth. Here, evolutionary naturalists are purposefully experiencing physical sacrifices to encounter truths about life and death. See George Levine, *Dying to Know: Scientific Epistemology and Narrative in Victorian England* (Chicago: University of Chicago Press, 2002).

26. Stephen, *Playground of Europe*, 108.

27. Bruce Hevly, "The Heroic Science of Glacier Motion," *Osiris*, 2nd ser., 11 (1996): 66, 84.

28. Tyndall's Journals, 4 August 1857.

29. Ibid., 13 July 1858.

30. Stephen, *Playground of Europe*, 121.

31. John Tyndall, *Mountaineering in 1861: A Vacation Tour* (London: Longman, Green, Longman, and Roberts, 1862), 33.

32. Tyndall, *Hours of Exercise in the Alps* (London: Longmans, Green, 1871), 370.

33. Leslie Stephen, trans., *The Alps; or, Sketches of Life and Nature in the Mountains by H. Berlepsch* (London: Longman, Green, Longman, and Roberts, 1861), 16.

34. This was especially true for biogeographers, from Hooker to Darwin to Alfred Russel Wallace. See Michael S. Reidy, "From the Oceans to the Mountains: Spatial Science in an Age of Empire," in *Knowing Global Environments: New Perspectives on the Field Sciences*, ed. Jeremy Vetter, 21–48 (Piscataway, NJ: Rutgers University Press, 2010).

35. Stephen, *Playground of Europe*, 295.

36. Tyndall's Journals, 4 August 1861.

37. Ibid., 9 August 1857.

38. John Tyndall, *Faraday as a Discoverer* (New York: D. Appleton, 1868), 185.

39. Michael S. Reidy, "John Tyndall's Vertical Physics"; Ruth Barton, "John Tyndall (1820–1893)," in *Dictionary of Nineteenth Century British Philosophers*, ed. W. J. Mander and Alan P. F. Sell, 2 vols. (Bristol, UK: Thoemmes Press, 2002), 2:1137; and "Professor Tyndall and the Scientific Movement," *Nature* 36 (1887): 217.

40. John Tyndall, "On the Study of Physics," in *Fragments of Science*, 6th ed., 2 vols. (London: Longmans, Green, 1892), 1:282.

41. Tyndall's Journals, 18 August 1861. The entry, though dated 18 August, was written on 21 August about his experiences on 19 August, the day of the first ascent of the Weisshorn.

42. Bell, "Stephen," 447.

43. Stephen represents merely one example of a student who purposefully mingled athleticism and mathematical studies in Cambridge. For the intense connection between rigorous athletic training and the similarly "tough regime of disciplined learning" involved in the Mathematical Tripos Exams, see Andrew Warwick, "Exercising the Student Body: Mathematics and Athleticism in Victorian Cambridge," in *Science Incarnate: Historical Embodiments of Natural Knowledge*, ed. Christopher Lawrence and Steven Shapin, 288–323 (Chicago: University of Chicago Press, 1998).

44. For Leslie Stephen's life, see Noel Annan, *Leslie Stephen: The Godless Victorian* (Chicago: University of Chicago Press, 1984); Maitland, *Life and Letters*; and Bell, "Stephen."

45. Maitland, *Life and Letters*, 80–81. They succeeded in their second attempt, having been rebuffed earlier by bad weather.

46. Lightman, "Robert Elsmere and the Agnostic Crisis of Faith," 283.

47. Annan, *Godless Victorian*, 2.

48. Stephen, *Alps*, 263.

49. Ibid., 263–64.

50. Stephen, *Playground of Europe*, 38–39.

51. Ruth Barton, "'An Influential Set of Chaps': The X-Club and Royal Society Politics, 1864–85," *British Journal for the History of Science* 23 (1990): 53–81.

52. Stephen, *Playground of Europe*, 45.

53. Ibid., 66, 65.

54. Ibid., 68.

55. Ibid., 180.

56. Ibid.

57. Ibid., 93.

58. Ibid., 104.

59. Ibid., 108.

60. Ibid., 196.

61. Ibid., 276.

62. Ibid., 277.

63. Ibid., 292.

64. Ibid., 282.

65. Ibid., 292.

66. Annan, *Godless Victorian*, 92.

67. Maitland, *Life and Letters*, 97.

68. The "bold man" was John Tyndall. In July 1872, Tyndall became embroiled (anonymously at first) in what was soon referred to as the "Prayer-Gauge Debate." In defending the need for a rational and experimental verification of prayer, he poked fun at the religious fervor of the day. Here, Stephen is poking fun at Tyndall. Leslie Stephen, "A Bad Five Minutes in the Alps," in *Essays on Freethinking and Plainspeaking* (New York: G. P. Putman's Sons, 1905), 179.

69. Ibid., 184.

70. Ibid., 186.

71. Ibid., 189.

72. Ibid., 196.

73. Ibid., 200.

74. Ibid., 201.

75. Ibid., 208.

76. Ibid., 210. Here, and elsewhere, Stephen referred to Darwin as he contemplated death. In a letter to Charles Norton, for instance, Stephen noted that while Montaigne had taken consolation from Lucretius when contemplating the end of his own existence, he (Stephen) took his own consolation from Darwin. As David Amigoni noted, "Stephen was claiming Darwin as a guide to action, or rather resignation, in the present, and situating him authoritatively in a tradition of rational thought about death." David Amigoni, "Proliferation and Its Discontents: Max Muller, Leslie Stephen, George Eliot and *The Origin of Species* as Representation," in *Charles Darwin's "The Origin of Species": New Interdisciplinary Essays*, ed. David Amigoni and Jeff Wallace (Manchester, UK: Manchester University Press, 1995), 124.

77. Stephen, "Bad Five Minutes," 214.

78. Ibid.

79. Ibid., 225. In case there was any doubt, Stephen added an endnote to his text: "It may be

as well to say, for the credit of the noble science of mountaineering, that the foregoing narrative is without even a foundation in fact." Ibid.

80. Annan, *Godless Victorian*, 91.

81. As quoted in Maitland, *Life and Letters*, 89.

82. Stephen, *Playground of Europe*, 107.

83. Past scholars have argued that these words led Tyndall to resign from the Alpine Club. See Annan, *Godless Victorian*, 91; Eve and Creasey, *Life and Work*, 389; and Fergus Fleming, *Killing Dragons: The Conquest of the Alps* (New York: Grove Press, 2000), 212. Yet, Catherine Hollis has recently shown that Stephen first made these remarks in 1865, three years after Tyndall resigned. See Catherine W. Hollis, *Leslie Stephen as Mountaineer: "Where does Mont Blanc end, and where do I begin?"* (London: Cecil Woolf, 2010), 35.

84. Lightman, "Robert Elsmere and the Agnostic Crisis of Faith," 299.

85. Lightman, *Origins of Agnosticism*, 149–50.

86. As quoted in Maitland, *Life and Letters*, 79; and Bell, "Stephen," 448.

87. Barton, "John Tyndall: Pantheist," 129–30; and Lightman, *Origins of Agnosticism*, 150.

88. Lightman, *Origins of Agnosticism*, 150.

89. As quoted in Maitland, *Life and Letters*, 104.

90. Tyndall's Journals, 18 July 1857.

91. Lightman, "Robert Elsmere and the Agnostic Crisis of Faith," 300.

92. Stephen and Pollock, *Lectures and Essays*, 404–5.

93. "Cosmic Emotion," *New York Times*, 28 August 1881. This is a short summary of Frederic Harrison, "Pantheism and Cosmic Emotion," *Nineteenth Century* 10 (1881): 284. Harrison wrote it as a reply to "a most interesting paper contributed by the late Professor W. K. Clifford and discussed at the Metaphysical Society." See Frederic Harrison, *The Creed of a Layman: Apologia Pro Fide Mea* (London: Macmillan, 1907), 194.

94. Tyndall, *Hours of Exercise*, 123.

95. As quoted in Eve and Creasey, *Life and Work*, 384; and Tyndall, *Hours of Exercise*, 292.

96. Leslie Stephen, "Sunset on Mont Blanc," *Cornhill Magazine* 28 (1873): 458. For an enlightening analysis of this and other passages from Stephen's mountaineering publications, see Hollis, *Leslie Stephen as Mountaineer*.

3

Paradox: The Art of Scientific Naturalism

GEORGE LEVINE

In his famous "Belfast Address," John Tyndall celebrated the literary quality of scientific writing:

> It has been said by its opponents that science divorces itself from literature; but the statement, like so many others, arises from a lack of knowledge. A glance at the less technical writings of its leaders—of its Helmholtz, its Huxley, and its Du Bois-Reymond—would show what breadth of literary culture they command. Where among modern writers can you find their superiors in clearness and vigor of literary style? Science desires not isolation, but freely combines with every effort towards the bettering of man's estate. Single-handed, and supported not by outward sympathy, but by inward force, it has built at least one great wing of the many-mansioned home which man in his totality demands.[1]

Tyndall's praise of scientific writing is itself part of the scientific naturalists' epic enterprise to change the way we see and the way we feel about the world. It was a project that we usually try to, and should, read in the context of social and cultural history, but I want to shift the focus and take what might seem a retrograde step—that is, to take the naturalists' expressed effort seriously and to a large degree sympathetically, and read it as a project of art as well as of science. Tyndall knew that his enterprise required more than narrow focus on humans' rational powers. If he were to make his readers see nature turned upside down, he needed both the power of rhetorical control and a mythic power to reshape the world. The ambition was epic.[2] A full understanding and a just evaluation of the work of the naturalists entails some direct engagement with their intended objectives, however necessary it is to recognize the social forces that impelled them and that they were trying to influence. There is a literary context that usefully supplements the social one.

Today, we tend no longer to be open to the kinds of progress narratives that Tyndall and his colleagues envisioned, and historians of science are distrustful of heroic rhetoric. Adrian Desmond, in several remarkable books, has thrown down the gauntlet. Without understanding the special interests of the writers, and the detailed social context of their writings, Desmond argues, our sense of their project and the full significance of their language is lost: "Surely," he wrote, "some . . . larger picture," which would include, for example, a recognition of the literal significance of Huxley's war metaphors, "will ultimately prove more satisfying than simply accepting Science's 'War' with Theology as an inevitable development of the rational mind." As Desmond puts it, Huxley "was refracting the light of science through an ideological lens," making science "part of the State apparatus."[3]

In addition, thanks to the work, among others, of Bernard Lightman and Frank Turner, historians and philosophers of science have recognized fundamental intellectual inconsistencies in the naturalists' program. Lightman has demonstrated the inherent contradictions in naturalist thought and its vulnerability to philosophically strenuous questioning of its metaphysical bases.[4] Epistemologically, scientific naturalism seems not to have had a leg to stand on. Turner has talked of "the existential, intellectual, and moral bankruptcy of scientific naturalism,"[5] echoing judgments made by many of the naturalists' contemporaries. Nevertheless, some years ago when I cited the phrase at a conference Turner was attending, he blanched at his own rhetoric and asked, "Did I write that?" It is now, alas, too late to find out how he might have wanted to diminish the severity of the dismissal.

The naturalists' rebellious and belligerent rhetoric, given their great successes, might seem disingenuous. But one needs to remember that despite the rise of secularism in the nineteenth century, the Victorian period was one of the most actively religious in the history of Britain; religion was not going away and its secular power remained great. Most of the naturalists, although fiercely keeping religion out of cosmology, sustained in their language strong traces of a religious inheritance. Lightman points out how the naturalists retained their own kind of religiosity and deliberately turned naturalism into an alternative creed, with ambitions to do the work traditionally associated with religion.[6] But these religion-like ambitions were the other side of the epic and often visionary quality of their writing.

Something of the nature of that ambition and the confidence with which they developed it emerges shortly after the Tyndall passage with which this essay opens: "The impregnable position of science," he says, "may be described in a few words. All religious theories, schemes, and systems, which embrace notions of cosmogony, or which otherwise reach into its domain, must, in so far as they do this, submit to the control of science, and relinquish all thought of controlling it."[7] This is no-nonsense prose. It is declarative, definitive, imperial, and defiant, its supreme confidence joined to a sense of righteous superiority. Tyndall simply displaces religion from its traditional position of intellectual authority, while at the same time suggesting and demonstrating (he was no cultural slouch either) that science gave space to those elements of feeling and spiritual aspiration that had been claimed as exclusively religious. Science can succeed in the high moral project of "the bettering of man's estate" where traditional high culture has failed because—the irony is implicit—*that* culture is guilty of not being high enough, of "lack of knowledge." Against the narrowness that cannot see the important connections between science and literature, Tyndall implies his own depth of literary culture. Implicitly, he makes the claim that in a culture dominated by literary figures like Carlyle, Dickens, George Eliot, Tennyson, and Browning, science writing should take an equal place. Tyndall's prose projects a scientific culture positively Arnoldian.

The naturalists' goal was no less than the transformation of common sense. Their prose is veined with a risk-taking determination to jolt the reader out of normal ways of seeing. "It is not right to be proper," W. K. Clifford concludes one of his strongest essays.[8] The risk taking is part of the great Romantic project to see the world in a grain of sand—or a piece of chalk—transforming the slightest object of our ordinary lives into something wonderful. Describing the growth to understanding from the infant to the boy, Tyndall makes the point: "The boy finds the simple and homely fact which addressed his senses to be the outcome and flower of the deepest laws."[9] Moving from the most ordinary and obvious experiences, the falling of the dew, the whiteness of the snow, the blueness of the sky, he makes the invisible natural order in which we live visible: the most disparate and ordinary things are united—nothing, as he puts it, is without "brotherhood in Nature." From Wordsworth's celebration of the ordinary, the naturalists move to shock readers into recognizing that their "common sense" is sustained by myths. No wonder Tyndall was so successful in his lectures, replete with the fireworks of real experiments, making science a popular entertainment and changing the world at the same time.

But such epic efforts toward the creation of a brave new world inevitably encounter contradictions. Romantically, but with the instruments of science, they evoke the world in progress narratives as a place of wonder, but, with Huxley, insist that science is only "organized common sense."[10] The project was itself paradoxical, representing as rationally comprehensible a world that contradicted itself. Ethical and epistemological crises follow even from the most fundamental naturalist assumptions—that, for example, the transcendental must be excluded from all explanations, or that nature is uniform. Here is Huxley, in an early essay, almost writing a manifesto for the myth of progress:

> If these ideas be destined, as I believe they are, to be more and more firmly established as the world grows older; if that spirit be fated, as I believe it is, to extend itself into all departments of human thought, and to become co-extensive with the range of knowledge; if, as our race approaches its maturity, it discovers, as I believe it will, that there is but one kind of knowledge and but one method of acquiring it; then we, who are still children, may justly feel it our highest duty to recognize the advisableness of improving natural knowledge, and so to aid ourselves and our successors in our course towards the noble goal which lies before mankind.[11]

The incantatory quality of this impressive periodic sentence mixes epistemology with ethics, in a way characteristic of the naturalists; it is an emotively charged call to intellectual arms. Suggesting a steady and slow progressive transformation toward its period in "mankind," it yet urges rhythmically a unification and triumph that is as much moral as intellectual. The "noble goal" is Tyndall's "bettering of man's estate." In the name of displacing religion, it opens a religiously charged crusade for science.

Tyndall's aggressive "Belfast Address" is a yet more famous instance of this kind of paradox. Delivered provocatively in the heart of a fiercely Protestant city in an intensely Catholic country, it might well be taken as a long spelling out of the progress myth by way of consideration of science's greatest heroes and their work. Among its many literary devices is a fictional dialogue staged with Bishop Butler designed to demonstrate Tyndall's sympathetic openness to intelligent opposition and the compatibility of his scientific perspective with true religious belief. But the sustaining frame of the address is the ethically charged and happy if difficult story of the inevitable growth of knowledge to its current climax in naturalist thought. Although some of the naturalists did speak of the likelihood or inevitability of cosmic disaster in the very distant future, they affirmed from a secular perspective that knowledge increases, morality develops, and despite bumps and setbacks, science works for the growing good of the world, if usually without that touch of complex sadness that marks the "Finale" of George Eliot's *Middlemarch*. Tyndall's voice at the end sounds a note of spiritual need for which he was often mocked, but it is not, I think, a mere bone thrown to religious resisters. For many of the naturalists, at critical points, when they get into the deep waters of beginnings and ultimate meaning, their sense of the limits of science grows with a rhetoric rising to mysticism.

As he moves to consider "the *creative* faculties of man," Tyndall arrives at one such place. The human mind, he says, may not be satisfied even with the greatest work it has produced, and, "with the yearning of a pilgrim for his distant home, will still turn to the Mystery from which it has emerged." "Here, however, I touch a theme too great for me to handle, but which will assuredly be handled by the loftiest minds, when you and I, like streaks of morning cloud, shall have melted into the infinite azure of the past."[12] Tyndall seems always ready to invoke the limits of science, and when he does, he calls for something more than simple rationality. A crude empiricism, like bad poetry, sees only surfaces, but the "scientific imagination" penetrates beyond the visible and the world of common sense, even as far as such world-defining ideas as evolution by natural selection, or such extraordinary perceptions of significance as we find in *Middlemarch*, where in the portrait of the researching doctor Lydgate, George Eliot imagines an ethically charged scientific vocation equivalent to that projected by Tyndall.

Here she too, in an intense imagination of scientific labor, rises to poetic language. Lydgate's scientific imagination "reveals subtle actions inaccessible by any sort of lens, but tracked in that outer darkness through long pathways of necessary sequence by the inward light which is the last refinement of Energy, capable of bathing even the ethereal atoms in its ideally illuminated space."[13] I know that it is cheating, in my effort to credit the art and imagination of the naturalists, to quote so great a writer as George Eliot, who though intellectually and personally close to the naturalists is not one of them. But her scientific imagination also aspires to that transformation of the material into the spiritual that Tyndall implies, and she manages it in the space of a sentence.

George Eliot would have found intimations of Lydgate's vision in Tyndall's essay "The Scientific Uses of the Imagination," written about the time she was in the midst of *Middlemarch*. There, in his discussion of the blueness of the sky, are George Eliot's ethereal atoms, particles too small—"infinitesimals"—to be observed individually even under a microscope. These "particles of our sky may be inferred from the continuity of light. It is not in broken patches, nor at scattered points, that the heavenly azure is revealed." Transforming the common sense of things, he concludes, "In the atmosphere, we have particles which defy both the microscope and the balance, which do not darken the air, and which exist, nevertheless, in multitudes sufficient to reduce to insignificance the Israelitish hyperbole regarding the sands upon the seashore."[14] The rhetoric soars as the senses fail, and invisible infinitesimal particles swarm in the most common elements of our world. At the limits of science common sense yields to imagination and art.

Without Tyndall's sometimes excessive mysticism, Darwin's arguments are also constructed this way. In virtually every example, he demonstrates the failure of common sense to recognize what nature is up to—in the birds' song in spring, in the structure of the eye, or in the bees' hive. "Paradox" is one of the conditions of the naturalists' arguments everywhere.[15] We do nothing all day, Clifford tells us, but change our mind.[16] We don't *really* see what we think we see, but supplement what the eye literally receives with what experience and inference provide.

The very name *positivism*, which Huxley contemptuously rejects (but which certainly fits most of the positions he takes), encourages paradox because it is always more consistently negative than positive. It is an oppositional stance, a position of distrust and skepticism. It is *anti*metaphysical. It dismisses all knowledge not achievable rationally and empirically, and it turns experience into a series of intermediary energies that leave us with universal subjectivity. The most complete empiricism, as in Ernst Mach or Karl Pearson, is one that recognizes that it never engages "reality." We are all in Plato's cave, getting indirect signals of what's really going on. Arthur Balfour, in his sweeping critique of "naturalism," rightly calls "startling" the account of "experience" that empiricist scientists offer.[17]

The world looks to be one thing; astonishingly, it is really another. Ironically, even naturalism's antimetaphysical epistemology is built on fundamental metaphysical assumptions, and the deep ethical urgency of the naturalists' project is bogged down in the impossibility of deducing an "ought" from an "is." To write like a scientific naturalist is to create a world that is fundamentally self-contradictory.

Ethics and Epistemology

One of the paradoxical implications of the naturalists' position is that if they follow their own logic, adhere to a rigorous empiricism, reject metaphysical assumptions, confront, as Huxley was to do, the absolute wall between description and prescription, they would have no grounds for their progress narrative or for naturalism itself. In the face of such contradictions, the naturalists, so forcefully insistent on the development of natural knowledge, and so ready to make the leap from that to authority over the "totality" of human experience, fall back regularly on the impossibility of knowing, draw blanks in the face of the big questions of ethics and epistemology, confess ignorance, and descend, or perhaps ascend, into Tyndallian mysticism of the sort that W. H. Mallock mocked. Mallock parodies Tyndall as saying, "Let us beware then of not considering religion noble; but let us beware still more of considering it true."[18] Settling for the humility of facing the limitations of what they knew, the naturalists more than occasionally get rather pretentious and self-righteous about their own honesty, but the genuine struggle to confront the limits is part of their great literary legacy.

Paradox upon paradox, the naturalists were fully aware that their anti-metaphysical stance, and the work of science, were dependent on metaphysically unprovable assumptions. Early in his career, Huxley had already confronted the paradox unapologetically. "All physical science," he admits,

> starts from certain postulates. One of them is the objective existence of a material world. It is assumed that the phenomena which are comprehended under this name have a "substratum" of extended, impenetrable, mobile substance, which exhibits the quality known as inertia, and is termed matter. Another postulate is the universality of the law of causation; that nothing happens without a cause (that is, a necessary precedent condition), and that the state of the physical universe, at any given moment, is the consequence of the state at any preceding moment. Another is that any of the rules, or so-called "laws of Nature," by which the relation of phenomena is truly defined, is true for all time. The validity of these postulates is a problem of metaphysics; they are neither self-evident nor are they, strictly speaking, demonstrable. The justification of their employment, as axioms of physical philosophy, lies in the circumstances that expectations logically based upon them are verified, or, at any rate, not contradicted, whenever they can be tested by experience.[19]

The justification is perfunctory, but the rhetoric triumphs because of Huxley's boldness in admitting the contradiction. He does not so much try to reason himself out of the paradox, as make it part of an attack. Accepting that much that we do and think depends upon a kind of faith, he goes on, "It is surely plain that faith is not necessarily entitled to dispense with ratiocination because ratiocination cannot dispense with faith as a starting-point; and that because we are often obliged, by the pressure of events, to act on very bad evidence, it does not follow that it is proper to act on such evidence when the pressure is absent."[20]

The problem can lead, in its crooked way, to the transformation of empiricism into idealism. To overcome empiricism's limits, Karl Pearson, critical of science's failure to recognize the philosophical weakness of its assertions, developed a statistical system that made scientific truth depend on probability rather than on the ontological status of the object studied.

Huxley's word *agnosticism* is perhaps the most famous expression of unknowing among the naturalists. He registers its negative force first: "It came into my head as suggestively antithetic to the 'gnostic' of Church history, who professed to know so much about the very things of which I was ignorant."[21] Thus, like its older cousin, positivism, agnosticism is aimed primarily at ridding the world of nonsense statements, but it does imply that something that resists normal scientific observation might be out there.

It resonates with a feeling that few critics have noticed. George Santayana regards the word as "Romantic," in that it alludes indirectly to a cultural condition and a state of feeling rather far from the rational optimism of the naturalists' narratives of progress. As Santayana puts it,

> The agnostic was haunted by ghosts of substance, filling his whole experience with a sense of discomfort, ignorance, and defeat. Those substances were real but elusive; and though he never saw them, the agnostic remembered only too well the tales once told concerning them, and secretly desired to have assurance of their truth; only he thought such assurance was eternally denied him by his psychological constitution.[22]

Santayana's description applies with painful precision to Leslie Stephen, whose dark and sometimes angry broodings in defense of agnosticism, and moral outrage against pious critics of science and naturalism, implied a deep sense of loss to which he was unable to reconcile himself. In "An Agnostic's Apology," Stephen's melancholy is on display almost melodramatically. We are, he writes,

> a company of ignorant beings, feeling our way through mists and darkness, learning only by incessantly repeated blunders, obtaining a glimmering of truth, by falling into every conceivable error, dimly discerning light enough for our daily needs, but hopelessly differing whenever we attempt to describe the ultimate origin or end of our paths; and yet, when one of us ventures to declare that we don't know the map of the universe as well as the map of our infinitesimal parish he is hooted, reviled, and perhaps told that he will be damned to all eternity for his faithlessness.[23]

Feeling the nostalgia, Stephen is angered that, in his heroic self-abnegation in the cause of truth, he is blamed for having lost the certainty for which he mourns. There are, of course, personal forces at work in Stephen's angry sadness. Stephen was writing this shortly after the death of his first wife. Just as Darwin's loss of his beloved daughter Annie seems to have intensified his feelings about Christianity's inadequacy,[24] so Stephen's inability to find consolation in Christianity for the loss of his wife seems to have pushed him yet further away from religion. Pride in loss and in the defense of truth at all costs produces a tone that is characteristic of the naturalists' efforts to reimagine the world.

Spencer, the inventor of "the Unknowable," was the great naturalist formulator of the naturalists' epistemological crisis. Not quite the materialist-reductivist that Huxley and Tyndall had almost become, and not hostile to religion, Spencer was nevertheless the most influential philosopher for the naturalists, and he initiates his system by plunging immediately into empiricist mystery and running into the same paradoxical problems. He was not adept at artful rhetoric, and I don't propose that his writing is a delight to read or has anything like the vitality, playfulness, suppleness, and richness of meaning that we find in Huxley, but his full engagement with the naturalists' myth of progress and his more than epic ambition are perhaps the best demonstrations of the paradoxical nature of the naturalist enterprise.

Santayana attempts to salvage from Spencer's philosophy, which had already in 1923 fallen out of fashion, some of the epic qualities that, I have been arguing, are part of the naturalists' vision. "Herbert Spencer," he says,

> dutifully gathered reports from every quarter and let them settle as they would in the broad levels of his system, as in geological strata; whence that Homeric sweep with which he pictures progress and decay, not in aversion from the severities of natural existence, but as the mechanical sediment of the tides of matter and motion, perpetually surging. Of course this epic movement, as Spencer describes it, is but a human perspective; he instinctively imposes his grandiloquent rhythms on things as he does his ponderous Latin vocabulary, or as Empedocles or Lucretius imposed their hexameters.[25]

Santayana's perspective is particularly apt for my point, not least because he is so much better a writer than Spencer. He traces with some sympathy Spencer's maneuvers through the contradictions inherent in the naturalists' efforts to construct a world rationally and empirically satisfying, and coherent with the really Christian moral ideals they imported into the material world; he hears even in Spencer's usually dry and systematic language a quality that derives from an unspoken obsession with lost substance and epistemological limit.

Spencer's early essay "Progress: Its Law and Cause" (1857) makes all the recognizable naturalist moves: he does not aim at the "solution of the great questions with which Philosophy in all ages has perplexed itself. Let none thus deceive themselves," he warns. "The ultimate mystery remains just as it was." In his terms at least, all "fearless inquiry," implicitly like his own heroic efforts, "tends continually to give a firmer basis to all true religion."[26] Even with the freethinking naturalists' usual contempt for those he calls "timid sectarians," he leaves an enormous mysterious gap in his theory into which religion is invited to enter.

For, as he puts it, when one traces back the evolution of things, one cannot ever understand how the initial conditions out of which all of nature has evolved came to be. There will always remain the question: and what preceded? Looking inward, he finds that "both terminations of the thread of consciousness" are beyond grasp; similarly, one cannot ever know things in "their essential nature"; and trying to understand the "sensations" that are the source of empirical knowledge, he finds that one "cannot in the least comprehend sensation," or "how sensation is possible." And so, like Tyndall when he spins away from what can be known empirically, he becomes dramatic, or as dramatic as Spencer can be. His rhetoric rises as the Unknowable looms. "In all directions," the scientist's investigations

> eventually bring him face to face with the unknowable; and he ever more clearly perceives it to be the unknowable. He learns at once the greatness and the littleness of human intellect—its power in dealing with all that comes within the range of experience; its impotence in dealing with all that transcends experience. He feels, with a vividness which no others can, the incomprehensibleness of the simplest fact, considered in itself. He alone truly *sees* that absolute knowledge is impossible. He alone *knows* that under all things there lies an impenetrable mystery.[27]

When this essay became part of Spencer's larger enterprise, in his *First Principles* (1862), he claims, in an aphoristic formulation that might be appropriate for a Mallock or even an Oscar Wilde, that the scientist, "more than any other, truly *knows* that in its ultimate essence nothing can be known."[28] Science, then, is nescience.

The effects certainly are startling: the naturalists' narrative, as it confronts deep positivist skepticism, leans toward religion, for what cannot be known is religion's province. The great tradition of English philosophy, which always carries with it that burden of skepticism that Hume developed so radically, becomes in Spencer's system the support for religion. In this respect, the empiricist argument of the scientists matches closely the pattern of thought we find in the writing of John Henry Newman making his case in *The Grammar of Assent* (1870). Newman, however, occupies that dark space with the positive force of faith, which takes priority over the rationality that confirmed the reality of the space in the first place.[29] Spencer begins with no faith, and it is impossible even to conjecture about the Unknowable. It is simply what could explain everything if we could only know it. Thus, behind Spencer's natural evolutionary system, there is something that might be described as "spiritual," or a "God." In the Unknowable lies the hidden, lost substance, after all.

In facing the question of the relations between science and religion, as he does in the first chapters of his *First Principles*, Spencer begins by seeking some underlying matrix of truth that unites them. Along the way, he gives nothing of the naturalist ground away, and he offers a narrative of progress through which science gradually wrests authority away from religion at stage after stage of the long growth to knowledge. In this respect, religion retreats before the new truths with which science belies traditional beliefs about the way the world is. Nevertheless, Spencer concedes, given the persistence and the virtual universality of religion, it must have at its foundation some truth, and since truth is also the fundamental quest of science, there must be some space for conciliation. We hear again the motif: "Positive knowledge does not, and never can, fill the whole region of possible thought." So "there must ever remain two . . . antithetical modes of mental action."[30]

Spencer's final reconciliation of these two antithetical modes wins, from me at least, a prize in paradoxicality, if I may use a noun that my computer program rejects with red underlining. All that underlies the naturalist project seems here to be paradox. I quote this also in part as contrast to what I take to be the greater literary skills of many of the other naturalists:

> Common sense asserts the existence of a reality; objective science proves that this reality cannot be what we think it; subjective science shows why we cannot think of it as it is and yet are compelled to think of it as existing; and in this assertion of a reality utterly inscrutable in nature, religions find an assertion essentially coinciding with her own. We are obliged to regard every phenomenon as a manifestation of some power by which we are acted upon; though omnipresence is unthinkable, yet as experience discloses no bounds to the diffusion of phenomena, we are unable to think of limits to the presence of this power; while the criticisms of science teach us that this power is incomprehensible. And this consciousness of an incomprehensible power, called omnipresent from inability to assign it its limits, is just that consciousness on which religion dwells.[31]

This exercise in paradox is verbally bewildering; it has not the literary power that some of the similar mysticism and paradoxicality of the other great naturalists display, and it leaves us with a religion that exists just where we haven't a clue.

As Santayana puts it critically in his discussion of Spencer's "unknowable,"

> Genuine religion professes to have positive knowledge and to bring positive benefits: it is an art; and to ask it to be satisfied with knowing that no knowledge can penetrate to the heart of things is sheer mockery: the opposite is what religion instinctively asserts.[32]

That is to say, Spencer's plunge into the paradoxical "unknowable" is another form of Huxley's more direct confrontation with paradox in *Evolution and*

Ethics (1893), and it does even less to satisfy the religious spirit than does Huxley, in his forceful and unreasoned assertion of the need for morality to do battle with nature. Spencer's long story of the evolutionary development of everything is, finally, built on a grand paradox that he, no more than Huxley, pretends to be able to resolve with the naturalistic methods available to him.

Original Sin

No text better exemplifies this argument than Huxley's "Prolegomena" to *Evolution and Ethics*. Despite the deserved demystification to which Desmond and others have subjected it, it survives its moment and still has the literary power to challenge and reshape our imaginations. I had evidence of that when, in an undergraduate course I recently taught, with readings including Darwin, Lyell, Wallace, Thomas Hardy, and Joseph Conrad, among others, it was the big hit. Students found the allegory of the garden rich in possible meanings as it pointed in one direction to the organization of our own societies, in another to the conditions of empire, and in yet another to the processes of nature. They were taken by the lucidity of Huxley's prose and the richness of his metaphorical imagination, and by the clarity with which he laid out crucial and still living issues and challenged their own assumptions. They certainly felt something of the shock Huxley sought to provoke in its radically paradoxical argument.

Their immediate sense of the literary power of the essay was intensified by the way its vision was translated shortly after in a yet more famous book, Conrad's *Heart of Darkness* (1899), where the garden metaphor explodes into full-scale narrative and the opening lines of Huxley's essay seem to be re-echoed.[33] Huxley's vision from the windows of his modern study of a prehistoric England in a "state of nature," trampled on by "prehistoric savages,"[34] is echoed almost precisely in Marlow's first speech as he stares at the darkened Thames, with the city of London all around: "This also . . . has been one of the dark places of the earth."[35] Mr. Kurtz looks very much like Conrad's version of Huxley's "perfect administrator," both figures with strong moral missions, the administrator of Huxley's allegorical garden to resist the "cosmic process," Kurtz, as ideal of European culture, the redeemer of the imperialist looting of Africa. Students found satisfying Huxley's ironically formulated rejection of the idea that "human society is competent to furnish, from its own resources, an administrator," who by "systematic extirpation, or exclusion, of the superfluous" could ensure that the allegorical garden of the "ethical process" would not be destroyed by the "cosmic process." "I doubt," wrote Huxley,

> whether even the keenest judge of character, if he had before him a hundred boys and girls under fourteen, could pick out, with the least chance of success, those who should be kept, as certain to be serviceable members of the polity, and those who should be chloroformed, as equally sure to be stupid, idle, or vicious. The points of a good or of a bad citizen are really far harder to discern than those of a puppy or a short-horn calf.[36]

The Swiftian satire intensifies the horrific implications of eugenics and of preservation of the "garden." The garden ideal suddenly transforms into a butcher's nightmare. Like Huxley's "perfect administrator," Kurtz is shown to have absorbed the very best of Western education; as Huxley's administrator, charged with eliminating the worst, cannot discriminate them from the best, so Kurtz ends up savagely scrawling the unforgettably chilling line, "Exterminate all the brutes."[37] Huxley's perfect administrator enters, "extirpating and excluding the native rivals, whether men, beasts, or plants." "When the colony reached the limit of possible expansion, the surplus population must be disposed of somehow."[38] This businesslike prose, following the lines of logic and moving toward extermination, is almost more chilling than Kurtz's sudden emotional exclamation. The allegory resonates with affective power beyond its literal terms, as the possible application expands to eugenics' programs, to imperialist ventures, to totalitarian governments. Huxley's satiric denigration of the perfect administrator is almost surely, as Desmond argues, a conservative defense against socialism, but it was for the students, beyond ideology, a powerful dramatization of human and scientific limitation.

Most startling and shocking in this essay of paradox is Huxley's cool explanation of the inadequacy of the Golden Rule. "If," Huxley puts it, in the prose of dry understatement fit again for Swiftian satire,

> I put myself in the place of a man who has robbed me, I find that I am possessed by an exceeding desire not to be fined or imprisoned; if in that of the man who has smitten me on one cheek, I contemplate with satisfaction the absence of any worse result than the turning of the other cheek for like treatment.

With a power still to shake and disturb, Huxley argues with cynical coolness that the consequences of the Golden Rule "are incompatible with the existence of a civil state, under any circumstances of this world which have obtained, or, so far as one can see, are likely to come to pass."[39]

And yet, in what might be taken as a further paradox, Huxley's imagination of the entirely secular world of the garden is governed by a Christian vision, intensified, in fact, just because, with Calvinist austerity, he disconnects nature from all the values on which our civilization is built. The epistemological, ethical, and biological limits that his dry scientific vision reveals leave a world that excludes not only utopianism, but pleasure itself. Ironically, Huxley's is the fallen world of Christian doctrine, for he claims, as much preacher as evolutionist, "Every child born into the world will bring with him the instinct of unlimited self-assertion."[40] It is original sin. And original sin entails the struggle for existence, so that the garden, culture itself, can only be sustained by repressing—although Huxley believes that we can never entirely succeed—what we are all born with. The traditional elements of Christianity are subjected to this rationalist, secular translation and become dark and abrasive, not so much evil as amoral.

Huxley's connection of a naturalist perspective with an austere Calvinist vision of a sinful world threatens, in the essay's progress, to subvert the naturalists' progress narrative, as it moves from the small-scale "three or four years have elapsed," to the ultimate destruction through the cosmic process. The garden, a mere "patch," thrives until the birds, and blown seeds, and blight and mildew and overpopulation do their work. The parable of the garden is a parable of what Huxley insists on, "the distinction . . . between the works of nature and those of man."[41]

Natural beings like us must act *against* nature in order for the ethical to survive at all. That "instinct of unlimited self-assertion," which Huxley also believes is a condition for "success in the war with the state of nature outside," threatens every horticultural move; and thus, the garden, the ideal organization of human community, can survive only if we repress our fundamental natures.[42] Freud looms. Christian renunciation of the world is a not illogical inference. The naturalists, who recognize no forces in the processes of this world but natural ones, have to recognize that nature has no intrinsic moral qualities: "In sober truth," John Stuart Mill had famously put it, "nearly all the things which men are hanged or imprisoned for doing to one another are nature's everyday performances."[43] So the "ethical process," which produces a "state of art," wards off the cosmic force that produced it. Human society, the "garden" of Huxley's remarkable and beautifully sustained metaphor, is bent on "the creation of conditions more favorable than those of the state of nature" for its survival and happiness. But Huxley renounces happiness as well: we should, he insists, "cast aside the notion that the escape from pain and sorrow is the proper object of life." The "ethical process" works "in opposition to the state of Nature, the State of Art." The climax of such a vision can only be a paradox, as it is in Huxley's famous assertion in the "Prolegomena": if you say that "the cosmic process cannot be in antagonism with that horticultural process which is part of itself—I can only reply that if the conclusion that the two are antagonistic is logically absurd, I am sorry for logic, because as we have seen, the fact is so."[44] This is perhaps the climactic paradox of the nineteenth century.

It is not only Huxley who has to wrestle with this paradox looming for all naturalists. Their position, usually so confidently affirmed, requires them to wrestle with the contradictions that brought Huxley at last to this point. *Evolution and Ethics* reflects a state of mind quite different from that intimated in Huxley's early optimism. The potentially subversive and revolutionary program implied in the optimistic prose of so much of the naturalists' work, not least Huxley's own, becomes a kind of social conservatism often connected to the Carlylean call to work and renunciation. Huxley's highly moralistic insistence on resistance to the brutalities of nature is not a form of evolutionary ethics but a rejection of their possibility. Huxley's nostalgia for the world that his whole career as a proselytizer for science was bent on disrupting gives to *Evolution and Ethics* a contradictory but almost elegiac note. The cosmic process has produced the humans whose art produces the garden, art, law, ethics, that fight the cosmic process. The "Prolegomena," in effect, is an argument that paradox is the inevitable form of the full naturalist program.

Conclusion

I have emphasized the paradoxical nature of the naturalists' vision not because I want to demonstrate that there are self-contradictions in their enterprise, but for almost a reverse reason: because the strategy of the paradox can be a powerful, disturbing, and attractive force in engaging readers imaginatively in alternative ways to see the world. It is a strategy of art, and at their best, the naturalists might be thought of as artists.

In emphasizing the naturalists' penchant for paradox in relation to the intellectual contradictions that inhere in their way of looking at the world, I have not emphasized sufficiently the creative power of that paradoxical mode, its capacity to change the way we look at the natural world and think about the world of human culture. I know it is comparing great things to small, but it is useful to remind ourselves that it is possible to come away deeply moved from the epics of Dante or Milton without believing in the literal realities or accepting the social and moral implications of their narratives, so it should be possible to come away from reading the best prose of the scientific naturalists. That prose often provides a thrilling vision, or many visions. "We on the earth's surface," Tyndall claimed in his essay on "The Constitution of Nature," "live night and day in the midst of ethereal commotion."[45] It does not matter now that we know that the Victorians got it wrong about ether: that sense of "commotion" as we look at the silent serenity and dignity of the movement of the stars or the still and apparently inanimate stolidity of the most ordinary stone infuses the world with energy, changes the way we can think about those stars and those stones, about our own powers of perception, about our relation to the natural world we think we know. Tyndall's description of "the idea of an all-pervading ether which transmits a tingle, so to speak, to the finger ends of the universe every time a street lamp is lighted" quietly plays with internal rhyme, and assonance, and alliteration as it animates the world and registers better than most poetry might do the vision of the universal connection of things. It tries to make us feel to our fingertips the experience of what is not visible.

I have dealt only partially with the profound ethical contradictions in the naturalist program, although their ethical struggles were probably more important in the controversies than the epistemological ones. In that context, it is important to recognize how their essays strain, and often very successfully, to evoke a sense of the author's honor, and of the not disingenuous labor toward the new gospel that they all believed needs to be broadcast. Clifford, intellectual bad boy though he was, often seems the most purist of the purist moralists, more rigorous and demanding than any clergy, as he shouts that "truth is a thing to be shouted from the housetops, not to be whispered over rosewater after dinner when the ladies are gone away."[46] I love William James's complication of this way of thinking,[47] but there remains the refreshing boldness of Clifford's "robustious" manner, the wonderful contradiction of the bad boy being bad for a greater goodness. His no-nonsense prose seems justified: truth requires directness, not careful decorous disguise, just because for Clifford (and the scientific culture for which he proselytized) so much was at stake. A refusal to stand on ceremony or accept conventional protocols in the pursuit of the truth is a condition of personal moral integrity and of a viable community as well. We all know Clifford's famous edict, quoted most notably and negatively by James: "It is wrong always, everywhere, for any one, to believe anything upon insufficient evidence."[48] The rhythmic and alliterative urgency of Clifford's assertion suggests something of the moral energy that drives it and allows it to retain its impracticable power in spite of James's wonderful and subtle critique. Truth is not only a goal in the naturalists' drama of progressive enlightenment; it is heroic moral drama as well.

But Clifford's noisy insistence on truth is a reminder that the naturalists are always on the verge of the irrational they abjured; they are always, in their aspiration to detachment and objectivity, passionately committed to values that take them beyond their science and link epistemology and ethics in what

Stephen Jay Gould would have called a single magisterium after all. This epic aspiration emerges strongly from their best writing, in tensions that paradoxes best express, and with a power beyond the dominant rational mode they insisted on. Paradox is the dramatic force of the naturalists' epic of the march to truth over the corpses of dead beliefs in transcendental power. It has startling and disturbing power, even as it dramatizes, once one steps back from it, the theoretical incoherence of the great naturalist project. The literature of the naturalists is full of an intensity and vitality that at the same time expose its contradictions. It is a daring prose they write, and it is they who, recognizing the paradoxical nature of their enterprise, help us to critique it.

Notes

1. John Tyndall, "The Belfast Address," in *Fragments of Science*, 2 vols. (New York: Appleton, 1899), 2:198–99.

2. John Holloway's classic argument applies to Tyndall as well as to the more literary of the Victorian sages. Holloway argued that the distinctive characteristic of their prose was that it never relied exclusively on rational argument, but depended for its power on invocation of experience and feeling. John Holloway, *The Victorian Sage* (New York: W. W. Norton, 1965).

3. Adrian Desmond, *Huxley: Evolution's High Priest* (London: Michael Joseph, 1997), 637, 642.

4. Bernard Lightman, "'Fighting Even with Death': Balfour, Scientific Naturalism, and Thomas Henry Huxley's Final Battle," in *Thomas Henry Huxley's Place in Science and Letters: Centenary Essays*, ed. Alan P. Barr, 323–50 (Athens, GA, and London: University of Georgia Press, 1997).

5. Frank Miller Turner, *Between Science and Religion: The Reaction to Scientific Naturalism in Late Victorian England* (New Haven, CT: Yale University Press, 1974), 36.

6. See, among others, Bernard Lightman, *The Origins of Agnosticism: Victorian Unbelief and the Limits of Knowledge* (Baltimore: Johns Hopkins University Press, 1987).

7. Tyndall, "Belfast Address," 197.

8. W. K. Clifford, "On Some of the Conditions of Mental Development," in *Lectures and Essays*, ed. Leslie Stephen and Frederick Pollock, 2 vols. (London: Macmillan, 1901), 1:117.

9. John Tyndall, "On the Study of Physics," in *Fragments of Science*, 1:290.

10. T. H. Huxley, "On the Educational Value of the Natural Historical Sciences," in *Science and Education* (London: Macmillan, 1893), 45.

11. T. H. Huxley, "On the Advisableness of Improving Natural Knowledge," in *Methods and Results* (London: Macmillan, 1893), 41.

12. Tyndall, "Belfast Address," 201.

13. George Eliot, *Middlemarch*, ed. David Carroll (1871; repr., Oxford: Oxford University Press, 1998), chap. 16, 162–63.

14. John Tyndall, "The Scientific Uses of the Imagination," in *Fragments of Science*, 2:124.

15. For an extensive discussion of paradox in Darwin's prose, see my *Darwin the Writer* (Oxford: Oxford University Press, 2011), particularly chaps. 4 and 5.

16. Clifford, "On Some of the Conditions of Mental Development," in *Lectures and Essays*, 1:79.

17. Arthur Balfour, *The Foundations of Belief: Notes Introductory to the Study of Theology* (New York: Longmans, Green, 1895), 107.

18. W. H. Mallock, *The New Republic*, ed. J. Max Patrick (1877; repr., Gainesville: University of Florida Press, 1950), 42.

19. T. H. Huxley, "The Progress of Science," in *Method and Results*, 60–61.

20. T. H. Huxley, "Agnosticism," in *Science and Christian Tradition* (London: Macmillan, 1895), 243.

21. Ibid., 239.

22. George Santayana, *The Unknowable: The Herbert Spencer Lecture* (Oxford: Clarendon Press, 1923), 11.

23. Leslie Stephen, *An Agnostic's Apology and Other Essays* (London: Smith, Elder, 1893), 40.

24. But see John van Wyhe, "The Annie Hypothesis: Did the Death of His Daughter Cause Darwin to 'Give Up Christianity'?," *Centaur* 54 (2012): 105–23. This is not the place to enter the renewed controversy about whether Annie's death led to Darwin giving up Christianity. As Wyhe reminds the reader, Darwin contended that his progress toward disbelief was slow and ultimately, therefore, painless. There is much evidence that this process was in operation before Annie's death and van Wyhe musters strong negative evidence that the death did not lead to a sudden deconversion. Positive evidence in writing or no, it is hard to imagine that Darwin's already keen sense of suffering throughout the biological world that made belief in a universally loving and omnipotent god impossible was not exacerbated by Annie's death.

25. Santayana, *Unknowable*, 3.

26. Herbert Spencer, "Progress: Its Law and Cause," in *Illustrations of Universal Progress: A Series of Discussions* (New York: Appleton, 1864), 58.

27. Ibid., 60.

28. Herbert Spencer, *First Principles* (New York: A. L. Burt, 1904), 57.

29. In *The Origins of Agnosticism*, Lightman points out the agnostic strategy of using an eminent religious thinker with fideist sympathies as an accomplice. Although Huxley, Stephen, and Spencer found Mansel somewhat more useful than Newman in this respect, they "referred to Newman as a sceptic, but Mansel, as one of their own, an agnostic," 115.

30. Spencer, *First Principles*, 13.

31. Ibid., 85.

32. Santayana, *Unknowable*, 19.

33. Conrad scholars have long noted the likely connection between Huxley's *Evolution and Ethics* and *Heart of Darkness*, among others Ian Watt, *Conrad in the Nineteenth Century* (Berkeley: University of California Press, 1979), 162–65; and Zdzislaw Najder, *Joseph Conrad: A Chronicle* (New Brunswick, NJ: Rutgers University Press, 1984), 249–50. Both scholars point to the similarity of ideas and language, and Watt recognizes in Kurtz the manifestation of primitive origins.

34. T. H. Huxley, "Prolegomena," in *Evolution and Ethics* (London: Macmillan, 1911), 1.

35. Joseph Conrad, "Heart of Darkness," in *Youth and Two Other Stories* (Garden City, NY: Doubleday, Page, 1924), 48.

36. Huxley, "Prolegomena," 23.

37. Conrad, "Heart of Darkness," 118.

38. Huxley, "Prolegomena," 18, 21.

39. Ibid., 32, 90.

40. Ibid., 44.

41. Ibid., 11.

42. Ibid., 27.

43. John Stuart Mill, "Nature," in *Three Essays on Religion* (London: Longmans, Green, 1874), 28.

44. Huxley, "Prolegomena," 86, 12.

45. John Tyndall, "The Constitution of Nature," in *Fragments of Science* (New York: D. Appleton, 1872), 8.

46. W. K. Clifford, "Right and Wrong," in *Lectures and Essays*, 2:175–76.

47. See, among other essays, William James, "Is Life Worth Living?," in *The Will to Believe and Other Essays* (London: Longmans Green, 1897), 32–62.

48. W. K. Clifford, "The Ethics of Belief," in *Lectures and Essays*, 2:179.

Institutional Politics

4

Huxley and the Devonshire Commission

BERNARD LIGHTMAN

For Thomas Henry Huxley, 1871 was an unusually hectic year. Apologizing in a letter of 7 July to Anton Dohrn, a German marine biologist, for being a bad correspondent, he explained, "I have been frightfully hard-worked with two Royal Commissions and the School Board all sitting at once, but I am none the worse, and things are getting into shape."[1] But the hectic pace took its toll. By December, plagued by incessant nausea, he was unable to give lectures and struggled to write letters. Huxley was close to a complete collapse.[2] Writing to Dohrn on 3 January 1872, he admitted, "Though beyond general weariness, incapacity and disgust with things in general, I do not precisely know what is the matter with me."[3] After securing a leave of absence from the School of Mines, he booked passage to Egypt, a popular tourist destination during the winter. He left on 11 January, and returned to London on 6 April, after exploring Cairo and traveling up the Nile in a houseboat. Huxley's work on the Royal Commission on Science Instruction and the Advancement of Science, also known as the Devonshire Commission, was undoubtedly a factor contributing to his illness, though in 1871 he was also busy working on the London School Board and the Royal Commission on Contagious Diseases. From 14 June 1870 up to 11 July 1871, the Devonshire Commission had met forty-one times to examine witnesses. It was demanding work, and completely unpaid. The Royal Commission on Contagious Diseases completed its work by the end of 1871. Before Huxley left for Egypt, he resigned from the London School Board.[4] But he held on to his appointment to the Devonshire Commission after he returned from Egypt. Why was Huxley so committed to the work of the Devonshire Commission, to the point where he was willing to risk his fragile health?

There were two reasons. First, Huxley and some of his X Club allies had lobbied to have the Royal Commission established.[5] In August of 1868, Alexander Strange, an army officer and astronomer, delivered a paper "On the Necessity of State Intervention to Secure the Progress of Physical Science" to the British Association, calling for a consistent government policy for science. Huxley, along with John Tyndall, Edward Frankland, and Thomas Hirst, was appointed to a British Association committee in November 1868, charged with the task of determining whether or not there was sufficient national provision for scientific research. The committee issued a questionnaire to men of science in March 1869, and later concluded, based on the replies, that substantial sums of money were required to fund individuals prepared to undertake scientific research full-time.[6] In late 1869 the British Association approached the government about establishing a formal inquiry, and, in February 1870, shortly after his election as prime minister, William Gladstone agreed to appoint a Royal Commission under William Cavendish, the seventh Duke of Devonshire, with Norman Lockyer serving as secretary. The commission was charged with the task of studying the existing national provision for scientific instruction and the advancement of science.[7] Huxley was not about to resign from the commission after the work that he and his friends had put into establishing it. Second, there seemed to be real potential for having a significant impact on the future of British science through the work of the commission. Whereas his work on the London School Board was limited in scope to London elementary schools, the Devonshire Commission had the mandate to deal with all levels of education throughout Britain, as well as the crucial issue of state support of science in general. Virtually all aspects of British science were to be put under the microscope of the commission. As Adrian Desmond has observed, in his role as commissioner, Huxley "was tasting real power."[8]

Although the Devonshire Commission was an important nineteenth-century government commission charged with investigating the state of British science, scholars have paid relatively little attention to it. There is no book-length study. Much of the scholarship that does deal with the commission is over thirty years old. Peter Alter's *The Reluctant Patron* contains scattered references to the commission, while half of Roy MacLeod's article on the endowment of science movement is devoted to an account of the establishment of the commission and its reports.[9] A. J. Meadows dedicates one chapter to the Devonshire Commission in his biography of Norman Lockyer, while D. S. L. Cardwell devotes a short section within a chapter of his work on the organization of English science.[10] Huxley's role on the commission, which also deserves more attention from scholars, is, unsurprisingly, little understood, despite the interest in Huxley as an educator.[11] In her forthcoming book on the X Club, Ruth Barton has analyzed the views of the members of this illustrious club on state support for science, focusing in particular on Joseph Dalton Hooker and Tyndall. I second her call for further research on the opinions of Huxley, Hooker, and Frankland through a comparison of the evidence they gave before the Devonshire Commission.[12]

In this chapter I offer a study of Huxley's role on the Devonshire Commission. It is not always easy to disentangle Huxley's main concerns from those of the rest of the commissioners. Desmond assumes that the recommendations of the committee fully reflected Huxley's wishes. He refers to the Devonshire Commission as Huxley's "political machine."[13] But Meadows, who treats the commission from Lockyer's point of view, maintains that its final recommendations represented an overwhelming endorsement of the views that Lockyer had been advocating.[14] By analyzing the questions that Huxley put to the various witnesses before the commission, as well as the evidence he gave, we can discern what Huxley hoped to accomplish. I will argue that Huxley's main concerns were the research grant to the Royal Society, the fate of the Royal School of Mines, defending his friend Hooker from the imperialist aims of Richard Owen, and the establishment of a national system of scientific education that began in elementary schools. An assessment of Huxley's role on the Devonshire Commission reveals what, for him, constituted the main priorities of scientific naturalism in the early 1870s.

The Royal Commission and the Institutions of Scientific Naturalism

The power of the leading evolutionary naturalists was based, in large part, on their positions at important scientific institutions. When the Devonshire Commission began to meet in 1870, Tyndall was already well ensconced at the Royal Institution. He had been appointed professor of natural philosophy there in 1853. In 1867 he succeeded Faraday as superintendent. Similarly, Hooker had recently been put in charge of an important scientific institution. In 1865 he had succeeded his father as director of Kew Gardens. Frankland's home base was the Royal College of Chemistry. He had been appointed to a permanent professorship in 1868 when A. W. Hofman decided to stay in Germany. Huxley was located at the Royal School of Mines, where he was professor of natural history. He had been working at the School of Mines since 1854. For Huxley, the Devonshire Commission provided a means of protecting, or even strengthening, the institutional structures that supported scientific naturalism. The composition of the nine-man Devonshire Commission ensured that he had sympathetic allies who would support this goal.

William Cavendish, the seventh Duke of Devonshire, chaired the commission. Cavendish, an industrialist, politician, and landowner, had been the first chancellor of the University of London, from 1836 to 1856, and the first president of the Iron and Steel Institute when it was founded in 1868. Cavendish had the reputation of being a model for the industrious, public-spirited mid-Victorian aristocrat. He later offered sixty-three hundred pounds toward the establishment of the Cavendish Laboratory at Cambridge in 1874.[15] The fifth Marquess of Landsdowne, Henry Charles Keith Petty-Fitzmaurice, politician and landowner, also came from the nobility. He had just graduated from Oxford, where he had been tutored by Benjamin Jowett. Since he showed promise as a Liberal politician, he was appointed as a junior whip by Gladstone in 1869 and then to the commission. George Gabriel Stokes, a physicist, had been Lucasian Professor at Cambridge since 1849. He held conservative religious views uncongenial to scientific naturalism.

From Huxley's perspective, the five other members of the commission were more likely to be allies. Henry John Stephen Smith was a mathematician, who, since 1861, had held the Savilian Professorship of Geometry at Oxford, as well as a mathematical lectureship at Balliol College.[16] Though he respected the classical learning so highly valued at Oxford, he was also known to harbor liberal sympathies due to his support of the promotion of the natural sciences within the university. James Kay-Shuttleworth, civil servant and educationist, had been a longtime supporter of more scientific education in the public schools. Nearing the end of his career, and no longer a major player in educational politics by 1870, he offered "strong support" to Huxley.[17] Bernhard Samuelson, ironmaster and advocator of technical education, had served on both the Acland Committee in 1867 and the Society of Arts Committee on Technical Education in 1868.[18] William Sharpey had long held the chair of anatomy and physiology at University College London, and served as secretary of the Royal Society from 1854 to 1872. He was seen as the father of modern physiology in Britain. Sharpey knew Huxley from long ago—as the student to whom he had awarded the gold medal for anatomy and physiology for his work on part one of the Bachelor of Medicine exam in 1845.[19] Another potential ally on the commission for Huxley was fellow X Clubber John Lubbock, banker, politician, and scientific writer on archaeology and entomology. Huxley also had an "ace in the hole," the secretary of the commission, Norman Lockyer, civil servant, astronomer, and journal editor. Huxley and Lockyer had worked together previously on a failed periodical, the *Reader*. As Melinda Baldwin points out in her chapter, when Lockyer became editor of *Nature* in 1869, he drew on the pens of Huxley and other scientific naturalists in the early years of the journal.[20] Including himself and Lockyer, Huxley had a majority on the commission, which ensured that issues of concern to scientific naturalism were seriously considered and that individuals who shared those concerns were called as witnesses. Frankland, Hooker, and Spottiswoode, as well as Huxley himself, were among the witnesses who gave testimony.

The very first issue dealt with by the commission, an inquiry into the state of the Royal School of Mines, the Geological Survey of Great Britain and Ireland, the Mining Record Office, and the Museum of Practical Geology, all located in Jermyn Street, and the Royal College of Chemistry on Oxford Street, fit perfectly into Huxley's agenda. For some time he had been touting an ambitious plan to create a general Science School that, not by chance, also provided him the opportunity to escape his unhappy situation at the School of Mines. The plan involved forming the new school under one administration and at South Kensington, bringing together the pure science part of the School of Mines, chemistry from the Royal College of Chemistry, and mathematics from the School of Naval Architecture.[21] It effectively separated the School of Mines from the Geological Survey and its museum, the latter remaining on Jermyn Street. Despite Huxley's growing public reputation throughout the 1860s in the wake of the controversies over evolution, his career had remained static. As naturalist to the Geological Survey in charge of fossil collections in an institution designed to be a museum, Huxley could not pursue his dream of creating a cadre of biologists who would bring laboratory methods to British schools and universities.[22] Just before the Royal Commission began to call witnesses, a letter to the editor in *Nature* indignantly pointed to Huxley's intolerable situation. "Is it not a crying shame," the anonymous contributor wrote, "that at the present time such a man as Huxley is completely isolated from the younger biological workers, and instead of, like Cuvier, having a large laboratory manned by an enthusiastic body of scholars, ready to dissect everything after its kind, is penned up in an abominable den in Jermyn Street, and . . . has, in fact, to work upon the world through the bars of a prison cage?"[23]

Huxley wasted no time building a case for his plan. On 14 June 1870, he questioned the very first witness to appear before the commission, Henry Cole, secretary of the Science and Art Department of the Committee of the Council on Education, about the desirability of establishing a new Science School. Since Cole was one of those working with Huxley to make the plan a reality, he was an extremely cooperative witness. Through questioning Cole, Huxley established that when the School of Mines was created in 1851, the original aim was to institute chairs of general science, as the founders believed that "technical science must be made subordinate to general science." To Huxley's leading question if it was desirable to permit only the mining students to attend the general courses, Cole merely answered, "I think not."[24] This line of questioning established that Huxley's plans for the School of Mines were more in keeping with the original intentions of the founders.

The following day Huxley was called as a witness. Responding to questions about the problems with the current examination system, Huxley declared that the "ideal system" would be to bring all science teachers to London to a "normal" or "proper training school." Huxley told the commission that he believed that "if you get a body of trained teachers all over the country, they will do your technical work for you very much better than you can do it in any other way." When asked if the School of Mines could house a training school of the kind he envisioned, Huxley answered in the negative. In the School of Mines there was "no means of teaching several of the subjects practically." The lack of a biological laboratory meant to Huxley, "I cannot teach in the proper sense of the word." All that could be said about the School of Mines is that it could provide "a nucleus of an efficient body."[25]

When other witnesses were called in June, Huxley pressed them for details about the shortcomings of the School of Mines and pushed for the establishment of a new Science School. Huxley summed up the testimony of Trenham Reeks, registrar of the School of Mines, with the rhetorical question, "So that the whole School of Mines has been grafted, so to speak, and not a very clever graft, upon a very different plan?" Through his questioning of Andrew Crombie Ramsay, an examiner in the Science Department of the Committee of the Privy Council on Education, Huxley tried to demonstrate that the increase in the number of surveyors for the Geological Survey of Great Britain had caused space problems at the School of Mines. In the course of questioning C. W. Merrifield, principal of the Royal School of Naval Architecture at South Kensington, Huxley led him to acknowledge that no school existed for students to obtain a general education combined with a preliminary course of study including mathematics and general physics.[26]

Roderick Murchison, the director of the Geological Survey and of the Royal School of Mines, and a fervent opponent of Huxley's scheme, was not called as a witness until 8 July 1870. In his late seventies, Murchison was still suffering from the effects of a stroke. The Duke of Devonshire inquired if he had been asked his opinion as to the proposed move of the educational branches of the Royal School of Mines from Jermyn Street to South Kensington. Murchison responded angrily, "Never; and I wish now to place on record my protest against the scheme for breaking up the Royal School of Mines, and to express my surprise that such a plan should have been contemplated without consulting myself and the council of lecturers in our establishment."[27] Compared to other witnesses, Murchison's appearance before the commission was quite brief.

Unsurprisingly, when the *First Report* of the commission appeared on 9 March 1871, the recommendations were fully in line with Huxley's vision. Denying that there was a "necessary connexion" between the Geological Survey and the Royal School of Mines, the commissioners urged that the two be separated, and that the School of Mines and the Royal College of Chemistry be combined to form a new institution called the "Science School," to be "accommodated in the buildings at South Kensington, now nearly completed." The Science School was to be provided with "sufficient Laboratories and Assistance for giving Practical Instruction in Physics, Chemistry, and Biology," and, in addition to offering theoretical instruction to students in the Royal School of Naval Architecture and Marine Engineering, the new school was also to be made available "for the instruction of many Science Teachers throughout the country."[28]

The *First Report* touched off a wave of protest against the scheme to create a new school, and its opponents blamed Huxley. An anonymous letter to the *Times* on 6 April 1870 referred to "Professor Huxley's plan for founding an imposing National College of Science" as "a somewhat puerile passion for symmetrical arrangement."[29] Huxley responded in a letter to the editor that appeared on 11 April denying that the commission's proposal was "in some especial sense, mine." He objected against the insinuation that the proposal was the "product of a far-reaching ambition to 'found an imposing National College of Science' which fires my soul." On the contrary, Huxley declared, he had advised his colleagues on the "real difficulties in the way of carrying out the change" and of the probability that their recommendation would be met by an "unreasonable outcry." Nevertheless, all of the commissioners had supported it after a careful consideration of the issue.[30] Huxley did not want his plan to be seen merely as a scheme that would benefit him personally by furthering his career. It cast aspersions on his character and raised questions about the validity of the plan.

That "abominable leader in the Times of April 6th," Huxley told Lockyer on 20 April, was likely was written by John Percy, lecturer on metallurgy at the Royal School of Mines. He had decided that if Percy and his allies requested to appear before the commission, he would not ask them any questions. But he thought it would be desirable to "put on record all that the opposition has to say" so that the commissioners could state that they had heard the objections from "their side without effect."[31] In the end, Murchison, Percy, and other opponents of the plan to separate the School of Mines from the Geological Survey submitted letters to the commission on 22 April 1871. Members of the commission formally responded. Huxley did not sign his name to the written response by the commissioners. Instead, he wrote to the commission stating that as a professor in the Royal School of Mines, he could not take part in the discussion, and proposed to absent himself from the meetings of the commission dealing with the subject in question.[32] After the appearance of the letter in the *Times*, Huxley had decided that it was best to let his fellow commissioners deal with his opponents in order to fend off charges that his own personal ambitions were driving the agenda of the commission.

The fate of the Royal School of Mines was not the only issue that Huxley dealt with that involved important consequences for the institutional power of scientific naturalism. During the early 1870s the status of Kew Gardens—and therefore Hooker's personal standing in the scientific world—had become a reason for concern to Huxley and his allies. Acton Smee Aryton, the first commissioner of the Office of Works, and Hooker's boss, clashed with Hooker several times when he tried to exercise his authority over Kew. Egged on by Owen, Aryton proposed that Kew's collections be transferred to Owen's new natural history museum, to be located at South Kensington. This was one move to South Kensington that Huxley and his friends vigorously resisted. It would have destroyed Kew's claim to be a center of botanical research and reduced it to a public park. Since the state of national scientific museums, and the scientific portions of national museums, was considered by the commissioners to be part of their mandate, Huxley had the opportunity to help bolster Hooker's position at Kew.

One of Huxley's strategies for helping Hooker depended on criticizing the scientific value of Owen's natural history collections at the British Museum. While questioning John Phillips, professor of geology at Oxford and keeper of the University Museum, on 15 July 1870, he drew from his witness a condemnation of the way Owen displayed his collection. Instead of providing a well-arranged and indexed collection, the British Museum confronted the student with a huge number of disorganized specimens. Sarcastically, Huxley asked, "You do not think that the public taste or knowledge of natural history would be improved by being able to inspect a mile of beetles?"[33] Huxley questioned representatives from the British Museum carefully on the organization of their collections. John Winter Jones, principal librarian and secretary of the British Museum, was asked on 13 March 1871 why the museum was not divided in half so that students could examine the collections without being disturbed by the public.[34] When Owen appeared before the commission on the same day, Huxley's questions were on the principle of selection used in the display of specimens.[35] Hooker gave his testimony on the following day.

Huxley asked how specimens could be displayed so that both the public and the men of science could have uninterrupted access. Hooker's solution was to have spacious galleries in which the specimens were closed in front toward the public and accessible to working naturalists at the back.[36]

Huxley also asked questions about the comparative value of the collections at Kew and the British Museum. He maintained that the Kew collection was superior. In the process of examining William Carruthers, keeper of the Botanical Department at the British Museum on 28 April 1871, Huxley asserted that the collection at Kew "is the only great scientific herbarium at present." He insisted, "The extent of accommodation and working is far greater than anything that you have at the British Museum." Carruthers vehemently disagreed.[37] Huxley could go to extreme lengths to defend the reputation of Kew when faced with a hostile witness. When the engineer Captain Douglas Galton, Francis Galton's cousin, seemed to side with Owen concerning the future of Kew, Huxley began to act as if he were a prosecuting attorney in a court trial. First, he demonstrated that although Galton had an interest in biology, he did not know systematic botany or zoology. If he did, he would be aware "that in the proper naming of a plant or an animal it is necessary to have a complete systematic collection of plants and animals." Then Huxley tried to show that Galton was unsympathetic to the entire purpose of the commission. Did Galton approve of the action taken by the government to support science in the same way that it supported art and literature? Galton replied, "I accept that action of the Government as a fact." Huxley persisted: "May we not have the advantage of knowing whether you approve of the fact or not?"[38] Huxley aimed to discredit Galton as a witness before his fellow commissioners.

As the commission was gathering evidence on the scientific role of museums, another issue came up that Huxley had to respond to in order to protect Hooker, undermine Owen, and defend some of his own schemes. Owen and his allies had proposed that elementary instruction in science could be undertaken by museums through the presentation of lecture courses. Huxley realized that this proposal expanded Owen's role in scientific education while threatening his own plans to direct the teaching of elementary science. Through his questioning on 14 March 1871 of William Henry Flower, Hunterian Professor of Comparative Anatomy and Physiology, and conservator of the Museum of the Royal College of Surgeons of England, Huxley made the point that the museum was not the proper place for giving "elementary instruction." Prompted by Huxley, Flower stated that he saw no advantage in it.[39] Flower was a former colleague of Owen's, and his successor as Hunterian Professor. But in the early 1860s he had supported Huxley during the

"Hippocampus controversy" with Owen. In the spring of 1872 Huxley was so confident that he had defeated Owen's proposals that he planned lunch with Tyndall rather than taking his place on the commission when Owen was next scheduled to appear.[40]

Much to his surprise, Owen's proposal concerning lectures at museums surfaced again at the commission on 14 November 1873, when a group of witnesses, including Edwin Chadwick and Lieutenant Colonel Strange, appeared, purporting to represent the Council of the Society of Arts. They recommended that scientific instruction be given at museums, not just the Science Schools at South Kensington. Huxley pointed out that the curator of a collection was too busy to lecture and that it was not necessary to have access to a large collection to provide elementary education. When Chadwick stated that some professors at the British Museum thought there was an important connection between the instruction to be given to schoolmasters and the collections at the British Museum, Huxley angrily interrupted him. The opinions of the professors at the British Museum, Huxley asserted, were already in evidence. Huxley wanted to know the position of the Council of the Society of Arts. Had the issue been discussed by the Council of the Society? Was the deputation reporting their "deliberate judgment"? Strange was forced to back down. He responded that the deputation's position represented their individual opinions, not the collective opinions of the council.[41] By questioning the authority of the deputation to speak on behalf of the Council of the Society of Arts, Huxley was able to blunt the effect of their support for Owen's proposal.

The *Fourth Report* of the commission, on museums, published in 1874, reflected all of Huxley's views on the British Museum and Kew. The report did not recommend the institution of systematic courses of lectures in the British Museum or the transfer of Kew's collections to Owen's domain. It did recommend that the British Museum build cases open at the back so as to be accessible to students and scientists (in line with Hooker's suggestions), and that the British Museum should only display a selection of its specimens and reserve the rest for the purposes of scientific investigation.[42] Just as Huxley, aided by Cole and other witnesses, had outmaneuvered Murchison on the issue of the future of the School of Mines, with the help of key witnesses he had stymied Owen on the fate of the Kew Herbarium.

Educating the Nation to Be Scientific Naturalists

On 20 March 1871, William Grylls Adams, professor of natural history at King's College London, was one of the witnesses who came before the commission. Adams explained why it was almost impossible to undertake scientific work at a sophisticated level in his classes. Huxley took the opportunity to dig deeper into the problem pointed to by Adams. It was one of many occasions when Huxley asked questions that led in this direction. "I understood from you just now," Huxley declared, "that you consider that the defective state of secondary education in England has been a very great obstacle to giving thorough scientific instruction in King's College?" Adams replied that it was a "very great obstacle." Even the best students only knew "a little of trigonometry and something of mechanics." This set up Huxley perfectly for his next question: "Do you yourself think that there is the least difficulty in introducing a very much greater quantity of scientific instruction into secondary schools, so as to enable young men to take advantage of your higher instruction?" Huxley received the answer he wanted. Adams did not think that there would be any difficulty introducing a greater quantity of scientific instruction.[43]

But for Huxley the problem was not just at the secondary school level. Students were ill prepared to cope with scientific instruction in secondary schools because they did not receive the proper instruction in elementary schools. Throughout the meetings of the commission, Huxley constantly pushed for a system of scientific education that began at an early age. The end result would be the production of students who entered university with a firm grasp of the general principles of science. Some of these students would even have the ability to do scientific research at an advanced level. Only then would the British educational system have the capability of developing a cadre of world-class scientists that could compete with such scientific powerhouses as Germany. As Barton has pointed out in her forthcoming X Club book, Huxley and his allies were active across the entire range of educational institutions and initiatives, from elementary schools to universities, and from evening classes for working-class adults to public schools for rich youths.[44] Expanding the teaching of science was a key component of their efforts to convert Britain to scientific naturalism. This was especially important during a period when, thanks to the Education Act of 1870, educational reform was the focus of much debate, and when the scientific education of the masses had been left to popularizers of science like John George Wood and Ebenezer Cobham Brewer, who perpetuated the natural theology tradition.[45] Huxley's involvement in new projects in the 1870s—including the International Scientific Series, Macmillan's Science Primers, and his *Physiography* (1877)—that were geared toward disseminating scientific naturalism to a popular audience is all of a piece with his championing of a new system of scientific education.[46]

Huxley raised the idea of starting scientific education at an early age when he appeared as a witness before the commission on the second day of hearings. He testified that math and physical geography were taught in most elementary schools, but no other branch of science. In addition to those subjects, he recommended the introduction of elementary physics, chemistry, botany, and human physiology into the standard curriculum.[47] Later, when witnesses with any connection to elementary schools came before the commission, Huxley often took the opportunity to ask them about their opinions on the question: what is the earliest age that children can be taught scientific subjects? He fought to correct the view that children younger than twelve were incapable of understanding elementary science. Thomas Coomber, headmaster of the Bristol Trade School, asserted during questioning that boys could not be taught science before the age of twelve. Could not they be taught elementary botany? Huxley asked. Coomber pleaded ignorance—his teaching was confined to the physical sciences. But Huxley persisted. Couldn't a boy much younger than twelve be taught the principles behind the thermometer, the barometer, and the air pump?[48] Huxley tried to demonstrate to Henry H. Sales, visiting agent of the Yorkshire Union of Mechanics' Institutes, and honorary secretary of the Yorkshire Board of Education, that children at the age of ten were quite capable of taking instruction in physical science.[49]

In his interaction with the Reverend James Fraser, lord bishop of Manchester, Huxley lowered the age for scientific instruction even further. He explained to Fraser that introducing physical science into elementary schools did not require "something very grand and abstract and gigantic." It merely meant giving children "an elementary acquaintance with the phenomena of nature, and with plain and obvious natural laws, which may be made intelligible even to children." Then he asked if Fraser was familiar with a German work, Sandmerer's *Lehrbuch der Nature Kunde*, which laid out three courses, one for those seven to nine years of age, another from nine to eleven, and a third for those eleven to thirteen. Fraser's mention of Brewer's *Knowledge of Common Things* as a text that had been popular in many schools drew from Huxley an extremely negative reaction. Brewer's book, Huxley asserted, was unsuitable, as it merely provided information instead of leading "the mind of a child to what may be called purely scientific considerations."[50] Huxley's criticism of Brewer echoed his general concerns about the low quality of science books for the public. They complement his depiction of most popularizers of science as literary hacks, starting in 1854 with his disparagement of George Henry Lewes's *Comte's Philosophy of the Sciences* and the tenth edition of the *Vestiges of the Natural History of Creation*.[51]

In addition to the problem that more scientific instruction needed to be introduced into the elementary schools, there was also the difficulty of locating qualified science teachers. Huxley's questioning of his fellow X Clubber Edward Frankland on 14 February 1871 helped him to draw attention to that issue. How much training, he asked Frankland, was needed to ensure that the "teaching of elementary science in the country is to be what it ought to be, . . . something equivalent to that nine months' practical training?" Frankland cooperatively replied, "I should like it, certainly." Then Huxley asked another leading question: "Is it not the case that from the method which has been pursued by elementary teachers in this country they are almost entirely incompetent to teach science, and that they have not understood in the least degree what the teaching of science is?" Again, Frankland agreeably answered, "Certainly, that is the case."[52]

If there were no qualified science teachers, then who would teach the science classes that Huxley desperately desired to include in the elementary and secondary school curriculum? Here is where Huxley's plan to move part of the School of Mines to Kensington to establish a new Science School came into play. This was to be the educational institution that was to train a new cadre of science teachers. Huxley had his friends endorse the idea when they came before the commission. Henry Cole, secretary of the Science and Arts Department of the Committee of the Council on Education, the first witness to appear before the commission on 14 June 1870, agreed that training would avoid cramming, one of the pitfalls of the current examination system, which financially rewarded teachers for every student who passed. Huxley lashed out at those teachers who took a number of "very small boys, and cram[med] them like turkeys, with just what they can put out again when the examiner asks them questions." If teachers had to pass an exam of their own, Huxley believed, it would cut down on incompetent instructors who crammed their students.[53] Henry Roscoe, professor of chemistry at Owens College, was summoned to appear before the commission on 3 March 1871. Roscoe was a close ally of Huxley in education matters. Huxley engaged him in a discussion of the quality of science teachers currently in the educational system. Roscoe believed that teachers were as a rule "exceedingly ill educated" for teaching science. He enthusiastically approved of a small-scale program that brought "men up to London to be trained in teaching" and expressed his desire to see "the system carried out very much more fully, and classes of a similar character set up in all localities in which science colleges exist."[54]

Huxley insisted that laboratory work had to be an integral component of the training of science teachers, so that they, in turn, could teach laboratory methods to their students. The laboratory was that crucial "truth-spot" of scientific naturalism—the privileged site of the discovery of new scientific knowledge as well as the best school space for students to learn in. Instead of learning about nature from books, Huxley wanted students to observe for themselves in a laboratory setting. During his testimony as witness before the commission, Huxley contended that the "great blunder that our people make, I think, is attempting to teach from books; our school-masters have largely been taught from books and nothing but books, and a great many of them understand nothing but book teaching." When they tried to teach science, "they make nothing of it." Huxley argued that a child needed to be taught "through its eyes, and its hands, and its senses."[55] During the proceedings of the commission on 21 June 1872, Huxley won the support of J. S. Burdon Sanderson, professor of practical physiology at University College, for the notion that institutions providing teacher training should be equipped with a laboratory in which to conduct original research.[56] When J. F. D. Donnelly, inspector for science in the Science and Art Department at South Kensington, made his appearance before the commission on 9 March 1871, Huxley stated that "we hear on all sides" that "there is a defect in the existing scientific teaching from want of practical laboratory instruction." He asked Donnelly if there was any way of examining laboratory work. Donnelly replied that the current system only allowed questions about the laboratory on the written exam. Would the department, Huxley inquired, be capable of organizing a "*vivá voce* practical examination" focusing on what was learned in the laboratory?[57]

Huxley's ideas for overhauling the curriculum in elementary and secondary schools were closely tied to his vision of the increased role of science at the universities and colleges. Students would arrive at university with a far better preparation in the sciences, allowing for more advanced-level work. But, if the ancient universities were to become hospitable to modern science, significant changes would have to be made. An article in *Nature*, which appeared shortly after the commission had questioned several teachers from Oxford, made that clear. "It is known to all the world that science is all but dead in England," the anonymous author declared. "It is also known that science is perhaps deadest of all at our Universities. Let any one compare Cambridge, for instance, with any German university."[58] Huxley's questions during the commission on science in the universities and colleges were framed to achieve two main goals. First, he wanted science to be recognized as an essential component of a university and college education, on a par with classical literature. Second, he desired to make it possible for students who wanted to concentrate on science to have the opportunity to receive a thorough training, including laboratory research.

Huxley's questions during the summer and fall of 1870 to Oxford professors were often barbed with implicit criticisms of the emphasis on classical languages and literature. He asked the Reverend Bartholomew Price, professor of natural philosophy at Oxford, if it was possible "for a man to obtain the highest honors which Oxford has to give at present without even having heard whether the earth goes round the sun or the sun round the earth." When Price asserted that a student taking a scientific degree should have preliminary training in literature, Huxley inquired if it was equally desirable "that persons taking a literary degree should have had an equivalent preliminary training in science."[59] Price responded in the affirmative. Huxley was also aggressive in his exchanges with the Reverend John Prideaux Lightfoot, rector of Exeter College. Huxley asserted that anyone taking a university degree should be compelled to study both literature and science in order to be well grounded in the elements of both. Then, if they like, they could "specialize themselves in particular departments." Lightfoot responded by saying that any person who specialized in science would be one-sided, to which Huxley replied that any person who had an exclusively literary culture would also be one-sided. When Lightfoot appealed to tradition, exclaiming, "History would hardly support such a view," Huxley did not back down. Appealing to the immense changes in the new industrial age, he asked, "Do not you think that, considering the enormous place which science now occupies, not only in the practical world, but in the speculative world, a person who has only literary culture must be regarded as one-sided?"[60] Lightfoot doubted that Oxford and Cambridge undergraduates were equal to the task of cultivating both. Huxley made a similar argument in the early 1880s in his famous debate with Matthew Arnold about the desired components of a modern education. During the hearings of the commission, Huxley maintained that universities and colleges had a responsibility to ensure that all students possessed a minimal level of scientific knowledge upon graduation.

Providing proper teaching methods and financial opportunities for those university students who wished to specialize in science were also important issues for Huxley. The appearance of an Oxford professor again offered the occasion for Huxley to express his low opinion of the state of science in the universities and colleges. Huxley expressed astonishment when Marmaduke Lawson, professor of botany and curator of the Botanic Garden, gave an account of the biology course currently offered at Oxford. "Did I rightly understand you to say," Huxley exclaimed, "that a man might pass through the whole of his biological course without having had his attention directed to the correspondencies [*sic*] between the structure of plants and the structure of animals, as laid down, for example, in Schwann's famous memoir?"[61] Huxley wanted his fellow commissioners to share his shock that essential morphological facts were not covered in a biology course. Huxley also drew the commissioners' attention to the lack of laboratories at British universities. During his examination of Robert O. Cunningham, professor of natural history in Queen's College Belfast, he established that there were no dissecting rooms.[62] In the course of questioning the Reverend Thomas Fowler, fellow and tutor of Lincoln College, Oxford, on 11 March 1873, Huxley spelled out the best methods of teaching physical science. Ideally, each university would have three classes of science teachers, including a demonstrator (who guided students in laboratory dissections); a *répétiteur*, or examiner (who gave subsidiary instruction); and a professor in chief (who lectured).[63] But many university professors had no assistants and no laboratory facilities.

Huxley objected to the current system of fellowships in universities. Few were given to science students. Questioning Benjamin Collins Brodie, professor of chemistry at Oxford, Huxley asked if the "mechanism of fellowships" might not be used to support an able man "who devoted himself to some special branch of science." Brodie did not believe that the existing system would allow it.[64] Huxley asked the Reverend Thomas George Bonney, fellow and tutor of St. John's College, Cambridge, a geologist, about the rewards offered to students proficient in the natural sciences. "Is not the working of the present system," he inquired, "rather in the opposite direction, that good men are rather bribed away from natural science by finding that rewards are only obtainable in other directions?" Bonney replied that up until very lately this was the case. But currently good natural science students in St. John's had no reason to fear that they would "injure their interests by following natural science."[65] Huxley was not convinced. He maintained that financial incentives to study science at the university and college level were lacking.

In cases where the quality of science teaching at the university level could be compromised by religious bias, Huxley opposed the provision of government funding. When Peter Martin Duncan, professor of geology and paleontology at King's College, confirmed that professors at his institution had to be members of the Church of England, Huxley asked if that would interfere with making the best possible appointments. "That is rather a serious objection, is it not," Huxley asked, "to bringing in the principle of endowment there?" "Quite so," Duncan replied.[66] Huxley also made religious exclusion (of both students and faculty) an issue when Robert Dyer Lyons, dean of the faculty of medicine at the Catholic University of Ireland, appeared before the commission. Lyons stated that Protestants could attend and receive degrees, except in theology. Huxley pointed out that the first official document describing the rules of the university contradicted Lyons, as it seemed to exclude Protestants.[67]

Just as Huxley's plan to move the School of Mines to South Kensington had been endorsed by the commission's *First Report*, his views on the teaching of science were approved in the five reports dealing specifically with education. The *Second Report* (1872) covered scientific instruction in elementary day schools under the Education Department and science classes under the Science and Art Department. The recommendations included the teaching of the rudiments of physical science for older children in elementary schools (beyond what was required in the regulations of the new code) and elementary lessons to younger children needed to prepare them for more advanced instruction to follow; the creation of a higher grade science teacher; and the establishment of recognized Science Schools with proper funding.[68] The *Sixth Report* (1875) focused on scientific instruction in secondary schools, and it drew on the answers to a questionnaire sent out to public and endowed schools, and on the information gathered through personal visits by Lockyer.[69] In these institutions the present state of scientific instruction was "extremely unsatisfactory." The omission "from a Liberal Education of a great branch of Intellectual Culture is of itself a matter for serious regret; and, considering the increasing importance of Science to the Material Interests of the Country, we cannot but regard its almost total exclusion from the training of the upper and middle classes as little less than a national misfortune." The commissioners recommended that at least six hours a week should be devoted to the study of natural science and that not less than one-sixth of the marks in general school exams be allotted to science.[70] Study of books alone was criticized as insufficient, and oral teaching, accompanied by appropriate illustrations and experiments, was deemed to be necessary. But even oral teaching supplemented by reading books was not enough. The report insisted on the need to practice the methods of observation and experiment in the field or in the laboratory.[71]

The *Sixth Report* made explicit the link that Huxley perceived between elementary and secondary schools and the universities and colleges. It looked back to earlier reports on science at the universities and colleges, which had recommended that students have the freedom to choose their "principal lines of study" without being compelled to pass examinations in subjects with no direct bearing on their careers. This recommendation could be made only if students were well grounded in the principal branches of knowledge before entering university. Literary culture, up to a certain point, was "indispensable for the Scientific Student," but "in like manner, evidence of corresponding Scientific Culture should be required from the Student of Classical Literature or of Theology."[72]

The *Third Report* (1873), concentrating on Oxford and Cambridge, argued for the principle that universities should be "centres of Scientific Education" and "centres of Original Research." Toward that end the report recommended the creation of university scholarships in natural science, in addition to the college scholarships, comparable to those existing for classical learning; increasing the scientific professoriat, as well as increasing their salaries in order to attract the most eminent scientific men; the appropriation of fellowships for those engaged in original research; and the establishment of laboratories for research.[73] The *Seventh Report* (1875) was on the University of London and the universities of Scotland and Ireland. The commissioners recommended boosting government assistance due to increases in students (in the case of Edinburgh), because of the need to hire assistants for professors (Edinburgh, Glasgow, St. Andrew's, and Queen's University in Ireland), or on account of the need to purchase equipment (Edinburgh, St. Andrew's, and Queen's University in Ireland).[74] These recommendations were in line with what these institutions had requested.

The *Fifth Report* (1874) dealt with institutions of recent voluntary origin that were mainly dependent on what was referred to as "voluntary support," including the two metropolitan colleges, University College and King's College, Owens College at Manchester, the College of Physical Science at Newcastle upon Tyne, and the Catholic University of Ireland. Increased government support was recommended for University College and Owens College. The College of Physical Science, the commissioners believed, would merit assistance in the future if it continued to develop its program. But the two universities with religious affiliations were denied funding. The religious restrictions imposed upon the selection of scientific professors and lecturers at the Catholic University of Ireland were mentioned as a major factor in ruling out a grant from public funds. Though the teaching staff of King's College were inadequately paid and the institution financially poor in general, the commissioners were only willing to consider a grant of public money if the institution applied for a new charter that abolished all religious restrictions on the selection of professors of science and on the privileges extended to science students.[75]

Huxley's plans for a radical reform of scientific education at every level, from primary schools through to universities, are reflected in the commission reports. Throughout the sessions of the commission, he asked questions designed to outline a detailed blueprint for the way British students were taught about science. He pushed for scientific instruction to be introduced much earlier in the standard curriculum in primary schools. He criticized the overemphasis in Oxford and Cambridge on classical languages and literature, and championed the notion that science was an essential subject for all university students. He recommended that programs be set up to enable students to concentrate on science and to do original laboratory research. In sum, as a member of the commission, he vigorously defended the idea that science was an integral part of a liberal education.

Science and the State

The issue of *Nature* for 15 June 1871 carried the abstract of a paper read at the Royal United Service Institution by Alexander Strange. Titled "On the Necessity for a Permanent Commission on State Scientific Questions," the paper had a significant impact on the hearings of the Royal Commission on Scientific Instruction and the Advancement of Science, although Strange denied that he was treading on its toes. Strange began by asserting, "The duty of the Government with respect to Science is one of the questions of the day. No question of equal importance has perhaps been more carelessly considered and more heedlessly postponed than this." Now that a "hearing had been obtained for it," Strange declared, neither the governing class nor the masses were qualified "to discuss it intelligently." The governing class had been educated over thirty years ago when science had no significant role in higher education, while the masses were "virtually destitute of scientific knowledge." Scientific men, some officers in the naval and military services, and professionals who were engaged in applying science practically, such as engineers and manufacturers, were the section of the community most qualified to deal with the issue.

But Strange then reversed gears, and claimed he was not going to answer the question How can the state aid science? That was the commission's question. Rather, his question was, How can science aid the state? Strange pointed out that there were innumerable questions involving science on which government had to decide, yet, for the most part, state scientific questions were dealt with "capriciously, inefficiently, irresponsibly," or even with neglect. Science could best aid the state "by means of a permanent scientific commission or council, constituted for the purpose of advising the Government on all State scientific questions." Strange insisted that the council should be purely consultative. Near the end of the article, Strange offered one more suggestion that he considered essential for the "efficient administration of scientific State affairs," the appointment of a minister of science. "When we have all Scientific National Institutions under one Minister of State," Strange affirmed, "advised by a permanent, independent, and highly qualified consultative body," then the advancement of science in Britain could make real progress.[76]

Strange's proposals put Huxley in an awkward position. A number of his fellow X Club members were not receptive to Strange's recommendations. Under examination, Hooker stated that a science council, as envisioned by Strange, would be too far-reaching. Its "magnitude" made him "doubt whether it will be possible, under the present condition of science in this country and of the scientific institutions, to get the Government to entertain the view of constituting a body so extensive, and armed with powers which must interfere with the authority and responsibility of the ministers of the Crown."[77] Barton, in her forthcoming book on the X Club, has persuasively argued that the X Clubbers were not the standard-bearers for more state support for science. Hooker and Tyndall, in particular, were not enthusiastic about the endowment of research schemes popular at the time. There was not agreement within the X Club as to what should be done and by whom.[78] Huxley's questions of commission witnesses reveal that his support for a science council and a minister of science was limited. In the area of state support for science, as compared to the issues surrounding the institutional base of scientific naturalism and scientific education, Huxley's path was less clear-cut. However, he was more receptive to the plan to increase the funding to the research grant administered by the Royal Society.

Strange appeared before the commission several times after the publication of his June 1871 *Nature* article where he laid out his ideas of appointing a minister for science and the establishment of a science council. On 8 May 1872, he was given the opportunity to review those ideas in front of the commissioners.[79] Strange's proposals were explored by the members of the commission through their questions of various witnesses, especially once the subject of the relation of the state and science became the focus of the meetings. If the number of questions asked on a specific topic are any gauge of a commissioner's interest, then Huxley's relatively few queries about Strange's proposals indicate that they did not appeal to him. The day that Strange reviewed his proposals for the commission, Huxley objected that the duties of the council members would be so onerous that no first-rate scientist could become involved without sacrificing valuable research time.[80] When Edward J. Reed, who had held the office of chief constructor of the navy in the Admiralty, was examined on 10 July 1872, Huxley put a series of questions to him about Strange's recommendations. "Do you contemplate that if such a council as you have been speaking of were established," he asked, "it should be empowered to take the initiative in any advice that it may give to the departments of the Government, or that it should merely act when called upon?" Huxley's question challenged Strange's insistence that the council adopt a merely consultative role by opening up the possibility that it take initiative. If there was to be a council, Huxley envisioned a more autonomous body. Whereas Reed preferred that the council be entirely subordinated to the minister, Huxley stated that he was "contemplating a scientific council which should be perfectly independent of the minister."[81] Huxley's preference may have been a reaction to the problems that Hooker had encountered with Ayrton. He may have wished to avoid setting up a situation where a government official had the power to limit the effectiveness of scientists with superior expertise. Later, Huxley's attitude toward Strange's proposals would have been affected by Strange's appearance before the commission on 14 November 1873, where he defended Owen's proposal to give museums an important role in disseminating scientific knowledge through lectures.

Aside from Strange's recommendations, the other major issue connected to state support of science concerned the increase of research funding. In 1850 a fund of one thousand pounds was allocated to the Royal Society by Parliament. Administered by the Royal Society's Government Grant Committee, the fund was used to subsidize publishing costs and to provide financial support for the more expensive research projects proposed by individual scientists.[82] Tyndall and Huxley had received grants from this fund in the early 1850s.[83] Although Huxley supported the increase of the allocation, he had reservations about one of the strings attached to it. The increase in funding was to be accompanied by a change in the constitution of the committee. Huxley seemed to be more comfortable leaving the Royal Society in full control of making the decisions as to which applications were successful. The current arrangement suited his purposes well. Since 1862 he and Tyndall had served as members of the Government Grant Committee numerous times, which allowed the X Club to have a significant impact on the decisions made about what kind of science was worthy of support.[84] Huxley also wanted to change the mandate of the committee however it was constituted. He believed that it should be more proactive in shaping the trajectory of British scientific research.

Huxley asked a number of witnesses about changes to the composition of the committee. That issue became entangled with the proposal to create a science council, which, along with its other duties, would adjudicate the grants. But there was also the proposal that whether or not a new science council was formed, the committee should include representatives from the major British scientific societies. To Edward Sabine, he asked, "Supposing the sum at the disposal of the Government Grant Committee were increased ten or twentyfold, would you be prepared to suggest any modification in the constitution of the committee?" Sabine answered, "Yes."[85] But Edward Henry Smith-Stanley, Earl of Derby, did not think that an increase necessitated a different administrative structure.[86] Robert Arthur Talbot Gascoigne-Cecil, Marquess of Salisbury, when asked if the Government Grant Committee "would be the best channel through which these funds should be administered," replied that "no better could be found. There is no body which is so thoroughly well constituted for the purpose of representing the scientific world as the Committee of the Royal Society." Huxley pushed further, asking, "You prefer that to any new fangled scientific council?" "Yes, certainly," Salisbury responded. "We know how it works, and it would be pretty sure to go on working as well as it does now."[87] Huxley's problem was that he supported the increase of the funding but was unhappy with the notion that other scientific societies would have representatives on the grant committee. His questioning of Strange on another issue made the point that the Royal Society was the supreme scientific society. He asserted that the Royal Society "has a right to be considered in an entirely different light from all the rest," not only "on account of its seniority," but also because of the conditions of membership. Whereas other societies merely required that an individual pay an entrance fee and subscription in order to join, you had to be elected to a fellowship in order to be a member of the Royal Society.[88]

The other issue for Huxley concerned the relatively reactive approach that constrained the activities of the Government Grant Committee. If the funding was to be increased, Huxley wanted the committee to have more leeway to encourage new lines of research rather than merely choosing from among the submitted applications. When Thomas Henry Farrer, permanent secretary of the Board of Trade, appeared before the commission, Huxley asserted that the Government Grant Committee took "no initiative." He then asked if the government had considered it desirable to have the new body, "in addition to its power of acceding to or refusing applications made to it by a person who wishes to pursue a particular line of inquiry, it should initiate inquiry?" Farrer believed that "it ought to be part of its duty to consider what are the lines of inquiry which deserve assistance, and to bring those to the notice of the Government."[89] But Alexander Williamson was not so obliging. Huxley inquired, "Do you contemplate entrusting any such council as this which might be formed with the origination of scientific inquiry, or would you limit its functions to the encouragement of original investigators who might apply?" Williamson asserted that the committee should stick to judging applications. If the committee took initiative, it could lead to abuse.[90]

The *Eighth Report*, the final one, published in 1875, focused on the advancement of science and on the relations of government to science. The report celebrated the achievements of past British scientists, such as Dalton, Davy, and Faraday, accomplished "without aid from the State." But their success, the existence of "numerous Learned Societies," and the "devotion of some few rich individuals to the current work of Science" should not blind those who recognized the importance of science to the crying need for state aid to research. It "must be admitted that at the present day Scientific Investigation is carried on abroad to an extent and with a completeness of organization to which this country can offer no parallel." Though the work done by private individuals was of great value, it did not provide the support "needed in the interests of the Science." Though the scientific societies promoted the discussion and publication of scientific facts, "these Societies do not consider it any part of their corporate functions to undertake or conduct Research." The commissioners concluded, "The Progress of Scientific Research must in a great degree depend upon the aid of Governments. As a Nation we ought to take our share of the current Scientific Work of the World."[91] National pride and responsibility justified an increase in state aid for science.

Huxley would have been pleased with the recommendation of the commission on the government grant to the Royal Society. Witnesses had unanimously supported enlarging the grant. Noting that the current grant had "contributed greatly to the Promotion of Research," the report suggested that "the amount of this Grant may with advantage be considerably increased." Moreover, the grants, previously restricted to research expenses, could, in certain circumstances and with proper safeguards, go toward remunerating investigators for their time and labor.[92] There is no indication that Huxley opposed the recommendation championed by Lockyer that state aid be provided to establish a new observatory devoted to astronomical physics and an organization charged with the task of observing tidal phenomena.[93]

The recommendations concerning the appointment of a minister of science and the creation of a Science Council represented a compromise with Huxley's position. Many witnesses had been attracted to Strange's idea of appointing a special minister of science to coordinate the functions of the government with regard to science. The commissioners recommended the creation of a Ministry of Science to deal with "the proper allocation of Funds for Research; the Establishment and Extension of Laboratories and Observatories; and generally, the Advancement of Science and the Promotion of Scientific Instruction as an essential part of Public Education." The staff of the ministry would need a council representing the "Scientific Knowledge of the Nation" to help them respond to scientific questions that came from the various government departments. The council would represent the chief scientific bodies in Britain. Huxley's views were incorporated in the assertion that "its composition need not differ very greatly from that of the present Government Grant Committee of the Royal Society." The Council of the Royal Society was to select the members, which would include representatives of other important scientific societies and some individuals nominated by the government. Although the committee would not have the power of initiating investigations, in exceptional cases it could offer suggestions to the minister.[94]

However, the relationships between the Government Grant Committee, the Science Council, and the minister were left somewhat vague. The issue of the autonomy of the Science Council was not discussed. To what extent did it report to the minister? And what about the Government Grant Committee? Was it to remain within the fold of the Royal Society or would it be run out of the Science Council? The report stated, "We think that the Functions at present exercised by the Government Grant Committee might be advantageously transferred to the proposed Council."[95] "Might" left it somewhat open. In sum, the desire to increase the grant for research to the Royal Society and to have a real scientific presence in the machinery of government led the commissioners to leave open the possibility of increased governmental control over both funding and scientific institutions. This was a danger that Huxley was sensitive to given the situation that his close friend Hooker had found himself in during the Ayrton controversy.

During the late 1860s and early 1870s, there were fears that Britain was falling behind the other industrialized nations. An article in *Nature* took the position that national industries depended primarily on original scientific research. The global position of England had faltered with the lack of scientific progress. "Whether we confess it or not, England, so far as the advancement of knowledge goes, is but a third or fourth-rate power."[96] Germany always seemed to supply the unflattering comparison. The superior laboratory capabilities in state-funded German universities were the envy of the scientific naturalists. But pushing for state funding from the British state could potentially result in a loss of autonomy. The government could set up a system to ensure accountability. A council under the thumb of a minister who was not a scientist might not be sympathetic to the goals of scientific naturalism, and, in Huxley's opinion, Britain could advance its knowledge only if it pursued those goals. The Ayrton controversy cast a huge shadow over the issue of state funding for science. An increase of funding to the Royal Society was another matter. If the current system were preserved in its essence, then scientific independence was not threatened.

Conclusion: Playing the Political Game

The Devonshire Commission was created to evaluate the state of science, and to determine whether or not it should become a national priority. It met over eighty times from 14 June 1870 until 12 May 1874. Its investigation into the state of science in Britain was exhaustive. The members of the commission examined over 130 witnesses, asking them a total of 14,553 questions, and produced eight reports. The reports contained extensive recommendations on improving scientific education at all levels, on increasing state support for research, on the creation of new scientific institutions, and on devising new ways for scientists to have more input into government science policy. Desmond maintains that the Devonshire Commission "detailed the needs of a new scientific culture—university research, Oxbridge laboratories, a Natural History Museum, schoolteaching and the creation of jobs."[97] But the British governments that received the reports over the years did not act upon the majority of the proposals put forward by the commission. Alter argues that the work of the commission "largely came to nothing, because very few of its recommendations were implemented in the years that followed."[98]

Although he had authorized the formation of the Devonshire Commission, Gladstone was opposed to a centralized Ministry of Science and he was against state intervention in science. A devout High Church Anglican, Gladstone also had grave personal doubts about scientific naturalism. When the Conservatives won the election of 1874, those who were a part of the endowment of research movement that had begun to coalesce around the commission's recommendations hoped that Benjamin Disraeli and his colleagues would be more receptive to providing state support. But they accepted only the least expensive of the commission's suggestions.[99] For example, in April of 1876 the government grant to the Royal Society was increased by four thousand pounds for five years.[100] However, Hooker, then president of the Royal Society, was told that the Government Grant Committee had to be reconstituted to ensure that the additional funding was administered fairly. The presidents of the fifteen major scientific societies of England, Scotland, and Ireland were to become ex officio members of the committee.[101] Hopes for implementing more of the commission's recommendations were dashed when Gladstone returned to power in 1880. The Liberal emphasis on retrenchment ruled out increasing the funding to science.[102] After 1870, the rising tide of opposition to scientific naturalism also complicated the push for state endowment of science.[103] Moreover, an organized opposition to state support emerged at the same time. The Society for Opposing the Endowment of Research was founded in 1880.[104] Significant state support for science did not materialize until after the turn of the century.[105]

In addition to significantly increasing the government grant to the Royal Society, the commission gave Huxley another important victory that had a tremendous impact on his personal career. The separation of the School of Mines and the creation of the new Science Schools in South Kensington was really, from Huxley's point of view, the crucial achievement. By approving what Cole, Donnelly, and Huxley had been planning in the *First Report* of March 1871, the commission removed a number of obstacles to reorganizing the School of Mines and the College of Chemistry. The new institution served Huxley's purposes. When Murchison died in October 1871, another obstacle was gone. In July 1872, the Council of the School of Mines agreed that the departments of chemistry, physics, and natural history should move to South Kensington while the mining sciences remained at Jermyn Street. The move to Kensington improved Huxley's personal position, as well as Frankland's, and it strengthened the power of scientific naturalism. At South Kensington, Huxley had the space, money, and supply of teachers to train. Much of his subsequent career was devoted to increasing these resources to make South Kensington a powerful site for scientific education.[106] The Science Schools became, as Desmond has put it, "the driving force of change" for state-funded science education untrammeled by commercial or religious interests.[107]

In the end, the Devonshire Commission served Huxley's objectives by increasing the government grant to the Royal Society and in clearing the way for the establishment of his Science Schools. Huxley must have been satisfied that his time had been well spent, for he agreed to serve on more commissions in the future. During his lifetime he served on ten royal commissions, despite the tremendous drain on his time and even though it was unpaid work.[108] The number of commissions is impressive, but it does not necessarily prove that Huxley managed to obtain through them the full extent of scientific reform that he and his scientific naturalist allies sought. In the case of the Devonshire Commission, he experienced limited success in achieving his goals, and he had to be willing to live with trade-offs to satisfy his fellow commissioners. Most of all, he had to be careful not to provide his enemies new opportunities to weaken the power of scientific naturalism through political means. Though state funding for science was desirable, if it came with strings attached that tied the hands of colleagues like Hooker, it was not worth the loss of autonomy.

Acknowledgments

I am grateful to Ruth Barton and Gowan Dawson, whose comments on earlier drafts helped me to strengthen the piece substantially.

Notes

1. Leonard Huxley, *Life and Letters of Thomas Henry Huxley*, 2 vols. (New York: D. Appleton, 1902), 1:389.

2. Adrian Desmond, *Huxley: From Devil's Disciple to Evolution's High Priest* (Reading, MA: Addison-Wesley, 1997).

3. Huxley, *Life and Letters of Thomas Henry Huxley*, 1:395.

4. Desmond, *Huxley*, 410.

5. Peter Alter, *The Reluctant Patron: Science and the State in Britain, 1850–1920* (Oxford, Hamburg, and New York: Berg, 1987), 84.

6. Roy M. MacLeod, "The Support of Victorian Science: The Endowment of Research Movement in Great Britain, 1868–1900," *Minerva* 9 (1971): 202–3.

7. Ibid., 204.

8. Desmond, *Huxley*, 388.

9. Alter, *Reluctant Patron*; MacLeod, "Support of Victorian Science," 197–230. Jones's brief summary of the main recommendations of the commission is Whiggish. See R. V. Jones, "Domesday Book of British Science," *New Scientist* 49 (1971): 481–83.

10. A. J. Meadows, *Science and Controversy: A Biography of Sir Norman Lockyer* (Cambridge, MA: MIT Press, 1972), 75–112; and D. S. L. Cardwell, *The Organisation of Science in England* (London: Heinemann, 1972), 119–26.

11. Bibby, for example, devotes fewer than three pages to it. See Cyril Bibby, *T. H. Huxley: Scientist, Humanist, and Educator* (London: Watts, 1959), 115–18.

12. Ruth Barton, *The X Club: Power and Authority in Victorian Science*, forthcoming.

13. Desmond, *Huxley*, 394.

14. Meadows, *Science and Controversy*, 95. Cardwell's account emphasizes Colonel Strange's role. See Cardwell, *Organisation of Science in England*, 119.

15. J. G. Crowther, *Statesmen of Science* (Chester Springs, PA: Dufour, 1966), 221.

16. Smith was appointed in December 1870, after an original member of the commission, William Allen Miller, professor of chemistry at King's College London, died earlier in September. The change was to Huxley's advantage, as Miller was deeply conservative.

17. R. J. W. Selleck, *James Kay-Shuttleworth: Journey of an Outsider* (Newbury Park, Ilford, Essex, UK: Woburn Press, 1994), 395.

18. MacLeod, "Support of Victorian Science," 205.

19. Desmond, *Huxley*, 34.

20. By 1874 Huxley and his friends became disillusioned with the journal when Lockyer refused to rein in critics of scientific naturalism. See Ruth Barton, "Scientific Authority and Scientific Controversy in *Nature*: North Britain against the X Club," in *Culture and Science in the Nineteenth-Century Media*, ed. Louise Henson, Geoffrey Cantor, Gowan Dawson, Richard Noakes, Sally Shuttleworth, and Jonathan R. Topham, 223–35 (Aldershot, UK: Ashgate, 2004).

21. Barton, *X Club*, chap. 5.

22. Sophie Forgan and Graeme Gooday, "Constructing South Kensington: The Buildings and Politics of T. H. Huxley's Working Environments," *British Journal for the History of Science* 29 (1996): 435–68.

23. In Sicco, "Letters to the Editor. Relations of the State to Scientific Research—II," *Nature* 2 (12 May 1870): 24.

24. "Second Report," *First, Supplementary, and Second Reports with Minutes of Evidence and Appendices, Volume I of the Royal Commission on Scientific Instruction and the Advancement of Science* (London: Printed by George Edward Eyre and William Spottiswoode, 1872), 9. (All references to the Royal Commission reports will use page numbers, not question numbers.)

25. Ibid., 21–22.

26. Ibid., 30, 40, 63.

27. Ibid., 151.

28. *First Report of the Royal Commission on Scientific Instruction and the Advancement of Science* (London: Printed by George Edward Eyre and William Spottiswoode, 1871), 1–2.

29. "London, Thursday, April 6, 1871," *Times*, 6 April 1871, 9.

30. Thomas H. Huxley, "The Royal School of Mines. To the Editor of the Times," *Times*, 11 April 1871, 5.

31. Huxley to Lockyer [20 April 1871], Huxley Papers, 21.270, Imperial College of Science and Technology, London.

32. *Copies of Correspondence Relating to the First Report of the Royal Commission on Scientific Instruction and the Advancement of Science and a Memorial Addressed to the Department of Science and Art by the Director of the Royal School of Mines* (London: George Edward Eyre and William Spottiswoode, 1871), 4, 8; and *Royal Commission on Scientific Instruction and the Advancement of Science, Minutes of Evidence, Appendices, and Analysis of Evidence*, vol. 2 (London: George Edward Eyre and William Spottiswoode, 1874), app. 3, 2–4.

33. "Minutes of Evidence Taken before the Royal Commission on Scientific Instruction and the Advancement of Science," *First, Supplementary, and Second Reports with Minutes of Evidence and Appendices*, 195.

34. Ibid., 423.

35. Ibid., 432.

36. Ibid., 437.

37. Ibid., 530.

38. *Minutes of Evidence, Appendices, and Analysis of Evidence*, vol. 2, 298–99.

39. *First, Supplementary, and Second Reports with Minutes of Evidence and Appendices*, 446.

40. Huxley to Tyndall, 4 June 1872, Huxley Papers, 8.121.

41. *Minutes of Evidence, Appendices, and Analysis of Evidence*, vol. 2, 370–71.

42. *Fourth Report of the Royal Commission on Scientific Instruction and the Advancement of Science* (London: George Edward Eyre and William Spottiswoode, 1874), 3–4, 8–9, 19–20.

43. *First, Supplementary, and Second Reports with Minutes of Evidence and Appendices*, 450.

44. Barton, introduction to *X Club*.

45. Bernard Lightman, *Victorian Popularizers of Science: Designing Nature for New Audiences* (Chicago: University of Chicago Press, 2007), 364–69.

46. Ibid., 370–97.

47. *First, Supplementary, and Second Reports with Minutes of Evidence and Appendices*, 26.

48. Ibid., 410.

49. Ibid., 398.

50. Ibid., 575–76.

51. Lightman, *Victorian Popularizers of Science*, 359–64.

52. *First, Supplementary, and Second Reports with Minutes of Evidence and Appendices*, 370.

53. Ibid., 12–13.

54. Ibid., 513–14.

55. Ibid., 23.

56. *Minutes of Evidence, Appendices, and Analyses of Evidence*, vol. 2, 242.

57. *First, Supplementary, and Second Reports with Minutes of Evidence and Appendices*, 419.

58. "A Voice from Cambridge," *Nature* 8 (8 May 1873): 21.

59. *First, Supplementary, and Second Reports with Minutes of Evidence and Appendices*, 217.

60. Ibid., 259–60.

61. *Minutes of Evidence, Appendices, and Analyses of Evidence*, vol. 2, 363.

62. Ibid., 324.

63. Ibid., 358.

64. *First, Supplementary, and Second Reports with Minutes of Evidence and Appendices*, 233.

65. Ibid., 337.

66. Ibid., 529.

67. *Minutes of Evidence, Appendices, and Analyses of Evidence*, vol. 2, 329.

68. "Second Report," *First, Supplementary, and Second Reports with Minutes of Evidence and Appendices*, xix, xxix–xxx.

69. The commissioners agreed the large number of secondary schools precluded the examination of witnesses. See *Sixth Report of the Royal Commission on Scientific Instruction and the Advancement of Science* (London: George Edward Eyre and William Spottiswoode, 1875), 1.

70. Ibid., 10.

71. Ibid., 5.

72. Ibid., 1–2.

73. *Third Report of the Royal Commission on Scientific Instruction and the Advancement of Science* (London: George Edward Eyre and William Spottiswoode, 1873), vii, xiii, xix, xxxiv, lvii.

74. *Seventh Report of the Royal Commission on Scientific Instruction and the Advancement of Science* (London: George Edward Eyre and William Spottiswoode, 1875), 12, 14, 20, 26, 40.

75. *Fifth Report of the Royal Commission on Scientific Instruction and the Advancement of Science* (London: George Edward Eyre and William Spottiswoode, 1874), 12, 21, 24, 28.

76. A. Strange, "On the Necessity for a Permanent Commission on State Scientific Questions," *Nature* 4 (15 June 1871): 130–33.

77. *Minutes of Evidence, Appendices, and Analyses of Evidence*, vol. 2, 231.

78. Barton, *X Club*, chap. 5.

79. *Minutes of Evidence, Appendices, and Analyses of Evidence*, vol. 2, 127.

80. Ibid., 131.

81. Ibid., 283.

82. Alter, *Reluctant Patron*, 21.

83. R. M. MacLeod, "The Royal Society and the Government Grant: Notes on the Administration of Scientific Research, 1849–1914," *Historical Journal* 14 (1971): 330.

84. Ibid., 334.

85. *Minutes of Evidence, Appendices, and Analyses of Evidence*, vol. 2, 142.

86. Ibid., 341.

87. Ibid., 349.

88. Ibid., 92.

89. Ibid., 276–77.

90. Ibid., 280.

91. *Eighth Report of the Royal Commission on Scientific Instruction and the Advancement of Science* (London: George Edward Eyre and William Spottiswoode, 1875), 24.

92. Ibid., 22, 47.

93. Ibid., 47. There is good reason to believe that Huxley supported the establishment of a new solar observatory. Meadows asserts that Huxley was behind the lengthy maneuvering needed to appoint Lockyer in 1882 to an astronomy post at the Normal School of Science, where

he could also guide the Solar Physics Observatory. See Meadows, *Science and Controversy,* 114 – 15.

94. *Eighth Report of the Royal Commission on Scientific Instruction and the Advancement of Science,* 46 – 47.

95. Ibid., 47.

96. "Our National Industries," *Nature* 6 (6 June 1872): 97–99.

97. Desmond, *Huxley,* 456.

98. Alter, *Reluctant Patron,* 90, 102.

99. MacLeod, "Support of Victorian Science," 214, 220.

100. In 1882 the total amount of the grant was reduced to four thousand pounds and stayed at this level until 1920. See Alter, *Reluctant Patron,* 21.

101. MacLeod, "Royal Society and the Government Grant," 338.

102. MacLeod, "Support of Victorian Science," 220. Gladstone's Liberals also won the elections of 1886 and 1892.

103. Ibid., 221–22. MacLeod states that there was a growing resentment against Tyndall's alleged "materialism," particularly after the "Belfast Address" (1874). By the early 1880s, he asserts, science had cut itself off from popular sympathy and support, and scientific naturalists were seen as arrogant. The reaction against vivisection reinforced the negative perceptions.

104. Ibid., 224.

105. Ibid., 229.

106. Forgan and Gooday, "Constructing South Kensington," 464.

107. Desmond, *Huxley,* 418. See also Barton's *X Club,* chap. 5.

108. In addition to the Devonshire Commission, Huxley sat on the Royal Commission on Scottish Herring Trawler Acts (1862), the Royal Commission on Sea Fisheries of the United Kingdom (1865), the Royal Commission on the Royal College of Science for Ireland (1866), the Royal Commission on Science and Arts Instruction in Ireland (1868), the Royal Commission upon the Contagious Diseases Acts (1870 –71), the Royal Commission on the Practice of Subjecting Live Animals to Experiments (1876), the Royal Commission to Inquire into the Universities of Scotland (1876–78), the Royal Commission on the Medical Acts (1881– 82), and the Royal Commission on Trawl, Net, and Beam-Trawl Fishing (1884).

5

Economies of Scales: Evolutionary Naturalists and the Victorian Examination System

JAMES ELWICK

One May morning in 1870, Thomas Henry Huxley must have sighed in resignation. We can only guess at this sigh, as no record of it exists. Yet it is a plausible inference to make, for such a reaction is the least that can be expected when 3,705 animal physiology exams show up at the door, all having to be marked in three weeks' time.[1] Throughout the rest of the month of May, large packets of Department of Science and Art tests appeared at other examiners' doors. John Tyndall received 2,613 exams in magnetism and electricity, and another 2,021 in acoustics, light, and heat; Thomas Archer Hirst got 3,995 exams in pure maths; Edward Frankland was given 2,694 exams in inorganic chemistry. The largest subject, physical geography, had 5,435 exams.[2] As standardized scales of scientific learning, these exams were being deployed in industrial quantities.

This chapter looks at scientific and evolutionary naturalism from an exam-centered perspective. Its focus is written exams between the early 1850s and the late 1870s: mostly science tests, especially Huxley's, although the points made about Huxley's exams can easily be extended to other scientific naturalists who also ran examinations. Huxley is given extra attention because of his reputation as a pioneering Victorian educationist, someone reforming everything from pedagogical techniques to bureaucracies.[3]

Accounts of Huxley's work in education, and indeed histories of Victorian education in general, tend to equate education with teaching. In such accounts, exams must always be a handmaiden to teaching: necessary evils, but not nearly as important.[4] There is also an overwhelming focus on formal schools, with educational "reform" being equated with the improvement of such institutions (or with external factors such as greater access to these institutions). But it is really just a recent assumption that education only occurs in formal schools, and that educational improvement entails the reform of such schools. There is no necessary connection.[5]

This emphasis on teaching over exams is generally and theoretically problematic because it ignores one important motivation for attending formal schools: to earn trustworthy credentials.[6] This perspective is specifically and historically problematic because it reads the situation backward: Victorians such as Huxley believed that exams were far more important than teaching, because the examiner's syllabus acted as a de facto curriculum for many Victorian teachers and students, with the later exam then making it evident whether the material on that curriculum had been properly learned.[7]

Once Huxley's educational career is reframed as being mainly driven by examinations, situated within and informed by this Victorian orthodoxy, then a different picture emerges. For instance, the only way a student could experience Huxley's much-vaunted new teaching laboratory practices at South Kensington was to perform highly in one of his May exams, for these tests determined who was admitted to Huxley's courses. We shall see that H. G. Wells, for instance, only got to study at South Kensington because of his skill at "getting up" different subjects and taking good exams in them.

But Huxley's exams did more than just recruit students to his teaching labs. Those hoping to succeed in his exams were obliged to learn the terminology and the methods of what was coming to be known as "biology." Examinations are tools with which to ensure some level of epistemological conformity, and through exams Huxley had the authority to determine who *properly* knew biology and who did not. Yet what he considered the proper understanding of biology was associated with metaphysical commitments to scientific naturalism and methodological materialism (the living body seen as a machine, for instance). Candidates wishing to pass their exam had a compelling incentive to follow Huxley's lead and associate such metaphysical commitments with biology. It is highly unlikely that someone answering a question about, say, eyesight by using the language of natural theology would have been given any marks by Huxley or one of his examiners. Due to the sheer numbers of candidates taking his tests, Huxley's exams and syllabi must have been at least as effective as any of his speeches or articles at strengthening the cultural authority of scientific naturalism. Or more effective: unlike casual audiences for speeches or articles, examinees were compelled to closely study Huxley's lessons if they wished to further their educational careers.

This chapter first gives a brief history of written exams in England,[8] both before and during the "mania" for these devices in the 1850s. Because this examination culture is best grasped by studying mundane routines, paperwork, and monies earned by individuals, we then look at some specific exam work done by Huxley's friend Tyndall. After laying out the workings of the Department of Science and Art (hereafter DSA), the chapter then reconstructs just what Huxley did with those 3,705 physiology examinations that arrived on that May morning in 1870. By focusing on seemingly dull things such as bureaucratic forms, one can supplement the Foucauldian view that exams were disciplinary tools used to maintain "surveillance" over students; for we see that examiners were themselves also subject to elaborate rules and routines.[9]

Indeed, what emerges is a picture of Victorian exams as parts of large, rationalized, systems of educational standards. The conceptual tools developed to understand standardization can be easily deployed to understand exams. For instance, it will be noted how the elaborate rules and routines that emerged in the late 1850s to govern what were known as "Local" exams made it possible for the examiner and examinee to transcend local context and particularity. They permitted not only trust in the results, but also uniformity: the conditions under which an exam was conducted in Yorkshire were deemed equivalent to those conducted in Kent. This allowed the *form* of the written examination to be ignored, making it possible to focus solely on its *content*—the answers therein, whether penned in Yorkshire, Kent, or Dublin.[10] It is this ability to transcend the local that made exams powerful in Huxley's time—and in our own.

The "General Mania" and the Standardized Routines of "Local" Exams

In the 1850s England went through what the *Athenaeum* called a "mania for examining everybody by written answers to printed questions"—using "the *examination-paper*, a list of miscellaneous questions, the answering of which requires book-knowledge positively remembered, rather than habits of useful learning ready to be shown."[11] This decade saw oral questions being replaced by written ones. This new materiality was important. Unlike the viva voce, or "living voice," format of oral questioning, written questions could be printed off identically and without limit. Meanwhile, a candidate's answers served as an abstraction of his or her knowledge, able to be taken away for further investigation and comparison. In 1845 the American educator Horace Mann compared the written exam answer to a "Daguerreotype likeness, as it were, of the state and condition of the pupils' minds . . . instead of perishing with the fleeting breath that gave it life, it remains a permanent record."[12]

To be sure, before the mania of the 1850s there had been written exams at other places in England, such as the Cambridge Maths Tripos. Indeed, these have been so well studied, most recently by William Clark, Christopher Stray, and Andrew Warwick, that it is not necessary to dwell on the

Tripos here.[13] Yet one point their research makes is worth rehearsing: by acting as a common rite of passage for young men who would go on to various
important positions, the Tripos experience made exams into self-evidently
valid measures of talent and industriousness, measures worth reproducing
in other contexts.[14] In other words, a high-scoring Tripos wrangler who later
became a vicar would likely support the testing of a local grammar school's
pupils with external written exams. While Cambridge was the "parent of the
examination-paper," according to that same *Athenaeum* reviewer, and likely
the earliest school to use written exams in the British Isles,[15] it was not the
only English school to use them before the 1850s. They were also employed
at military academies such as Woolwich and Sandhurst, and the East India
Company's military academy at Addiscombe and its college at Haileybury.[16]

In each of the above cases, however, written exams evolved *within* each
school, fit to the needs of the larger teaching system. At London University
it was the other way around: the school was built for the written exam from
its 1827 inception. In 1828 it was declared that the "strictness" of its various
exams, made intentionally difficult, raised the prestige of the university's credentials.[17] In 1836 a change was made to the university structure: the University of London became an examining board for King's and University College,
having nothing to do with teaching. From the beginning of its first examinations in November 1838, this remade University of London became utterly
dependent upon written tests. While the medical faculty retained some oral
questioning and practical tests, by March 1838 all of London's BA exams were
written, with viva voce questioning reserved only for unusual cases. One reason was for administrative convenience. Another reason was fairness: written
questions allowed honors candidates to be ranked more reliably, and published questions "exposed to scrutiny" the conduct of the examiners. Appeals
to fairness were again made when in July 1857 the University of London revised its charter to make the attainment of its BA conditional only on passing
its exams. This reform was supposed to give poorer students a chance at a
London BA by obviating the need to belong to one of its affiliated colleges
(and pay its dues). College membership was no longer necessary—and nor,
for that matter, was class attendance.[18] All that ultimately mattered for the
London BA student was to pass its tests, so confident was its Senate that exams alone could determine who had been properly educated.

By 1857 many other English exam systems had appeared. Some were internal, given by a single school to assess its candidates: these included the
exams of the Board of Trade for Masters and Mates of Merchantmen (1850),
the Oxford Honours Schools in Maths, Natural Science, History, Law and
Theology (1850), and the Inns of Court (1856). Others were external, in which

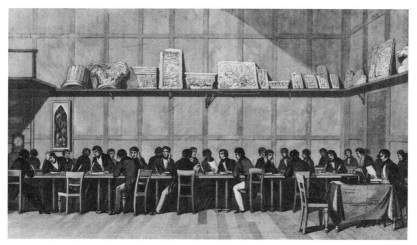

FIGURE 5.1. Candidates writing matriculation exam of the University of London, July 1842. University of London Archive, Senate House Library, University of London.

candidates from different schools showed up at a single place to take an examination: these included those of the College of Preceptors (1853–54), the Indian Civil Service (1855), and the Society of Arts (1856).[19]

While Scotland tended to have a more homogeneous education system, England's was more sectarian and decentralized, preventing any standard curriculum. As a result, external exams came to set English schools' curricula by default. Candidates at external exams were taught to syllabi published by examiners, then tested to see if they had learned that material. The effect was to crystallize the belief that teaching and examining must be kept separate; since examining set the standards to which teachers and students aspired, it was more important than teaching. Like most Victorians, Huxley took this division for granted. His involvement as an examiner at the University of London only strengthened his self-image of being an examiner first, then a teacher.

Arguably the most important English external exams were the Oxford Local Exams. They started in that watershed year of 1857, which in addition to the new charter of the University of London saw military and administrative reforms spurred by the Crimean War and Indian Rebellion. The significance of the Oxford Locals was the new organizational model of multiple "local" exam centers. The Oxford Locals were intended to strengthen private middle-class grammar schools (roughly equivalent to secondary schools) by setting explicit standards and then testing to them. In early 1857, Oxford scholars devised a syllabus of exam topics in subjects ranging from classics and languages to mathematics and chemistry, and sent that syllabus to participating grammar schools around Exeter. Those schools taught the subjects it listed. Then, on the morning of Tuesday 16 June, 107 fourteen- and sixteen-year-old boys showed up at an Exeter hotel, paid a fee, received an identifying number, and inside standardized booklets wrote answers to questions based on the syllabus. The exams ran for three days. Then the answer booklets were sent back to Oxford and were marked by July 1. The results were tabulated and published, and the highest-scoring candidates won prizes. All who passed received nondegree certificates with Oxford's imprimatur.[20]

One educational reformer later suggested that the exam method of the Oxford Locals might be "one of the great discoveries of our day" because it combined "economy of labour; absolute uniformity of standard, and all the security of the utmost publicity."[21] In modern-day terminology, "economy of labour" and "uniformity of standard" were rationalizing steps. Complicated tests intimately related to local settings—evanescent oral questions of each candidate, or different questions posed at a particular school, for instance—were distilled into *simpler tasks* and *standardized procedures* that transcended those local contexts. It was *simpler* for specialists to draw up questions in their field and delegate the actual running of the exam to a local committee, to organize the boys into age cohorts that answered identical questions, and for the papers to be marked all at once by those same specialists. It was *standardized* when a schedule of exams was drawn up, a set of specific rules was used to run the exams, and the identical questions printed off on uniform physical documents.[22] The third part of the discovery—"security of the utmost publicity"—referred to the eventual publication not only of the results but also of the questions, allowing anyone to see what had been asked at the exam. The public was thereby assured that the Oxford Locals were trustworthy and impartial indicators of a candidate's knowledge.

The rationalization of the exam process—coupled with a dawning awareness that examiners didn't have to be physically present at the exam itself—made it easy to add new test centers for the Oxford Locals. The assumption that examiners had to be physically present at an exam was likely a vestige of oral testing, and it had restricted just how many candidates could write a single exam. Removing this constraint made economies of scales possible: far more candidates could be tested with a single exam. Hence, one year later, in June 1858, the Oxford Locals had ten times as many boys as before (1,223) simultaneously sitting at eleven locations throughout England. More candidates made the Oxford Locals more important. In response, Cambridge started its own Local exams six months later, and London allowed its matriculation (entrance) exams to be taken at five provincial colleges—provided that extremely detailed rules were followed by candidates and by its own sub-examiners to maintain the matriculation exam's fairness and image of fairness.[23] One year later, in 1859, the exams of the DSA were set up in the same way, with satellite exam centers organized by local committees.

Exam centers throughout the British Isles, then overseas, were easily added thereafter. By 1864 colonial students were being tested in Trinidad (by Cambridge) and in Mauritius (by London), in exams ranging from classics to chemistry.[24] While this is to leave out many complications over the next thirty years, the procedures of Cambridge and London became more refined, and their results trusted enough, that it became unproblematic to compare examinees throughout the British Empire, regardless of ethnicity or gender. The point is obliquely made in one 1895 poem celebrating the Cambridge Local Examinations Syndicate.

> Though Roman legions ruled the world,
> Though Britain's thunderbolts are hurled
> At Monarchs in Ashanti Plains;
> The Locals Syndicate preside
> O'er realms more gloriously wide,
> Broad as the sky are their domains
> Black babes or yellow, brown or white,
> Cram manuals from morn to night
> No hue from culture now refrains;
> The infant startles from his cot,
> His bottle and his bed forgot,
> To moan aloud the name of K[eynes].[25]

The very standardization of various exams ultimately made it possible for each candidate to be removed from local custom, the "daguerreotype likenesses" of their minds compared with others. Like other systems of standards, the testing procedures had become routine and invisible, a testament to their success. We shall study some of these procedures in more detail with Huxley and his 1870 DSA physiology exam. But first we turn to the opportunities that exams offered London men of science in the 1850s, especially the scientific naturalists beginning their careers on the eve of mass examinations.

"These Horrid Examination Papers": Men of Science as Examiners

Many scientific naturalists began their careers during the mania of the 1850s, and so exams became a way for them both to shape science and to supplement meager incomes. Large sums of money were available. William Brock has estimated that about half of the £51,884 spent by the DSA between 1853 and 1870 on examining went to men of science, particularly members of the X Club such as Huxley, Tyndall, Frankland, and Hirst.[26]

Such large figures are easier to grasp by focusing on individual examiners. Consider the early career of John Tyndall. When he gained his position at the Royal Institution in May 1853 on a salary of £200 a year, promised raises did not always materialize, so he cast about for ways to supplement his income. In July 1855, as part of a modernization drive, the Royal Military Academy at Woolwich (which trained the Royal Artillery and Royal Engineers) instituted an entrance exam. Tyndall was offered its official examinership in science. He accepted it but grumbled about having to prepare questions not only in physics and chemistry, but also in metallurgy, geology, botany, zoology, and entomology. Tyndall's questions were worth only half as much as those in classics, which did not seem right for the training of "scientific" soldiers like engineers. It is unclear how much the position paid at first, but when in April 1857 he was offered just £25 to set exams for the next school year, Tyndall resigned.[27]

However, only nine months later, in January 1858, further reforms in military education following the Crimean War and Indian Rebellion got Tyndall a new position. He became examiner for the new Council of Military Education. His expanded duties included setting ten exams for students and candidates at Sandhurst as well as Woolwich. It paid £150 a year, and Tyndall told Hirst that the position might lead to something important.[28] Yet problems again arose: the military once more clawed back his salary by £50, and his questions in experimental science were still worth only one-third of those in Latin and one-half of those in Greek.[29]

Tyndall became disheartened, sometimes traveling to the wrong exam location by accident, and periodically complaining about "these horrid examination papers."[30] On both 20 December 1859 and 22 January 1861, he sent Huxley in his stead to watch over the Woolwich exam takers, in the 1859 case because Tyndall was packing for a Continental holiday. In the 1861 case, Tyndall did not mark those exams either—citing "unavoidable work," he gave them to Edward Frankland instead. But Frankland procrastinated. The exam was taken on a Tuesday and was due back on a Friday. When Tyndall received an official query as to why they were late, he paid a visit to Frankland's house on Sunday. There, he found that his panic-stricken friend had enlisted his wife, Sophie, to help him finish his marking; they did not complete the task until midnight. Tyndall ultimately resigned his examinership over this incident, although he rationalized it to his mentor Michael Faraday by saying he had not been given enough time in which to judiciously assess each candidate. Tyndall did not mention that he had delegated his marking.[31]

It took T. H. Huxley longer to become jaded about exams, partly because his early years were filled with exam successes. Indeed, he favored exams and a scholarship ladder to support poor but talented students because of his own history.[32] In September 1841, at age sixteen, he wrote to the University of London to ask about its matriculation exam. A year later Huxley spent nine hours taking a medal exam for the Society of Apothecaries. He had studied sixteen hours a day over the previous three months for it. He won its silver medal, which, when combined with a testimonial from John Henry Newman, gained Huxley and his brother free tuition at Charing Cross medical school. There, Huxley earned more prizes.[33] By August 1845 Huxley took London's first Bachelor of Medicine exam, being given the gold medal for anatomy and physiology by examiner William Sharpey. Although he could not complete his degree because his scholarship ended, such victories got Huxley talent-spotted as an assistant surgeon with a scientific bent, getting him posted to HMS *Rattlesnake* rather than to a more typical naval mission. This placement helped to make his career.[34]

If we jump ahead to July 1854, Huxley's first academic appointment at the School of Mines paid £200 a year. At the time, his friend Joseph Hooker earned £100 merely for setting two days' worth of botany exams for East India Company medical candidates. "I call that a decent wage," said Hooker coolly. Hooker would later suggest conspiratorially to Huxley that the cultural position of science would be helped when they had "sufficient command over the public, as examiners in London, and as confidential advisors and professors elsewhere."[35] While Huxley must have agreed with Hooker, more concretely the prospect of earning more money must have been just as attractive.

In the summer of 1856 Huxley got a tip from his friend William Benjamin Carpenter: because he had just been elected registrar of the University of London, Carpenter would have to give up his examinership in physiology and comparative anatomy. While proclaiming his impartiality, Carpenter encouraged Huxley to stand for his old position, and to canvass Senate members for the upcoming election. Thus prepared, Huxley ran against four other candidates, and won the position on 9 July. As Sharpey's "grandson," Huxley earned £125 a year to ask anatomy and physiology questions of medical students, sometimes in the very same MB exam that he himself had shone in eleven years before. When London's new BSc program began in 1859, four new exams were also added. To ease his examining load, Huxley was joined by George Busk (later another X Club member). Both men were paid £150 a year—a £25 raise for Huxley.[36]

The Department of Science and Art: Promoting "Free Trade in Education"

In 1859 Huxley also became an examiner for the Department of Science and Art, along with Tyndall. Frankland joined the DSA in 1868 or 1869, and Hirst in 1870.[37] Covering twenty-three subjects from machine construction to drawing, from "steam" to zoology, DSA exams followed the procedures of the Oxford Locals—a board of examiners set printed questions to be answered at satellite exam centers. But the DSA added something else: in its mission to produce new science teachers, it gave money to teachers based on their students' exam results. Good grades were literally monetized.

To discuss this shift, we must briefly venture into the larger world of English education. Following the Great Exhibition of 1851, the chemist–turned–civil servant Lyon Playfair pointed out that Continental participants who won prizes tended to be those who showed principles of science and art; he suggested that British industry might improve by following this example. Playfair's observation intrigued Prince Albert, who pushed for ways to make British industry more "scientific." In March 1853 Henry Cole's Department of Practical Art, which originally had been set up to encourage industrial design, had a "Science" section added to it, headed by Playfair; the combined group became the Department of Science and Art. The DSA bought land in South Kensington to establish museums and other institutions. When in 1856 the DSA became part of the Education Department, its mandate was made clear: increase the number of English art and science teachers, particularly those of the "Industrial Classes."[38]

But how? The DSA could not set up normal schools to train teachers— spending on education had exploded with James Kay Shuttleworth's new system of training pupil-teachers, and Chancellor Gladstone, eager to cut the government budget, was constantly pressing the DSA to reduce expenditures.[39] Nor could a standardized curriculum be imposed—the peculiarly decentralized, class-ridden, and sectarian nature of English education forbade it. It is worth recalling that in the 1850s there was no English education *system* to speak of, if by this word one denotes a graduated set of levels of formal schooling, usually sorted by age group. Instead, there were smaller branches—a poor student might attend a "British and Foreign school" for a few years; a middle-class student might head to a grammar school; an upper-class student might go to a public school; a working man might learn at a mechanics' institution.

Following the mid-Victorian passion for "free trade in education," rather than train new science teachers at a central location and thus increase their *supply*, the DSA decided to stimulate *demand* for becoming a science teacher.

Paying for exam successes clearly rewarded anyone who became a good science teacher.[40] Funding by exam success was an ideal policy for a civil servant because passing an exam was a clear signal that a student had properly acquired scientific knowledge: standards could be set without having to overburden education inspectors. The exam results were abstractions of knowledge gained—and these abstractions could be quantified and tabulated to defend DSA turf against encroaching political paymasters like Gladstone.[41] By tying payments and the production of new science teachers directly to exam successes, the DSA would be seen as exquisitely responsive to the improvement of science education. (Of course, to accept this notion, one would have to equate educational improvement with increased pass rates in exams—a point to be discussed below.)

The ingenious government mandarin Cole had already experimented with "payment on results" to increase the number of art teachers.[42] By January 1857, Playfair was moving in the same direction. Eager to cut funding for Irish public science lectures (he thought them of dubious value), but requiring justification, he had attendees take exams afterward. Playfair was stunned to learn that the Irish exam answers were "surprisingly good," presumably because the tests had "forced the pupils to read & study." The exam results showed Playfair that *how* science was taught was not as important as he had assumed. So why not leave teaching to the diverse methods of "private enterprise"? In a letter to Tyndall he declared that it was enough to stimulate science teaching by rewarding by exam results: his department cared not if science was taught in a school or a "garret, by books, oral demonstration, or Experiment."[43] Playfair's assumptions were identical to those informing the University of London's revised charter of that same year: passed exams by themselves were sufficiently demonstrative of good teaching.

In June 1858 Playfair left the DSA, and Cole became its sole head. Playfair's administrative duties were given to a twenty-four-year-old Royal Engineer and decorated Crimea veteran, Lieutenant John Donnelly. Like the successful wranglers and double firsts who equated education with exam successes, Donnelly was presumably also well disposed to exams, having entered Woolwich in 1849 as the top candidate.[44] By June 1859 he and Cole had turned Playfair's notions about funding by exam results into a formal process. The new DSA exams combined payment for exam results with the procedures of the Oxford Locals. Specialist examiners such as Huxley were hired to draw up and then publish syllabi of examinable topics. The first student exams were held in June 1860, and were taken by roughly 500 candidates.[45] Like the Oxford Local exams, the DSA's own exam system grew rapidly thereafter as new exam centers were added. In 1860, for Huxley's first animal physiology exam, there were about 100–200 candidates. In 1863 there were 349; by 1868, 1,182; and by 1870, to repeat, 3,705.[46]

Protocols of Examination

Just what did Huxley do with those 3,705 physiology exams of May 1870 with which this chapter began? His actions can be reconstructed from DSA reports, directories, and memos, as well as from Huxley's 1872 Devonshire Commission testimony, which is also discussed in Bernard Lightman's chapter in this volume. Such a method necessitates some caution—it sometimes means we must infer particular events from general rules. But this step is necessary to understand how such exams became both uniform (and thus rapidly expandable), yet impartial (and thus trustworthy) measures of science education.

The animal physiology exam was one of the most popular of the twenty-three different DSA topics. Broken up into three subexams, or "papers" in the British parlance, a candidate could only take one a year, and in succession from elementary, through advanced, to honors.[47] Candidates were both male and female and ranged from eleven to forty-five years of age.[48] To help candidates prepare, Huxley had issued a biannual syllabus with a list of testable topics since 1860: the more advanced the exam paper, the more specific and searching its anatomical and physiological questions. The syllabus also recommended various books for study, including Huxley's own *Lessons of Elementary Physiology*. Tyndall and Frankland also put their own textbooks on their syllabi, and even recommended each other's.[49] While it would be uncharitable to suggest that financial interest was the only motive, such recommendations to large and captive audiences cannot have harmed the sales of these books: Huxley's *Lessons*, for instance, went through four editions and twenty-nine printings between 1866 and 1895.[50]

As a DSA examiner in both animal physiology and zoology (a far less popular exam), Huxley received a one hundred–guinea retaining fee in addition to the other monies he earned.[51] A few months before the May exam date, Huxley wrote out the questions for the three papers. Each question was worth a certain number of points ("marks"), and only one hundred points' worth of questions could be answered in total. The intention was to create a uniform percentage scale, allowing quantitative comparisons of individual exams. Although the physiology syllabus changed fairly slowly over the preceding decade—its entire objective was to prevent surprises, after all—Huxley nonetheless had to ensure he was not repeating "hobby questions." For he was being closely watched by teachers, exam coaches, and students eager to divine what was to be in his upcoming exam. Since each student's exams would bring an average of thirteen shillings seven pence to his or her school, a great deal of money was at stake.[52]

The money involved made security important too. The leaking of exam proofs had already caused problems at the University of London. Thus, all DSA questions and proofs traveled between Huxley and the printers in special envelopes. Only certain printing houses could be used, and these firms had a single locked room in which all composition, printing, and proof checking was conducted. Finally, the proofs were watermarked so that if one was leaked, it could be traced back to a specific printer. By 1870, while there were some minor cases of cheating, no major cases of exam fraud had yet occurred.[53]

As Huxley and the printers finished their preparations, arrangements were made for candidates to take the exams. The May 1870 physiology exam was to be taken at 570 centers throughout the British Isles, with 65 in London alone. To set up a new class and assure the DSA that the exam proceedings would be trustworthy, a committee of local residents submitted a form attesting to their own respectability. Sometimes they were then visited by a department inspector—if he judged that no committee member held a position with enough "public responsibility," DSA acceptance was withheld until they added someone with such a job.[54] Meanwhile, exam candidates had until 31 March to preregister with their local committee. Anyone could take the exam, although only students of certificated teachers could earn DSA payments. New candidates arriving at the test center on the May exam day were given a unique identification number, partly for privacy and impartial treatment by the examiners, and partly so that the DSA could track each candidate over several years' worth of exams. At 6:55 p.m. in each center, the relevant sealed exam package was opened in the presence of at least three committee members.[55] At 7:00 all candidates throughout the British Isles began answering Huxley's questions. Over the next three hours, one of seventy-three specially appointed Royal Engineers might make a surprise inspection to ensure "uniformity of action"; after all, the DSA had two and a half pages of rules governing how exams were to be run.[56] When the exam ended, the committee filled out yet another form certifying that the rules had been followed, placed the exam papers in yet another security package, and mailed it back to South Kensington.

The speed of the Victorian post office meant that the 3,705 animal physiology exams were received by Huxley the following morning. Now he had three weeks to mark them. Thankfully, he had the help of one assistant examiner for every thousand papers. Donnelly seems to have arrived at this figure by taking 70 papers marked in a six-hour day—the tolerable daily limit—and multiplying it by fifteen business days.[57] Huxley had hired his assistants well in advance of the exam, insisting that each one be "able and distinguished" in physiology, capable of making quick yet accurate judgments on the candidates' answers. Each assistant received one pound for every 20 elementary papers he marked, and one pound ten shillings for every 20 advanced papers. The total amount works out to about fifty pounds per assistant: not a large sum, but a little more patronage that Huxley could dole out, giving him another tool with which to shape his field. Thus, Michael Foster—who been one of Huxley's laboratory demonstrators at South Kensington and then his choice as the new praelector of physiology at Cambridge in May 1870—eventually became physiology coexaminer; Foster selected various assistant examiners from among his physiology students.[58]

To secure marking uniformity, Huxley and his three assistants met as soon as he received the completed exams. They established their style of marking and defined various levels of answer quality. While Huxley's specific marking style is not known, he did think that the typical answer in the advanced physiology paper was equivalent to that of the average medical student.[59] For his part, Donnelly made suggestions to all DSA examiners: first-class papers should score over 69 percent in the elementary paper and 74 percent in the advanced: this mark was attainable by "a well-taught, clever, first year student, or an average second year student." Donnelly's system would give a first-class mark to about one-third of the elementary candidates and one-quarter of the advanced. Meanwhile, second-class papers should score over 29 percent in the elementary exam and 39 percent in the advanced: a mark open to "a fair, even a large, proportion of moderately stupid" candidates taught well for thirty to forty lessons. Huxley could mark the honors papers however he liked.[60]

At the end of this first meeting, each assistant carried away his batch of 1,000 papers, which would have left 705 papers for Huxley to mark himself (paying at least £35 5s, and taking up just over ten days in his active schedule). He had to mark the 28 honors papers himself (paying £4 4s). And to ensure that all papers were being marked uniformly, Huxley also had to look over at least 20 percent of his assistants' exams (735 papers, paying £1 for every 20 papers checked,[61] earning him £36 15s). After grading for about a week, all physiology examiners met again to further calibrate their marking: each of the assistants brought 50–60 representative and problem papers to the meeting, comparing them, asking Huxley questions, and revising their marking. Sometimes they would meet a third time. Impartiality was important to them: since the candidates were identified only by numbers, the examiners never knew their identities and did not want to.[62]

With the papers marked in three weeks, and the DSA having one week in which to prepare the lists and statistical tables, all exam results were published one month later.[63] Thus alerted, qualified teachers could apply to the DSA for any money they earned by sending in yet another form.[64] In 1870 a first-class elementary or advanced physiology paper paid £2; a second-class, £1; a first-class honors paper, £4. That year saw the average teacher earning £35 for all of his or her students' exams; the most paid out was £227 10s.[65] Although students themselves could not earn department money, they might win awards: medals, books or instruments, or scholarships both large and small. For Huxley's part, combining all of the payments above plus a few others, such as meeting fees and drawing up questions, he earned about £195 from his DSA examining in 1870.

The monies disbursed, the teaching and acquisition of scientific knowledge thus encouraged, the cycle would move again toward its May 1871 climax. Over the next decade the number of candidates continued to grow: 6,191 candidates took Huxley's physiology exam in 1883.[66] "To look over a thousand sets of answers to the same paper by people you don't care about, is next door to penal servitude," said one commentator of the DSA exams in that same year,[67] and it is hard not to shudder in agreement.

Huxley's Catechisms?

In the early years of the DSA's exams, Huxley was enthusiastic about them, telling Hooker in 1864 that they were "the most important engine yet invented for forcing Science into ordinary education." In July 1868 he testified before the first Samuelson Committee that the DSA's system was "one of the greatest steps ever made in this country towards spreading a knowledge of science among the people" because it disseminated science from "below upwards" rather than "from above downward."[68] But Huxley eventually followed Tyndall into disillusionment. His 1874 rectorial address at Aberdeen University looked back to his own days as an examinee: there, he confessed his shame at "how very little real knowledge underlay the torrent of stuff which I was able to pour out on paper." By 1877 Huxley described competitive exams as the "educational abomination of desolation of the present day."[69]

Huxley's disenchantment was not unique—by the mid-1870s, public opinion had shifted against exams, transforming them from near-magical devices of social reform into distasteful and overused tools. Even Playfair, who in 1857 had done so much to create payment by results, was by 1873 attacking the hegemony of exams.[70] This new resentment was partly caused by the massive expansion of the system of payment by results. In 1862 the DSA's system was taken by the department's Liberal political masters and deployed in the far larger realm of English primary education. Intended to reform and cut costs, the infamous "revised code" rewarded teachers at government-funded schools when their children did well in tests in the so-called three Rs. The format was seen as too mechanized. Matthew Arnold, who was a school inspector as well as a poet, thought the policy a disaster.[71] Yet the revised code, and more generally what Theodore M. Porter calls "the strange English prejudice that a new field is best established by creating an examination,"[72] was motivated by the classical Liberal belief in "free trade in education"—that it was better to reform education through *demand* and competition between teachers and students. It was thought that the only impartial way to determine success in these competitions was by written exams.

Unfortunately this verification was indirect. If good teaching was implied simply by exam successes, and funding depended on these successes, then it was in a teacher's interest to do whatever it took to get his or her students to pass exams. One option was to use rote memorization—a term more popularly known as *cramming*. The metaphor weds gastronomy with epistemology: facts are food to be consumed, but that sometimes remain undigested. For his part, Huxley complained of boys being "stood up in rows and crammed like turkeys," or "how those dogs of examinees return to their vomit"[73]—replacing the "folly" of Proverbs 26:11 with an incorrect or misunderstood point regurgitated in its entirety back onto an examination booklet. Teaching and learning by cramming was a rational, and rationalized, response to a rationalized exam system.

More specifically, cramming was a problem in science: the Rugby science teacher and former senior wrangler James Wilson remarked in 1867 that fields such as geology and chemistry were "frightfully crammable."[74] Books discussing science in a question-and-answer format abounded, such as Ebenezer Brewer's catechetical *Guide to the Scientific Knowledge of Things Familiar* (1847). Passing written science exams by rote memorization undermined Huxley's oft-stated belief that science was a mental habit, even an ethical stance, instilled by practice and direct observation. Huxley always thought that books alone could not teach science, and he especially disliked Brewer's *Guide*.[75]

Huxley's distaste for cramming was a major reason behind his 1869 development of laboratory teaching methods at South Kensington. Not only were his classes supposed to show science teachers how to better teach their subject—they were also supposed to get them to stop teaching through rote memorization.[76] Exams came first, *then* Huxley's teaching laboratory and Normal School of Science. The point is exemplified by the career of H. G. Wells, arguably the most famous product of Huxley's lab: Wells only got to South Kensington because he was so successful in DSA exams, and he wrote them mainly to earn money for an entrepreneurial grammar school teacher. Indeed, as an exam coach for the University Correspondence College, Wells would later teach students effective ways to pass the University of London biology exams. His cram course, which he called an "examiner defeating mechanism," was so successful that he condensed it into an 1893 biology text—his first book. Wells thus probably learned more about science from the exam process than he ever did from Huxley's laboratory.[77]

Moreover, there is a wonderful irony in Huxley's dislike of cramming and book learning. For Huxley was placed in the position of having thousands of people every year riding on his every word: not those published in journals or newspapers, but those appearing in his *syllabus*. Teachers' livelihoods and students' future ambitions depended upon how closely they paid attention to Huxley's examinable topics. And what were the subjects he wanted them to study? Many could not be learned inductively or with trained common sense, because they required prior naturalistic or even materialistic metaphysical commitments. Huxley's September 1863 syllabus, for instance, announces that examinable topics included "the general properties of living matter," "the living body considered as a machine," and "Hereditary transmission, and the modification of physical and mental characters by education, as the basis of a rational belief in the possibility of human progress."[78] Such principles cannot really be learned in a laboratory, no matter how well equipped. This point supports a later criticism made by Henry Edward Armstrong, the chemist who pushed for reforms in science teaching: contrary to the popular image of Huxley "as a master of education," Huxley was an overly didactic teacher who assumed that learning about science would automatically be followed by an embrace of its methods.[79]

While other examinable topics did not require metaphysical commitments, it was far easier to learn them from books, not experiments. For example, Huxley's February 1870 syllabus requires that students know the meaning of various anatomical terms; the chemical composition of air, water, ammonia, protein, and fat; the quantity of "dry solid and gaseous aliments" needed daily by an adult; and the composition of sweat, urine, or blood.[80] Such information could quickly be looked up in texts such as Huxley's own *Lessons in*

A TABLE OF ANATOMICAL AND PHYSIO-LOGICAL CONSTANTS.

THE average weight of the human body may be taken at 154 lbs.

I. GENERAL STATISTICS.

Such a body would be made up of—

	lbs.
Muscles and their appurtenances .	68
Skeleton	24
Skin	$10\frac{1}{2}$
Fat	28
Brain	3
Thoracic viscera	$2\frac{1}{2}$
Abdominal viscera	11

147*

	lbs.
Or of water	88
Solid matters	66

* The addition of 7 lbs. of blood, the quantity which will readily drain away from the body, will bring the total to 154 lbs. A considerable quantity of blood will, however, always remain in the capillaries and small blood vessels, and must be reckoned with the various tissues. The total quantity of blood in the body is now calculated at about 1-13th of the body weight, *i.e.*, about 12 lbs.

FIGURE 5.2. Huxley's "Table of Physiological Constants." T. H. Huxley, *Lessons in Elementary Physiology*, 2nd ed. (London: Macmillan, 1868), 331–32.

ANATOMICAL AND PHYSIOLOGICAL
CONSTANTS.

1. WHAT is the average weight of a man?

1. GENERAL STATISTICS.

2. How much of this weight will belong to muscles, skeleton, skin, fat, brain, thoracic viscera, and abdominal viscera respectively?

3. What additional weight must be allowed for the blood?

4. What proportion of the whole body-weight consists of blood?

5. What is the weight of water, and what of solids, which together form the body?

6. Of what elementary substances are the solids composed?

7. Under what four heads may the compounds they form be arranged?

8. What weight of matters of all kinds does the body lose in twenty-four hours; and of this weight how much is water, how much is carbon, how much nitrogen, and how much minerals?

9. What quantity of heat does the body part with in twenty-four hours?

10. What is meant by a foot-ton?

FIGURE 5.3. Alcock's questions to be answered by referring to Huxley's table. Thomas Alcock, *Questions on Huxley's Lessons in Elementary Physiology* (London: Macmillan, 1868), 66.

Elementary Physiology. In a second edition, published by Macmillan in 1868, some of the topics in the February 1870 syllabus are covered in *Lesson's* new "Table of Physiological Constants."

In the same year that Huxley's *Lessons* appeared, Macmillan (probably seeing a market opportunity) also published a companion book, *Questions on Huxley's Lessons in Elementary Physiology.* As the accompanying juxtaposition shows, *Questions* follows Huxley's format extremely closely, with the physiological constants being the authoritative text to be memorized. To be fair to Huxley, this is why he idealized the practical teaching of biology: initially used to teach religious principles, catechetical instruction—which often led to cramming—was a difficult format to do away with.

Unlike religious catechisms, however, the correct answers to Huxley's questions were those with naturalistic or materialist underpinnings. It is thus not surprising why Huxley described the DSA's exams as some of the most important measures ever taken to reduce "Parsonic influence" in schools.[81] He worked with what was at hand. The 3,705 candidates of his 1870 physiology exam—the ones who passed, anyway—had not only learned about physiology. They had also learned to associate scientific knowledge with a commitment to naturalism. Knowing this fact may have made Huxley's examining burden a little easier to bear; perhaps his sigh that May 1870 morning was followed by a slight smile.

Acknowledgments

I would like to thank Ruth Barton, Tansy Barton, Gillian Cooke, Angela Craft, Ruth Macleod, Julia Nguyen, Richard Temple, and David Zatzman. Figure 5.1 (ULC/PC26/14) and material RO1/2/3 and RC 40/21 appear with permission from the University of London Archive, Senate House Library, University of London; the Tyndall Correspondence and Tyndall Journal by courtesy of the Royal Institution; Huxley Papers, College Archives, Imperial College London; and Cambridge Assessment Archives Service, Papers of J. N. Keynes.

Notes

1. This combines the statistics of the Department of Science and Art's *Eighteenth Report* (London: Eyre and Spottiswoode, 1871), 49, 57, with the examination procedures outlined in Huxley's testimony for the Devonshire Commission, p. 19: Spencer Compton Cavendish, eighth Duke of Devonshire, *Royal Commission on Scientific Instruction and Advancement of Science: First and Second Report, Minutes of Evidence, Appendices* (London: Eyre and Spottiswoode, 1872), cited hereafter as *Devonshire Report.* Unfortunately, the precise date in May of Huxley's physiology exam has not been located.

2. Department of Science and Art, *Directory* (London: Eyre and Spottiswoode, 1870), iii; and DSA's *Eighteenth Report* (1871), 49. In addition to the physiology exams of that year, Huxley had to mark 114 in zoology.

3. One panegyric on Huxley and teaching is Cyril Bibby's *T. H. Huxley: Scientist, Humanist and Educator* (London: Watts, 1959). For an interesting counterargument—that Huxley was *not* a good teacher—see Henry Edward Armstrong, "Our Need to Honour Huxley's Will: Huxley Memorial Lecture," in *H. E. Armstrong and the Teaching of Science, 1880–1930*, ed. W. H. Brock, 55–73 (Cambridge: Cambridge University Press, 1973), 57–59. An excellent look at Huxley and his complicated educational context is Richard Jarrell's "Visionary or Bureaucrat? T. H. Huxley, the Science and Art Department and Science Teaching for the Working Class," *Annals of Science* 55 (1998): 219–40. The most important paper on Huxley's innovations with the teaching laboratory is Graeme Gooday's "'Nature' in the Laboratory—Domestication and Discipline with the Microscope in Victorian Life-Science," *British Journal for the History of Science* 24 (1991): 307–41.

4. The exception here, which *does* recognize the Victorian tendency to subordinate teaching to exams, is of course the collection of essays in Roy M. MacLeod, ed., *Days of Judgement: Science, Examinations, and the Organization of Knowledge in Late Victorian England* (Driffield, UK: Studies in Education, 1982).

5. Gillian Sutherland makes this exceptionally important point in her entry on "Education" in *The Cambridge Social History of Britain, 1750–1950*, ed. F. M. L. Thompson, 3 vols., 2:119–69 (Cambridge: Cambridge University Press, 1990).

6. See, for instance, Randall Collins's Weberian *The Credential Society: An Historical Sociology of Education and Stratification* (New York: Academic Press, 1979); Raymond Murphy, *Social Closure: The Theory of Monopolization and Exclusion* (Oxford: Oxford University Press, 1988), 163–65; Barry Barnes, *The Elements of Social Theory* (Princeton, NJ: Princeton University Press, 1995), 132–37; and David K. Brown, "The Social Sources of Educational Credentialism: Status Cultures, Labor Markets, and Organizations," *Sociology of Education* 74 (2001): 19–34. While this paper does not overtly discuss credentialism and the sciences, it is strongly informed by this perspective.

7. Teachers could not examine because of the inherent conflict of interest in assessing their own performance in the classroom. For this separation being "taken for granted," see, for instance, Robert Lowe, *Primary and Classical Education* (Edinburgh: Edmonston and Douglas, 1867), 5. Lowe was the politician on whose watch appeared the "payment by results" exams of both the Department of Science and Art (1858) and primary schools (1862's revised code). Karl Pearson's desire to reform the University of London in the early 1890s seems to be grounded in a rejection of the assumed split between examining and teaching. This would be in keeping with his idealization of all things German. See Theodore M. Porter, *Karl Pearson: The Scientific Life in a Statistical Age* (Princeton, NJ: Princeton University Press, 2004), 217–20.

8. Note that this focuses on *England*, and not Scotland or Ireland, although exams began to draw in non-English jurisdictions for comparison. For instance, the Scottish universities, with their wider-ranging arts degrees, were generally deemed superior to the English ones until the results of the new India civil service exams appeared in 1856; candidates taught at English universities did much better on these exams, illustrating the view that exams were unproblematic indicators of educational merit. See Michael Sanderson, *The Universities in the Nineteenth Century* (London: Routledge and Kegan Paul, 1975), 99.

9. Much scholarship on examinations depicts them as mainly surveillance tools: for instance,

see Keith W. Hoskin, "The Examination, Disciplinary Power and Rational Schooling," *History of Education* 8 (1979): 135–46; and Keith W. Hoskin and Richard H. Macve, "Accounting and the Examination: A Genealogy of Disciplinary Power," *Accounting, Organizations and Society* 11 (1986): 105–36. For a skeptical response, see Christopher Stray, "The Shift from Oral to Written Examination: Cambridge and Oxford, 1700–1900," *Assessment in Education* 8 (2001): 40–41.

10. This is to paraphrase Jon Agar, *The Government Machine* (Cambridge, MA: MIT Press, 2003), 1. This is also to take up Bruno Latour's pre–actor network theory work on immutable mobiles and inscriptions, particularly his "Visualization and Cognition: Thinking with Eyes and Hands," *Knowledge and Society* 6 (1986): 1–40.

11. [Augustus de Morgan], "Review of James Booth, *How to Learn and What to Learn*," *Athenaeum* 1520 (13 December 1856): 1531; the attribution to de Morgan is by D. S. L. Cardwell, *The Organisation of Science in England*, 2nd ed. (London: Heinemann, 1972), 85.

12. [Horace Mann], "Boston Grammar and Writing Schools," *Common School Journal* 7 (1845): 334.

13. Andrew Warwick, "A Mathematical World on Paper: Written Examinations in Early 19th Century Cambridge," *Studies in History and Philosophy of Modern Physics* 29 (1998): 295–319; Andrew Warwick, *Masters of Theory: Cambridge and the Rise of Mathematical Physics* (Chicago: University of Chicago Press, 2003); W. W. Rouse Ball, *A History of the Study of Mathematics at Cambridge* (Cambridge: Cambridge University Press, 1889); Michael Sadler, "The Scholarship System in England to 1890 and Some of Its Developments," in *Essays on Examinations*, ed. Michael Sadler et al. (London: Macmillan, 1936), 1–78; Sheldon Rothblatt, "The Student Sub-culture and the Examination System in Early 19th Century Oxbridge," in *The University in Society*, ed. Lawrence Stone, 2 vols., 1:247–303 (Princeton, NJ: Princeton University Press, 1974); John Gascoigne, "Mathematics and Meritocracy: The Emergence of the Cambridge Mathematical Tripos," *Social Studies of Science* 14 (1984): 547–84; Christopher Stray, *Classics Transformed : Schools, Universities, and Society in England, 1830–1960* (Oxford: Clarendon Press, 1998); Stray, "Shift from Oral to Written Examination," 33–50; and William Clark, *Academic Charisma and the Origins of the Research University* (Chicago: University of Chicago Press, 2006). On the Natural Science Tripos, see Roy M. MacLeod and Russell Moseley, "Breaking the Circle of the Sciences: The Natural Sciences Tripos and the 'Examination Revolution,'" in *Days of Judgement*, ed. Roy M. MacLeod, 189–212 (Driffield, UK: Studies in Education, 1982).

14. The most comprehensive list of Tripos men and their later careers can be found in Alex D. D. Craik, *Mr Hopkins' Men: Cambridge Reform and British Mathematics in the 19th Century* (London: Springer, 2007). Most of the leading proponents of civil service reform had previously been successful in exams: Stafford Northcote had been a scholarship student at Balliol College, while Charles Trevelyan had been one of the East India College at Haileybury's most successful students. In Parliament the measure's leading advocates were Trinity College Cambridge graduate T. B. Macaulay, who went through the Tripos, and the "double-starred first" William Gladstone, graduate of Christchurch College Oxford. The politician who first adopted exams for the India Civil Service, Robert Lowe, had been a successful Oxford classics coach. D. W. Sylvester, *Robert Lowe and Education* (Cambridge: Cambridge University Press, 1974), 4, 38; A. Patchett Martin, *Life and Letters of the Right Honourable Robert Lowe* (London: Longmans, Green, 1893), 66; and John Roach, *Public Examinations in England, 1850–1900* (Cambridge: Cambridge University Press, 1971), 210.

15. [de Morgan], "Review of *How to Learn and What to Learn*," 1531. The historical use of written exams at Trinity College Dublin is unclear: see Sadler, "Scholarship System in England to 1890," 53–54.

16. *General Half-Yearly Examinations of the Gentleman Cadets at the East-India Company's Military Seminary at Addiscombe* (London: J. & H. Cox, 1848), 3; Trevor Hearl, "Military Examinations and the Teaching of Science, 1857–1870," in *Days of Judgement*, ed. Roy M. MacLeod (Driffield, UK: Studies in Education, 1982), 110–12; and F. C. Danvers et al., *Memorials of Old Haileybury College* (London: Archibald Constable, 1894), 51–53.

17. *Second Statement by the Council of the University of London* (London: John Murray, 1828), 26–27.

18. On BA written exams, see N. B. Harte, *The University of London, 1836–1986: An Illustrated History* (London: Athlone Press, 1986), 91–95; University of London, *Minutes of the Committee of the Faculty of Arts [Printed]* (1837–38), 22, 24–25. "Scrutiny": *Minutes of the Senate*, vol. 1 (London: Taylor and Francis, 1837–43), 30. Revised charter of 1857: University of London, *Amended Draft Charter as Adopted by the Senate, June 4, 1857;* and University of London, *Minutes of the Senate* (London: Richard Taylor, 1853–58), 22 July 1857, 123, 129, 133, 137.

19. Cardwell, *Organisation of Science*, 85; Richard Willis, "Market Forces and State Intervention in Educational Enterprise: The Case of School Examinations from 1850 to 1917," *Journal of Educational Administration and History* 27 (1995): 100; Henry Trueman Wood, *A History of the Royal Society of Arts* (London: John Murray, 1913), 426–27; and P. S. King, *Guide to the Civil Service Examinations* (London: Parliamentary Paper Depot, 1856), vii. Note also the account of the civil service exams in Agar's *Government Machine*, 47–48, 52, 57–59.

20. *Middle Class Education: West of England Examination Papers, June, 1857* (Exeter, UK: W. H. Roberts, 1857), iii, viii, 49–50; and T. D. Acland, *Some Account of the Origin and Objects of the New Oxford Examinations*, 2nd ed. (London: J. Ridgway, 1858), 101, 75–76, 16.

21. J. P. Norris, "On the Proposed Examination of Girls of the Professional and Middle Classes," in *Transactions of the National Association for the Promotion of Social Science*, ed. George W. Hastings (London: Longman, Green, Longman, Roberts and Green, 1864), 408–9.

22. "Rationalization" denotes here the transformation of a system along certain principles so that more actions can be done with fewer inputs (labor, materials, or time). One kind of rationalization *analyzes*: breaking down complicated tasks into multiple independent and simpler ones, which facilitates specialization at each simpler task. Another form of rationalization *standardizes*: making diverse parts and tasks uniform, allowing substitution or repetition. The Oxford Local Examinations, as was the entire shift to written exams, were cases of rationalization by analysis and standardization. This point is informed by Elihu Gerson's "Reach, Bracket, and the Limits of Rationalized Coordination," in *Resources, Co-evolution and Artifacts: Theory in CSCW*, ed. M. S. Ackerman et al. (London: Springer, 2008), 9–10.

23. "Oxford Associate-in-Arts Examination," *Chambers's Journal* 246 (1858): 187; the rules are in University of London, *Minutes of Committees, 1853–1866*, 66–69.

24. University of London, *Minutes of Committees, 1853–1866*, 257–59; and Roach, *Public Examinations in England*, 145.

25. This poem is from the *Cambridge Review*, 28 November 1895. The target of the poem must have enjoyed it, as a copy can be found in John Neville Keynes's "Common-Place Book," Papers of Dr. J. N. Keynes, Cambridge Assessment Archives, PP/JNK/1/3. Keynes (father of John Maynard) was then head of the Cambridge Examinations Syndicate.

26. W. H. Brock, "The Spectrum of Science Patronage," in *The Patronage of Science in the 19th Century*, ed. Gerard L'Estrange Turner (Leiden, Netherlands: Noordhoff, 1976), 183–84.

27. Hearl, "Military Examinations," 114–15; and John Tyndall, "Journals, 1855–1872 [Typed]," Royal Institution of Great Britain, London, Journal VIa, 3 August 1855, 780; 1 May 1856, 838–40; and 18 April 1857, 927.

28. John Tyndall to Thomas Archer Hirst, 8 January 1858, Royal Institution MS JT/1/ HTYP/507.

29. *Papers Used at the Examinations for Direct Commissions, and for Admission to the Royal Military College, in July 1858* (London: Harrison, 1858), 13–16. The decline in salary is taken from John Tyndall's letter to Lyon Playfair, 1859, Royal Institution MS JT/1/TYP/3/986.

30. Tyndall, "Journals," Royal Institution, Journal VIIIa, 9 December 1859, 1166; and 1 December 1860, 1193.

31. Tyndall, "Journals," Royal Institution, Journal VIIIa, 20 December 1859, 1167; 1 January 1861, 1198; and 11 and 13 March 1861, 1203.

32. Jarrell, "Visionary or Bureaucrat?," 227.

33. Richard Wellesley Rothman to [T. H.] Huxley, 22 September 1841, University of London Archive, Senate House Library, University of London RO 1/2/3; and Adrian Desmond, *Huxley: From Devil's Disciple to Evolution's High Priest* (London: Penguin, 1997), 15, 18–19. Attempts to make contact with the copyright holders of the Rothman estate were unsuccessful.

34. Desmond, *Huxley*, 28, 33–35, 37.

35. J. D. Hooker to Thomas Henry Huxley, [12 September] 1854, Huxley Papers, College Archives, Imperial College London, 3.1; Hooker to Huxley, 1856, cited in David Layton, *Science for the People* (London: George Allen & Unwin, 1973), 71.

36. William B. Carpenter to Thomas Henry Huxley, [Summer] 1856, Huxley Papers 12.92–93. The four other candidates for the position were Hugh Carlile, Thomas Spencer Cobbold, Croker King, and Thomas Williams. University of London, *Minutes of the Senate, 1855–1858*, vol. 4 (London: Taylor and Francis, 1855–58), 47–48. The figure of £125 of 1856 is inferred by taking the 1865 examiner's wage of £150, then removing the £25 raise added when the BSc began in 1859: "Duties and Remuneration of Examiners [1866]," University of London Archive, Senate House Library, University of London RC 40/21; and University of London, *Minutes of Committees, 1853–1866*, 86–88.

37. *Scholastic Directory for 1861* (London: John Crockford, 1861), 8; DSA, *Directory* (1869), 3; and DSA, *Directory* (1870), iii.

38. Harry Butterworth, "The Science and Art Department Examinations: Origins and Achievements," in *Days of Judgement*, ed. Roy M. MacLeod (Driffield, UK: Studies in Education, 1982), 28. The "Industrial Classes" were defined as people not earning enough to pay income tax, those supporting themselves by manual labor, and the children of people in these two groups. DSA, *Eighteenth Report* (1871), 11.

39. Asher Tropp, *The School Teachers: The Growth of the Teaching Profession in England and Wales from 1800 to the Present Day* (Westport, CT: Greenwood Press, 1977), 18–19; and Elizabeth Bonython and Anthony Burton, *The Great Exhibitor: The Life and Work of Henry Cole* (London: V&A Publications, 2003), 198.

40. Harry Butterworth, "The Science and Art Department, 1853–1900" (PhD thesis, University of Sheffield, 1968), 33; and *Devonshire Report*, xix–xx. The phrase "free trade in education" can be found in many documents of the period, sometimes used supportively, sometimes critically. One disapproving example is in "The London University Calendar for 1857," *Quarterly Review* 9 (1857): 20.

41. This is to follow Theodore M. Porter's point in his *Trust in Numbers: The Pursuit of Objectivity in Science and Public Life* (Princeton, NJ: Princeton University Press, 1995), 8.

42. Butterworth, "Science and Art Department Examinations," 30.

43. Lyon Playfair to John Tyndall, 7 January 1857, Royal Institution MS JT/TYP/3/982–83.

44. Bonython and Burton, *Henry Cole*, 185; Margaret Reeks, *Register of the Associates and Old Students of the Royal School of Mines* (London: Royal School of Mines Old Students' Association, 1920), 126–30; and Butterworth, "Science and Art Department Examinations," 30. Butterworth notes that Cole was thinking about payment by results as early as 1857; Playfair's letter shows that he too was considering a similar kind of policy at about the same time.

45. Bernhard Samuelson, *Report from the Select Committee on Scientific Instruction* (London: HMSO, 1868), 4; Butterworth, "Science and Art Department Examinations," 32; and *Devonshire Report*, xix–xx.

46. Huxley's testimony is in Samuelson, *Report from the Select Committee on Scientific Instruction*, 401. 1863: DSA, *Tenth Report* (London: Eyre and Spottiswoode, 1863), 10.

47. It was policy to preserve only a few of the best exam papers and a few of the average once marking was completed, and even these papers did not last long—all were discarded after a year. *Devonshire Report*, 16.

48. On the age range, see Huxley testimony, in *Devonshire Report*, 25. On female examinees, see Margaret A. I. Macomish on the register (she was rated as one of the two best students in Huxley's 1871 biology course for teachers): DSA, *Nineteenth Report* (London: Eyre and Spottiswoode, 1872), 22.

49. DSA, *Directory* (1870), 75, 81, 82–83, 99–103.

50. Jarrell, "Visionary or Bureaucrat?," 232.

51. The rationale for the high wages was that they kept the best possible examiners performing a job "of a very laborious and repulsive nature." *Devonshire Report*, app., 8.

52. *Devonshire Report*, xxv–xxvi, app., 6–8; and H. S. Roscoe to Thomas Henry Huxley, 15 October 1871, Huxley Papers 25.269.

53. Allusions to fraud at London are taken from William Miller's questioning of Henry Cole, 14 June 1870; Miller himself mentioned this "considerable difficulty" about exam security. *Devonshire Report*, 15–16, 19. As a professor of chemistry at King's College and as a University of London Senate member, Miller may have been referring to leaks of medical exam proofs.

54. DSA, *Eighteenth Report* (1871), viii; and DSA, *Directory* (1870), 5, 30.

55. The DSA exams were held in the evening, as many of the science classes were already held at this time; the evening writings also made it easier for the volunteer members of the local committee to attend and supervise the exam.

56. DSA, *Eighteenth Report* (1871), x–xi; *Devonshire Report*, xxii–xxiii, 3; and DSA, *Directory* (1870), 8, 38–40.

57. *Devonshire Report*, app., 6–8.

58. Huxley testimony, in *Devonshire Report*, 19. T. H. Huxley to Michael Foster, 15 April 1872, Huxley Papers 4.38.

59. DSA, *Nineteenth Report* (1872), 43; and Huxley testimony, in *Devonshire Report*, 19.

60. *Devonshire Report*, app., 6–7.

61. Ibid., app., 6–8.

62. Huxley testimony, in *Devonshire Report*, 19, 21.

63. Ibid., 6–7.

64. In 1870 teachers were qualified to earn DSA monies in a subject by passing a special teachers' exam (before 1867), gaining a first- or second-class mark in the advanced paper of that subject, or graduating from any UK university. DSA, *Eighteenth Report* (1871), 10.

65. *Devonshire Report*, xxv–xxvi.

66. DSA, *Thirty First Report* (London: Eyre and Spottiswoode, 1884), 62.

67. H. W. Eve, "On Marking," in *Three Lectures on Subjects Connected with the Practice of Education*, ed. H. W. Eve (Cambridge: Cambridge University Press, 1883), 12–13.

68. Huxley to Hooker, 6 October 1864, cited in Butterworth, "Science and Art Department, 1853–1900," 221–22; and Huxley's testimony, in Samuelson, *Report from the Select Committee on Scientific Instruction*, 401.

69. Thomas Henry Huxley, *Science and Education* (London: Macmillan, 1895), 228–29, 410; the first quote is from "Universities: Actual and Ideal," 1874; the second is from "Technical Education," first published in 1877. See also Jarrell, "Visionary or Bureaucrat?," 232.

70. Lyon Playfair, *On Teaching Universities and Examining Bodies*, 3rd ed. (Dublin: Hodges, 1873), 26–27.

71. For Arnold's contemporary view of the revised code, see "The Code out of Danger," *London Review* 4 (1862): 429–30; the article is attributed to Arnold in its reprinted version in R. H. Super, ed., *Matthew Arnold: Democratic Education* (Ann Arbor: University of Michigan Press, 1962), 347–64.

72. Porter, *Karl Pearson*, 226.

73. Huxley's testimony, 28 October 1882, in Bernhard Samuelson, *Second Report of the Royal Commissioners on Technical Instruction*, 3 vols. (London: Eyre and Spottiswoode, 1884), 3:325; Huxley to Michael Foster, 16 September 1886, in Leonard Huxley, *Life and Letters of Thomas Henry Huxley*, 2nd ed., 3 vols. (London: Macmillan, 1908), 2:462.

74. J. M. Wilson, "On Teaching Natural Science in Schools," in *Essays on a Liberal Education*, ed. F. W. Farrar (London: Macmillan, 1867), 269–72; and Warwick, *Masters of Theory*, 188.

75. Layton, *Science for the People*, 112. Brewer's *Guide* was extremely popular, selling 113,000 copies in its thirty-one editions between 1848 and 1873. On Brewer, and Huxley's criticism of his *Guide*, see Bernard Lightman, *Victorian Popularizers of Science: Designing Nature for New Audiences* (Chicago: University of Chicago Press, 2007), 66–68, 367.

76. Henry Cole and Alan S. Cole, *Fifty Years of Public Work of Sir Henry Cole*, 2 vols. (London: George Bell, 1884), 1:311–12, says Huxley's "Normal School of Science" arose out of the exam system, a point corroborated by Huxley's 1868 testimony in Samuelson, *Report from the Select Committee on Scientific Instruction*, 401.

77. Wells even described his *Text-Book of Biology* (London: W. B. Clive, 1893) as "a cram book to be exact—on biology as it was understood by the University examiners." H. G. Wells, *Experiment in Autobiography* (Toronto: Macmillan, 1934), 84, 137–38, 282–85.

78. DSA, *Directory* (London: Eyre and Spottiswoode, 1863), 40.

79. Armstrong, "Our Need to Honour Huxley's Will," 70.

80. DSA, *Directory* (1870), 99–101.

81. Huxley to Henry Cole, Henry Cole Diary 7, 1861, cited in Bonython and Burton, *Henry Cole*, 199.

6

Odd Man Out: Was Joseph Hooker
an Evolutionary Naturalist?

JIM ENDERSBY

On 19 August 1868, Joseph Dalton Hooker gave the presidential address that opened that year's British Association for the Advancement of Science (BA) meeting in Norwich. He described recent advances in science, particularly the botanical works of his friend Charles Darwin, whose evolutionary ideas Hooker strongly endorsed. Then, toward the close of the address, he attacked natural theology, as being "to the scientific man a delusion."[1] Thomas Henry Huxley rose to give the vote of thanks and "expressed his entire concurrence in all that had fallen from Dr. Hooker, and observed that it would be improper, and, he thought, somewhat impertinent, that he should enlarge upon its merits." His sentiments were seconded by John Tyndall, who praised Hooker's "genius" and modesty.[2]

Hooker, like Huxley and Tyndall, was a proponent of scientific naturalism, and so his apparent hostility to religion seems unsurprising; reactions to his address in religious publications were equally predictable. For example, the Tory-sympathizing, High Church paper the *English Churchman* described the address as evidence of the "rank infidelity" that was the BA's "predominating hue." Daily newspapers joined the assault: the generally evangelical *Morning Advertiser* cited Hooker's speech as evidence that the "managers of the British Association proceed upon a settled plan, which is the result of a deep-seated enmity to revealed religion."[3] The conservative and High Church–affiliated *John Bull* called Hooker's speech a "melancholy exhibition of verbose mediocrity," little more than "an opportunity of puffing Mr. Darwin's latest hallucinations in transcendental anatomy."[4] And the *Nottinghamshire Guardian* reported it under the headline "Science Perverted to Infidelity," observing that "the infidel theory of Dr. Hooker is no new idea.

It is simply that of Dr. [*sic*] Darwin, and the learned infidels, sceptics, and rationalists who swarm throughout Christendom."[5]

At first glance, the content and reception of the Norwich address seem entirely predictable, apparently confirming much of what we think we already know about such overfamiliar topics as the Victorian "conflict" between science and religion. Although crude metaphors of warfare have long been abandoned by historians of both science and religion, analyzing scientific naturalism remains central to grasping the Victorian period. The tensions apparent in Hooker's address reflect a broad change in the ways natural laws were used to explain the workings of the physical universe during the nineteenth century. In 1800, no respectable man of science would have invoked nature's laws without mentioning (usually explicitly) that these laws had been designed and were maintained by God. *Respectable* is the key term here: many savants questioned both God's existence and his nature, perhaps offering deistic explanations instead, but speculations of this kind placed a British scientific man well beyond the pale of respectability. By contrast, in 1900 almost no one who wished to be taken seriously as a scientific figure would have considered mentioning God in a scientific paper or discussion. To do so would have been considered at best eccentric, at worst a serious breach of scientific etiquette. Among the questions this volume attempts to answer is when, why, and how did this change occur?

Both Hooker's speech and the criticisms it was subjected to illustrate the centrality of Charles Darwin's ideas within these wider changes. Bernard Lightman, following Frank Turner, has defined scientific naturalism as relying on explanations that excluded divine purpose and teleology in favor of the atomic theory of matter, the conservation of energy, and evolution. The last of these being so important in Lightman's view that he has occasionally described the new style of science as evolutionary naturalism.[6] Because of its links with Darwinism, naturalism is often connected to two other important changes that occurred at roughly the same time: secularization and professionalization.[7] As Turner wrote, in one of his last published papers, in mid-nineteenth-century British scientific communities,

> there arose a network of individuals who sought to define science within a narrower professional and naturalistic framework. This led to conflict within scientific communities themselves as well as between some scientists and religious figures over the character, goals, and cultural authority of natural knowledge.[8]

Turner ascribed a key role to these three connected movements, which gave men of science the cultural authority to pronounce on questions that had once been the exclusive terrain of churchmen and, in doing so, contributed to much broader changes that dramatically reduced the role of the churches and of religion in Victorian intellectual, cultural, and social life.

For convenience (and acknowledging the risk of oversimplification), I will refer to this style of argument as the "modernization" analysis, since it tends to stress the changes that made the Victorian world more recognizably like our own.[9] The historians who developed this analysis, including Turner and Lightman, do not of course present the changes as a simplistic triumph of rationality over superstition. Moreover, in recent decades each has contributed to a renewed scholarly emphasis on continuity—particularly in religion—during this period, an emphasis that provides a vital corrective to the earlier focus on disruption.[10] Nevertheless, the modernization analysis has retained its influence because it seems to capture something important about changing conceptions of nature and science within Victorian society.

Without disputing either the importance or nature of the changes that transformed Victorian Britain, this chapter will nevertheless seek to challenge aspects of this well-established interpretation.[11] The three "isms" (naturalism, secularism, and professionalism) are conventionally described as if they were aspects of a conscious movement, a deliberate program to reform and modernize Victorian Britain. For example, J. F. M. Clark describes the X Club as being so "effectively organized" that it was able to successfully "launch unified campaigns in favour of secularized science."[12] The loose cluster of sociological concepts that constitute modernization theory haunts these narratives, and as a result the narratives carry a residual teleology that leads to a confusion between outcomes and goals. Few historians would assume that the Victorians *strove* to lose their faith, yet we remain generally comfortable with the idea that men of science worked to become professional scientists. However, this second claim may apply no more widely than the first: undoubtedly, some Victorians worked to throw off what they saw as the shackles of religion, and some worked to become professionals (Huxley may perhaps serve as an example of both efforts).[13] However, for many Victorians the more secular society that gradually emerged in the nineteenth century was a somewhat unexpected outcome of compromises made for unrelated reasons.[14] Moreover, it was an unhappy outcome and might well have been avoided had it been foreseen. The same can be said of the emergence of a more meritocratic and democratic Britain.

Joseph Hooker exemplifies both the discomforts and (from his perspective) the undesirability of the compromises that transformed Britain. The complex role played by scientific naturalism in these changes only begins to make sense when we recognize that Hooker (like many men of science who

FIGURE 6.1. Young Hooker. Portrait by George Richmond (1855). Hooker complained to Darwin that "poor Richmond, who generally knocks off his chalk heads in two sittings, gave me eight I think, and grumbled all the time, and has turned me out a very lackadaisical young gentleman" (J. D. Hooker to C. Darwin, 17 March 1872, in Frederick Burkhardt et al., eds., *The Correspondence of Charles Darwin*, vol. 10, *1862* [Cambridge: Cambridge University Press, 1997], 119). (Courtesy of the Hooker family.)

were his contemporaries) was a gentleman first and a scientific naturalist second; there was more continuity between the early nineteenth-century gentlemen of science and their mid-Victorian successors than has perhaps been recognized. The meaning of the word *gentleman* was, of course, much debated in the nineteenth century, but its Victorian meaning usually included some (occasionally little more than residual) sense of being wellborn and well bred. However, a gentleman was also expected to possess one or more of the following: property (preferably land); education (preferably at a public school); a sense of honor; good manners; and, increasingly important during the Victorian period, a Christian faith (a broad, rather ill-defined faith being generally preferable to a doctrinaire one).[15] Hooker conformed, to varying degrees, to all these senses of the word, which would seem to make him an odd man out among the scientific naturalists, but may in fact make him typical of his contemporaries (or at least more representative than some of his more quotable contemporaries). To show why and how Hooker's example might lead us to revise our understanding of scientific naturalism, I propose to take each of the three "isms" associated with the modernization analysis in turn and demonstrate how Hooker's example leads us to reconsider their significance.

Professionalism

Hooker is often grouped with the "young guard" of British science, those whom Turner described as having taken up "the public championship of professionalized science."[16] Yet despite his close links to men like Tyndall and Huxley, Hooker was generally contemptuous of professions and professionals. His public service was prompted by a sense of noblesse oblige rather than any recognition of the public's rights, much less of his obligations to them. Hooker's career illustrates why we should reject the long-standing professionalization narrative, which in turn throws the presumed link between professionalization and scientific naturalism into doubt.

There are many reasons why *professionalization* is the wrong term with which to describe Hooker's career, but the simplest illustration is probably his conflict with Acton Smee-Ayrton, the government minister who oversaw the Royal Botanic Gardens at Kew in the early 1870s. The complexities of the "Ayrton affair" revolved around Ayrton's decision to treat Hooker as a paid government servant, whose salary was dependent on his abiding by normal civil service rules and protocols—for example, by putting government-funded building work out for competitive bids. Like his father, William Hooker (also director of Kew), Joseph Hooker employed a favorite firm of builders to carry out any work needed in the gardens. And, like his father, he used his network of contacts to find appropriate people to fill any vacancies. Ayrton decided that such practices were unacceptable; instead, all building work was to be overseen by the newly created position of director of public works, and would be put out for competitive bids before a contract was issued. Similarly, Ayrton's office would appoint Kew's clerical staff, after candidates had passed the relevant civil service examinations. Prior to Ayrton's appointment, both Hooker and his father had enjoyed a friendly, informal relationship with their masters in government. Ayrton's changes might have been acceptable to Hooker had Ayrton shown any willingness to treat the often-prickly Hooker with the respect he had become accustomed to. However, Ayrton was notoriously ill mannered and apparently delighted in using his political power to offend those who considered themselves his social superiors; he deliberately snubbed Hooker by offering his deputy an alternative post without consulting or even notifying Hooker of the proposed change. Hooker reacted by making a series of increasingly public (and increasingly intemperate) protests about the way he was being treated. These received considerable press coverage and, eventually, led to the matter being discussed in Parliament. Despite the widespread public support Hooker had enjoyed, he had called Ayrton a liar in a letter to the prime minister's secretary, an indiscretion for which he was forced to apologize. Ayrton neither resigned nor apologized, but lost his parliamentary seat shortly afterward, leaving Hooker with a distinctly Pyrrhic victory.

As Roy MacLeod has argued, the Ayrton incident was "one stage in the transformation of the amateur natural scientist into the professional 'scientific civil servant,' subject as any other civil servant to the rules of central departmental authority."[17] Indeed it was, but it is important to note that the professionalizer was Ayrton, not Hooker, who vigorously resisted the minister's attempts to introduce financial accountability and meritocratic standards. Although Hooker possessed greater scientific expertise, his belief that his rule should be unhindered in matters of choosing building contractors and hiring clerks reflects his confidence that a gentleman's genius was universal, in contrast to the specialized skills of the political (or other) professional. Hooker's disdain for public accountability and access can also be discerned in his later efforts to prevent the gardens opening earlier on public holidays.[18] Despite his hauteur, most of the newspapers took Hooker's side in the Ayrton affair, repeatedly describing him as a gentleman whose disinterested service to his country contrasted sharply with Ayrton's boorishness. One newspaper described Ayrton as "a man, whom the thick breath of a turbulent suburban democracy has blown for a moment into patronage and power," whereas Hooker was "a public servant whom all nations envy us, and whose loss to his country and to the interests of universal science would be absolutely irreparable."[19] Hooker undoubtedly shared this sense that his opponent was socially beneath him, writing privately (but not, of course, breaching gentlemanly etiquette in public) that "Ayrton was a Bombay attorney, & his mother a native woman which sufficiently accounts for his vulgarity."[20] Given Ayrton's ill-bred behavior, Hooker told Darwin that "I should lose caste altogether if I did not stand up & fight."[21]

The seeming difference in class between Ayrton and Hooker was apparent to most of the newspapers, yet their reporting of the affair not only revealed little sympathy for Ayrton's campaign for accountability, but also suggested considerable deference toward the kind of gentleman Hooker represented. The *Saturday Review* commented that Hooker "had been treated with systematic and persistent disrespect" as Ayrton violated "the most elementary rules of intercourse between gentlemen."[22] The *Times* reported the comments of William Francis Cowper-Temple, MP (a previous holder of Ayrton's position), who told Parliament that "the House should express an opinion *against* the treatment of scientific men as executive officers, and should insist upon their being treated as gentlemen." They had a right to be "treated with consideration, delicacy, refinement, and courtesy."[23] Clearly, Hooker was not alone in thinking that a scientific gentleman outranked an "executive officer," or any other professional.

In addition to criticizing Ayrton, many newspapers implied that Hooker was in some sense entitled to his position. For example, the widely read *Gardener's Chronicle*, in describing the background to the affair, noted that William Hooker

> came to Kew at a clerk's salary; he spent the best years of his life in his endeavours to raise the garden to its present position. He expended largely from his private fortune in the establishment and maintenance of the museum and herbaria. His zeal, his vast knowledge, his untiring labour, and, last not least, his courteous manners, were mainly instrumental in raising Kew to its present position.[24]

Clearly, Sir William Hooker (he was knighted in 1836) was a gentleman in all the senses discussed above: he possessed money (albeit not quite as much as the *Chronicle*'s writer implied); education; a work ethic combined with a disinterested sense of public duty; and, above all, good manners.[25] According to the *Chronicle*, Joseph had inherited these characteristics from his father and continued to enhance Kew's "efficiency as a great nursery," "its functions as a scientific establishment," and its role as "a huge pleasure garden."[26]

Joseph Hooker had inherited more from his father than good manners and a capacity for hard work. As many newspapers tacitly acknowledged, he also inherited his father's library and herbarium (collection of dried plants) and thus effectively inherited the directorship of Kew. Joseph clearly understood this; in a letter to his father he referred to the herbarium as "as much my future estates to be cared for by me, as if they were landed property."[27] Kew had no herbarium when William Hooker became director in 1841, so he began building one through a global network of correspondents and by encouraging his friends and relatives to leave their collections to Kew. The resultant collection was at the disposal of the nation, but much of it remained the private property of the Hooker family.[28] William planned to bequeath this substantial proportion of the collections to his son, but realized that Joseph would be able neither to house nor to maintain them. He therefore proposed that the government should consider "the propriety of purchasing the herbarium at a fair valuation and depositing it at Kew, as part of the Crown property attached to the Royal Gardens."[29] This suggestion was adopted: Kew got the herbarium, and a few weeks after his father's death, Joseph got the directorship. He was amply qualified to succeed his father; nevertheless, the government had been put in a position where it could hardly have made any other decision.[30]

William and Joseph Hooker were creating careers for themselves during a transitional period in British science, when the old world of patronage (exemplified in the career of Sir Joseph Banks, one of William's most important patrons) was slowly being replaced by the now more familiar one of formal scientific qualifications and open competition for positions. Judged by these later standards, the Hookers' maneuvers over the directorship seem nepotistic and even a little corrupt. While such anachronistic judgments are unhelpful, the recognition that the Hookers were improvising careers under rapidly changing circumstances should not lead us to assume that Joseph would have *preferred* to have obtained his position through an open, meritocratic process. There are good reasons to take the aristocratic implications of his phrase "my future estates" at face value. For example, during his first voyage, aboard HMS *Erebus*, he wrote to his father to discuss his future career prospects, concluding: "I am not independent, and must not be too proud; if I cannot be a naturalist with a fortune, I must not be too vain to take honourable compensation for my trouble."[31] Clearly, inheriting a fortune, rather than a herbarium, would have been his first choice.

The inheritance of estates was, of course, an aristocratic practice—the very opposite of meritocratic or democratic. The maneuvers over his "estates" at Kew are not the only evidence that Hooker saw himself as having more in common with the aristocracy than he had with members of the "turbulent suburban democracy." Hooker's sympathy for aristocratic principles went beyond his desire to be independently wealthy. In 1862 he wrote to Darwin, discussing the American Civil War and the difficulties each was finding in maintaining a collegial correspondence with their mutual friend Asa Gray, professor of botany at Harvard University. Hooker was disgusted by the war, which he regarded as the predictable result of the Americans being "a nation of upstarts," and was unable to share Gray's enthusiasm for the Union cause. He told Darwin that

> our Aristocracy may have been (& has been) a great draw back to civiliza-
> tion—but on the other hand it has had its advantages—has kept in check the
> uneducated & unreflecting—& has forced those who have intellect enough to
> rise to their own level, to use it all in the struggle— There is a deal in breeding
> & I do not think that any but high bred gentlemen are safe guides in Emergen-
> cies such as these.[32]

The dangers of giving political power to the "uneducated & unreflecting"
were clearly on Hooker's mind well before his brush with Ayrton.

Hooker argued that intelligent men like himself would be able to "rise"
by using their intellects "in the struggle" of life, but his comments capture
the complex compromise embodied in the Victorian gentleman, who was
paradoxically expected to be both wellborn and self-made. Hooker justified
his belief in the value of both nature *and* nurture by citing Darwin's work:
"Natural Selection," he asserted, must ensure that "the best trained, bred &
ablest man will be found in the higher walks of life," adding that "your 'Origin' has done more to enhance the value of the aristocracy in my eyes than
any social political or other argument."[33] Darwin was amused by this, reply-
ing that "your notion of the aristocrats being ken-speckle, & the best men of a
good lot being thus easily selected is new to me & striking. The Origin having
made you, in fact, a jolly old Tory, made us all laugh heartily." Nevertheless,
Darwin politely demurred from his friend's interpretation, noting that "pri-
mogeniture is dreadfully opposed to selection."[34]

Despite Darwin's teasing, Hooker described himself as "a philosophic
conservative, a strong Unionist, but not a Tory."[35] The phrase "philosophic
conservative" seems to capture key qualities of his character. These were
summarized by the naturalist Sir Charles James Fox Bunbury, after the two
men had known each other for many years. Bunbury recorded in his diary in
January 1868 (just a few months before the Norwich address):

> [Hooker] is, in natural science, a keen Darwinian, and in general a warm ad-
> vocate of what are called "liberal" and "progressive" doctrines, though not
> violent or extravagant; not a *subversionist* like Huxley. Eager to welcome new
> discoveries, and to follow up new thoughts and new suggestions, he is at the
> same time not at all deficient in veneration, and is able and willing to do full
> justice to the learned and the good of former times.[36]

To be a respected Victorian man of science, one's public behavior had to be
gentlemanly, and so it was vital not to appear as a "subversionist" (especially
of religion), nor to be too violent or extravagant. After the Norwich meeting, Hooker criticized Huxley's behavior in a letter to Darwin, commenting

that "poor Huxley made a sad mess of it by twice offending the clergy, totally without cause or warrant."[37] Hooker's "veneration" of earlier traditions—both scientific and, as we shall see, religious—was crucial to obtaining Bunbury's approval. This combination of progressive views, on scientific and social issues, and respect for tradition is captured in Hooker's self-description as being a "philosophic conservative." Reforms and change were needed, but they must be neither too rapid nor too radical if the "uneducated & unreflecting" were to be kept in check. Old-fashioned good breeding acted as a brake on the unsettlingly rapid change the Victorians were experiencing, hence its attraction for someone like Hooker, who wished to move up in the world without having that world changed beyond recognition in the process.

Nevertheless, if we accept that Hooker did indeed see himself as an aristocrat in some senses, like his friend the banker John Lubbock, it is important to remember that the British aristocracy had always been rather more open than its counterparts elsewhere—something that became more prominent in the late nineteenth century. Foreign visitors had long noted relatively high social mobility as a distinctive feature of British society, and this became more pronounced during Victoria's reign, as a complex compromise was being negotiated between the old elite (consisting largely of aristocratic and gentry families whose wealth came from land) and rising groups, including the manufacturing and merchant interests. This compromise between what might crudely be called the upper and middle classes was seen by many Britons as a distinctive feature of their country, which explained the absence of revolutionary unrest.[38] As the journalist Thomas Escott wrote, the great strength of English society was that

> the three rival elements—the aristocratic, the plutocratic and the democratic—are closely blended. The aristocratic principle is still paramount, it forms the foundation of our social structure, and has been strengthened and extended in its operation by the plutocratic, while the democratic instinct of the race has all the opportunities of assertion and gratification which it can find in a career open to talent.[39]

Again, the complexity of the Victorian gentlemanly ideal is apparent here: good breeding (the aristocratic principle) was "still paramount," but Escott acknowledged that the traditional ruling class had been bolstered by new money, coming from both industry and finance. Meanwhile, for those with neither blue blood nor large fortunes, careers were supposedly "open to talent." These are contradictory claims, and evaluating the economic and ideological underpinnings of nineteenth-century Britain's emerging class structure lies well beyond the scope of this chapter. My immediate concern is with the ways in which the gentlemanly ideal helped manage these contradictions, particularly for the men of science, who might be seen as parvenus amid Britain's intellectual leadership.

Several scholars have highlighted the ways in which the *gentleman*—especially as depicted in Victorian literature—facilitated a social and political accommodation between the aristocracy and the middle classes. The ambiguity of the term was precisely its attraction: those of noble birth and good family were gentlemen by right, as were such groups as Anglican clergymen, but as the title's scope was extended, it became increasingly ambiguous. In the Victorian period, it began to include a definite moral component, which enhanced the sense that anyone could become a gentleman (and that only gentlemen were fit to lead the country). In 1862 Fitzjames Stephen argued in the *Cornhill Magazine* that the term *gentleman* implied "the combination of a certain amount of social rank with a certain amount of the qualities which the possession of such rank ought to imply; but there is a constantly increasing disposition to insist more upon the moral and less upon the social element of the word."[40] Nevertheless, despite the comparative openness of Britain's ruling class, Robin Gilmour stresses that *gentleman* was *not* a democratic term and would not have been appealing to the Victorians if it had been. It was precisely the way the word might (or might not) bridge the gap between breeding and behavior that made it appealing.

Hooker has usually been lumped with the professionalizers because he was a prominent member of the X Club. Because of their support for Darwin and desire to reform British science, the group has been conventionally associated with professionalization, but Ruth Barton has argued persuasively that the club is better understood as a forum for gentlemanly networking; she offers the prominence of Anglican amateurs like the MP John Lubbock as evidence.[41] There are some senses in which the aristocratic Lubbock and Hooker had more in common than Huxley and Hooker did. For example, Hooker's grandfather Dawson Turner was (like Lubbock and his father before him) a banker, a member of the commercial class who generally saw themselves as slightly superior to the manufacturers and industrialists who are normally thought to epitomize the rising Victorian middle class. Moreover, Lubbock shared Hooker's mistrust of the lower classes, despite working to improve their condition.[42] Clark characterizes Lubbock as a Whig-liberal—a reformer, but a paternalistic rather than a democratic one. Education—devised and imposed by the middle class—was the vital tool with which to raise the moral standards of the lower classes.[43] This was a cause Hooker shared with his fellow members of the X Club, who worked to improve the examination system for men of science.[44]

FIGURE 6.2. Old Hooker. Every inch the gentleman. This is believed to be the last photograph ever taken of Sir J. D. Hooker, 3 June 1911. (Courtesy of the trustees of the Royal Botanic Gardens, Kew.)

The Lubbocks were part of the group that Noel Annan christened the "intellectual aristocracy," a network of Whig families (including the Darwins and Wedgwoods) who intermarried regularly and exercised a degree of cultural leadership in mid-Victorian Britain.[45] Hooker undoubtedly felt at home among such people, not least because they were in many senses an aristocracy of talent. Yet at the same time, he belonged to a distinguished family (the Hookers could trace their ancestry back to the sixteenth-century theologian Richard Hooker, "the Judicious"). Like many Victorians, Hooker's sense of "gentleman" included both inherited qualities and personal achievements: he effectively inherited Kew from his father and bequeathed the gardens to his son-in-law, William Thiselton-Dyer, yet at the same time worked hard both to enhance the value of his inheritance and to prove himself worthy of it.

In 1857, Hooker had written to Darwin about the relief he felt in "turning from the drudgery of my 'professional Botany' to your 'philosophical Botany.'"[46] Being a professional meant the "drudgery" of government bureaucracy, being answerable to the vulgar Ayrton, and making Kew accessible to "swarms of nursery maids and children" and responding to their "demands for luncheons, pic-nics & bands of music."[47] Given that Hooker had no desire to be a professional, it is implausible to see his advocacy of scientific naturalism as a means of becoming one.

Secularism

Hooker's career suggests that naturalism cannot necessarily be linked to secularism any more than to professionalization. Although *secular* simply means *non*religious, when applied to Victorian Britain it often carries the connotation of being *anti*religious; certainly, in the controversies around Darwinism, historians have tended to associate the term with a campaign to free science from church influence. It is therefore striking that Hooker's unshakable commitment to scientific naturalism never translated into public hostility toward religion (in contrast to Huxley's belligerence). Although, at Norwich, Hooker attacked natural theology as being "to the scientific man a delusion," he also condemned it because it was "to the religious man, a snare, leading too often to disordered intellects and to atheism."[48] This might be interpreted as politic rhetoric, designed to soften the force of his attack, but when he discussed the Norwich meeting in a letter to Darwin, Hooker commented that

> the Clergy throughout behaved **splendidly** like men & gentlemen. The Cathedral service was glorious, the anthem was chosen for *me* "What though I know each Herb & Flower" & brought tears into my eyes, & Dr Magees discourse was the grandest ever heard by Tyndall, Berkeley, Spottiswoode, Hirst or myself.[49]

Hooker's praise of the clergy, emotional response to the anthem, and enthusiastic description of William Connor Magee's sermon should be enough to make us at least consider taking Hooker's warnings against atheism at face value—not least because professing some form of Christian faith was another essential part of being a Victorian gentleman.[50]

Hooker was raised in a devout household, and as a young man he recorded his regular church attendances in letters to his father. He remained a member of the Anglican Church throughout his life and stipulated that he be buried alongside his father at St. Anne's Church on Kew Green.[51] Nevertheless, some aspects of Hooker's gentlemanly religious conformity appear rather superficial: when Huxley asked Hooker to stand as godparent to his son, Noel, Hooker responded that "in the abstract I hate and despise the spiritual element of the ceremony," but that "the pleasure of being in any recognised relationship to your child will sweeten any pill of doctrine that may be offered."[52] There is an evident contrast between the ways Hooker and Huxley present themselves in their private letters and their public behavior; as Turner noted, a degree of public religious observance was important to maintaining respectability, hence the decision of an agnostic like Huxley to have his child baptized in the first place.[53]

However, looked at more broadly, Hooker's comments on religious matters suggest more than simply keeping up appearances. In agreeing to act as godfather to Huxley's son, he acknowledged that he had done the same for other friends' children and "as the christening is to be done, it is a duty to see it done properly; 'devoutly, orderly and reverently,' and as I won't trust these parsons, I will go see it myself."[54] If Hooker really despised the ceremony, why would he care whether it was conducted properly? More seriously, Hooker's speech at Norwich and his reaction to the press comments on it suggest a degree of sympathy for religious, or at least spiritual, values. The *Pall Mall Gazette* reported the service in Norwich cathedral that closed the BA meeting, noting that the following anthem was sung, apparently "for the benefit of the 'botanical and Darwinian president'":

> What though I trace each herb and flower
> That drinks the morning dew
> Unless I own Jehovah's power
> How vain were all I knew[55]

The writer sardonically suggested that the anthem was a subtle but successful attack on Hooker's apparently antireligious sentiments: "Dr. Hooker could not sing his reply, and the Cathedral authorities thus cleverly stole a march upon him. A man cannot very well answer a clergyman in the pulpit, and to put one's views in the form of an anthem and chant it oneself is a still greater impossibility. Dr. Hooker was silenced, if not convinced."[56]

The newspaper's implication that he had been offended by the anthem caught Hooker's eye and he wrote to Darwin:

Have you seen the Pall Malls inuendo [*sic*] apropos of the Cathedral Anthem at
Norwich & me—it is all bosh—Dr Buck, the organist—, gave it out of a real
compliment—to me & I am sure (accepted it as such & enjoyed it extremely
though I must confess I at first thought myself a huge egotist to suppose it was
intended for me):—it was lovely and the Pall Mall may be d—d.[57]

Before the *Pall Mall Gazette*'s report had even appeared, Hooker had told
Darwin that the anthem "brought tears into my eyes." The earlier letter also
praised the sermon, by William Connor Magee, which was later published as
"The Christian Theory of the Origin of the Christian Life." Magee, a broadly
liberal Anglican, argued that both scientific and church men were teachers like
Christ, and "the final cause of all science and all philosophy is the enrichment
of human life."[58] This sentiment was not unusual; Lubbock also asserted that
"men of science, and not the clergy only, are ministers of religion."[59] Magee's
compliment to science was followed by a careful argument for the separation
of church and science, on the grounds that the truths of Christianity were susceptible neither to proof nor disproof because "the idea of a Saviour, whose
life, communicated to those who believe in Him, shall give them a new life, is
mysterious, inexplicable, supernatural."[60] Furthermore, "a demonstration of
the supernatural is an impossibility: it is a contradiction in terms. No amount
of evidence drawn from the world of nature can demonstrate the existence of
a world above nature."[61] It was therefore futile, he asserted, to try to prove the
supernatural reality of Christianity using natural evidence (as natural theology attempted to do); such efforts have "the effect of violently repelling those
who are not Christians."[62] Listening to these words, Hooker would surely
have recognized that he and Magee were in agreement about natural theology
and in acknowledging the independence of science and religion.[63]

However, while Magee acknowledged science's expertise in the natural
world, he asserted Christianity's reign over the supernatural realm. Christianity was much more than a philosophy or a set of moral guidelines; it was a life
in itself and "a life with its own conditions of existence, its own laws of development, its own peculiar phenomena, as real and distinct as those of any
other form of existence which science investigates and classifies." Magee asserted that the Christian life was outside natural explanation and thus marked
the limit of scientific competence; "before the mystery of life, as before its
twin mystery of death, science stands abashed and dumb."[64]

Magee's sermon was an acknowledged tour de force; Charles Kingsley
was in the audience at Norwich and described it as "the most glorious piece
of eloquence I ever heard." Its success helped ensure Magee's rapid eleva-
tion (he became bishop of Peterborough later that year, eventually rising to

archbishop of York).[65] Part of its success was that, as Hooker noted, Magee behaved like a gentleman by making no public attack on science or Darwin, declaring instead, "It is to Science then herself that we make our appeal for aid to Faith":

> We ask her to tell us whether she, too, has not her mysteries which cannot be defined, her dogmas which cannot be demonstrated. From the conceited half knowledge of the dabblers in science and the smatterers in theology—with their parrot-like cant about the unreasonableness of mystery and the absurdity of dogma, their solemn platitudes about the irreconcilable differences between science, of which they know little, and theology, of which they know less—we appeal to the true high priests of science, to those who in the inmost shrine of the temple stand ever reverently with bowed heads before a veil of mystery, which they know they can neither lift nor rend . . . behind which there is a light whose source they cannot reach to and yet whose rays are still the light of all their life.

Magee suggested that the leaders of the BA (some of whom might have been delighted to be described as the "high priests" of science) should consider whether "the very mysteriousness of religion . . . is its recommendation to a mind that has felt the mysteriousness of science."[66] It was precisely at the point of greatest ignorance, "where the supernatural seems almost natural and where the natural almost rises up into the supernatural," that science and faith must acknowledge their mutual ignorance: "there, where the man of faith exclaims 'I know!' and the man of science 'I believe!' there, where in the presence of life—that common miracle of science and religion—the disciple of each exclaims to the other 'Behold, I show you a mystery!' there it is that we men of faith feel our grasp to tighten on the hand of our too long estranged brother, the man of science."[67] Despite the incommensurability between their fields, Magee offered hope for a reconciliation between men of science and of faith, based on each acknowledging the limits of their expertise.

Given Magee's passionate and uncompromising religious message, Hooker's enthusiasm for this sermon might seem surprising, but only if we retain anachronistic expectations created by the modernization analysis.[68] Once we set these aside, it is clear that the Darwinian president of the BA was not advancing secularization by attacking religion, but in fact his "sermon" and Magee's echoed each other.

Apart from his criticism of natural theology (which Magee shared), Hooker did not attack religion; in fact, he lamented that science had been mentioned more frequently from the pulpits in recent years, "but too often with dislike or fear, rather than with trust and welcome."[69] This mistrust could be overcome, he asserted, if men of science recognized that questions of ultimate meaning and purpose lay outside their province and must be left to men of faith. He concluded his address with a quotation, not from Darwin or any other scientific man, but from a poem by his cousin Francis Turner Palgrave, "The Reign of Law."[70] The penultimate stanza asserted that "the sequences of law / We learn through mind alone," since we perceive only "outward forms" and our senses cannot comprehend the "One who brought us hither, / And holds the keys of whence and whither." The poet implies that the law-bound world of "visible things" serves only to direct our attentions to the "One who brought us hither," who made those laws and alone understands our ultimate purposes and fate.

As Palgrave expressed it, in the stanza with which Hooker ended his address: "He in His science plans, / What no known laws foretell," and concludes that "the seeming chance that cast us hither / Accomplishes His whence and whither."[71] Hooker used the poem to acknowledge the limits to scientific laws, but at the same time he rejected the subservience of science to religion that the Duke of Argyll had argued for in *The Reign of Law*, published the previous year.[72] Natural theology was rejected because it involved a fatal confusion between the proper domains of science and religion, but Hooker denied the value neither of religion nor of faith in general. On the contrary, he argued that "to search out the whence and whither of his existence, is an unquenchable instinct of the human mind," and in our search for answers, humans have "adopted creeds" and "eagerly accepted scientific truths that support the creeds," and this "unquenchable instinct" had advanced both religion and science. "Science has never in this search hindered the religious aspirations of good and earnest men; nor have pulpit cautions . . . ever turned inquiring minds from the revelations of science."[73]

Magee's sermon was preached on 23 August, just four days after Hooker's address, so while the preacher had probably heard the botanist, the sermon must have been written before Hooker spoke. The similarities between the two therefore suggest common purposes among the more liberal members of both the religious and scientific communities.

The press reactions to Hooker's address demonstrate that the hope for a reconciliation between science and religion was widespread; the hostile reactions quoted above seem to have been in the minority. Darwin congratulated Hooker on the speech and mentioned the reports in the *Times*, *Telegraph*, *Spectator*, and *Athenaeum*, and had heard of many "favourable newspapers," adding "there is a chorus of praise."[74]

Among the favorable reports in the provincial papers, the *Liverpool Mercury* praised Hooker for giving "the English working man" what he wants, "a just comprehension of the order of the universe," because such an understanding of "the laws which rule this complicated system of forces" must teach us "how to live rightly, and be rewarded with such welfare as the Author and Lord of life has appointed for the obedient." As a result, science "can never, as Dr. Hooker truly observes, be at variance with the aims of pure religion," as was exemplified in the work of "such men as Faraday, the physicist, and Darwin, the naturalist."[75] The gentlemanly restraint of Hooker's public pronouncements helped ensure a largely favorable reaction.

The complexity of the science and religion debate is exemplified in the reactions to Hooker's speech by two *Guardians*: the liberal, reforming *Manchester Guardian* and the London-based *Guardian*, which had High Church sympathies (albeit fairly broad-minded ones).[76] The London paper noted the predominance of Darwinism at the Norwich meeting, observing that in the biological section ("its own peculiar home") particularly, "Its reign was triumphant and almost unopposed." The meeting was, the paper concluded, "a true reflection of the general tone of thought in the younger part of the scientific world. The coming generation of naturalists are all more or less Darwinian." The paper was not uncritical of Darwinism, but mildly commented that it took "too little note of the Creative Will, which after all for anyone except an atheist—which Mr. Darwin certainly is not—must lie at the back of all visible phenomena." It was, the *Guardian* concluded, too soon to say what Darwinism's ultimate impact would be, but perhaps it would "suggest to theologians the modification of expressions by which, as the Dean of Cork [Magee] well reminded us, theological truth is apt to be clothed from time to time in the terms of a bygone philosophy."[77]

By contrast with its London-based namesake, the *Manchester Guardian* expressed some frustration over what it saw as the annual ritual embodied in the president's address. The rapid progress of the sciences "excite[s] uneasy apprehensions in many well-meaning but timid people." As a result,

> it appears to be almost a necessary part of the President's annual duty to make a sort of apology for science, and to assure his timorous hearers that she is by no means the dangerous person they suppose. Dr. Hooker accepted the inevitable task, and closed his address by vindicating her from the charge of being antagonistic to religion.

There was, the paper felt, a "loss of dignity, in these repeated protests on the part of science of her innocence of any designs against religion," not least because "they fail even to allay the apprehensions excited by unwelcome discoveries."[78]

In discussing the press reactions, Darwin had observed that the *Spectator* "pitches a little into you about Theology, in accordance with its usual spirit; for there is some writer in the Spectator who is the most ardent admirer of the Duke of Argyll."[79] Hooker replied, "What a pother these papers kick up about my mild theology!" He continued:

> An Aberdeen one calls me an Atheist & all that is bad: to me, who do not intend to answer their abuse, misquotations, garbled extracts, & blunders, it is all really very good fun. There were gentle disapproving allusions at Kew Church today I am told! I am beginning to feel quite a great man![80]

The generally positive response no doubt made the accusations of atheism and the "gentle" disapproval of his own parish easier to laugh off, especially in a private letter; one aspect of gentlemanliness was not to overreact to perceived slights. And in fact the *Spectator's* article was fairly measured; its writer admitted to sharing Hooker's dislike of "dogmatic theology," but accused men of science of indulging in dogmatism of their own, and cited Hooker's attack on natural theology as an example, but its criticisms were generally courteous and reasonable in tone. The article also highlighted an important change that helped science and religion become reconciled. Hooker had claimed that astronomers first broke the hold of dogma by demonstrating that "the sun does not go round the world." The *Spectator* demurred: "We doubt if that astronomical discovery did half so much to break the bonds of dogma as the discovery in morals called the right to private judgment."[81] As Lightman has shown, the idea that one could be an "honest doubter" in religious matters, a position encapsulated by Huxley's term *agnostic*, became increasingly respectable during the late Victorian period.[82] By contrast, in the early years of the century, those who were not active and public supporters of Christianity were assumed to be its enemies, or "infidels." What the *Spectator* called "the right to private judgment" was a widespread recognition that an individual's religious opinions were no longer the subject of public comment or condemnation.[83]

The ever more private nature of religious beliefs was an aspect of the increasing variety of both faiths and doubts in Victorian Britain. The well-known gradual removal of disabilities from Dissenters and other religions is probably best understood as both a cause and a symptom of this diversification of belief. As tolerance increased, so too did the difficulty of knowing which beliefs (if any) one might safely assume one held in common with an interlocutor. The debates around the *Essays and Reviews* (1860) and over Bishop John William Colenso's *Pentateuch and the Book of Joshua Critically Examined* (1862) exposed divisions within the Anglican Church that were often even sharper than those between believers and unbelievers.[84]

Hooker and the rest of the X Club supported liberal theologians in the controversies over both *Essays and Reviews* and Colenso, despite some disagreements about tactics, partly to defend freedom of thought and speech but also because they shared a principled opposition to reactionary theology and its role within the Anglican Church. In 1861, Lubbock organized a petition in support of the *Essays and Reviews* authors, but Hooker declined to sign, explaining that the names Lubbock had gathered represented only "the young progressionists in Science," whose "opinions are of no weight in religious matters." Hooker believed that unless accompanied by the signatures of older, more conservative men, the petition would "create a fission in the 'body politic' of scientific men." And, while he firmly believed there could be no compromise "between progression and non-progression" in scientific matters,

> I should be sorry to see anything done that would countenance a belief amongst the outsiders that our scientific differences influenced our religious views and this would be a very legitimate inference if your memorial was signed wholly or chiefly by men of one way of thinking, in such matters as "Origin of Species," "Age of Man," &c., &c.

While Hooker confessed to "an almost morbid aversion for clique or sectarianism," he suggested waiting for the "present excited state of the public mind" to subside, after which "I shall gladly sign a memorial addressed to the Essayists, thanking them for what they have done and requesting a Second Series of Essays."[85]

Nothing in this letter suggests any hostility to religion, but it again illustrates the distance between private and public attitudes. As Barton has argued, Hooker's refusal to sign was a disagreement over strategy; I would add that it was also an argument about good manners.[86] Hooker was generally averse to contentious or provocative language. He did not shun controversy as such, but restricted it to purely scientific debates, conducted courteously between qualified participants.[87] Just as Hooker criticized Huxley's gratuitously offending the clergy (see above), he wrote privately that although he liked and respected Colenso (Hooker was "pleased with his calmness, dignity and charity towards his opponents"), he nevertheless felt that the bishop had brought his difficulties upon himself: "He might in my opinion have said ten times as much as he has *in different language* and he would have created no sensation at all."[88]

At this stage in his life, was Hooker a believer and, if so, in what? The question is all but impossible to answer and that, surely, is as Hooker would have wanted it. In a private letter to his friend the naturalist James la Touche,

Hooker expressed some appreciation for Huxley's concept of "a religion of pure reason."[89] No mention here of agnosticism nor of hating and despising the "spiritual element" of sacraments like baptism, but Hooker was tailoring his message to his audience—la Touche was a clergyman and Hooker disliked causing unnecessary offense; once again, Hooker was conforming to gentlemanly norms.

Unlike Huxley or Tyndall, Hooker met this celebrated definition of a gentleman:

> If he is an unbeliever, he will be too profound and large-minded to ridicule religion or act against it; he is too wise to be a dogmatist or fanatic in his infidelity. He respects piety and devotion; he even supports institutions as venerable, beautiful, or useful, to which he does not assent; he honours the ministers of religion, and it contents him to decline its mysteries without assailing or denouncing them. He is a friend of religious toleration, and that, not only because his philosophy has taught him to look on all forms of faith with an impartial eye, but also from the gentleness and effeminacy of feeling, which is the attendant on civilization.[90]

This is, of course, from John Henry Newman's definition, which stresses the gentleman's unwillingness to cause offense. Newman did not regard gentlemanliness as any substitute for faith; indeed, his emphasis on the show and form of the gentleman was intended to highlight its hollowness, the lack of true Christian belief that the public, polished performance helped conceal.[91] For mid-Victorian gentlemen, Christianity became part of good manners; a writer in the *Contemporary Review* (1869) asserted that "Christianity is the revelation to us here of the Etiquette of Heaven."[92] The superficiality of this definition was repulsive to men like Newman, but his strong faith would have appalled many gentlemen—as much because it was strong as because it was Catholic. Many preferred an easy passage through society to contentious disputes over theological matters such as God's true nature; that, of course, was Newman's complaint, but it is precisely for that reason that his definition fits Hooker well (just as it fits Lubbock and even Darwin).

Conclusion: Naturalism

Hooker's rejection of deliberately controversial stances, such as Huxley's, illuminates his sense of how one became and remained respectable. For the men of science, scientific naturalism was, in part, a form of politeness, a way of avoiding unnecessary controversy—particularly over religion—and thus maintaining the equable sociability of a gentleman; this was, of course, one goal of Huxley's neologism *agnosticism*, which played such a crucial role in making private doubt respectable.[93]

To understand how naturalism could function as a form of etiquette, it is necessary to distinguish what might be called metaphysical naturalism from a purely methodological version of the same beliefs.[94] At its strongest, metaphysical naturalism asserts that the only entities that exist are those that the senses (refined and assisted by scientific method and instruments) can apprehend and measure; everything else, from God to transcendental anatomical archetypes, simply doesn't exist. Whether any major Victorian man of science really held to this strong view in its strongest sense is, of course, a moot point.[95] Nevertheless, opponents of scientific naturalism accused men like Tyndall and Huxley of holding this view in private, and they scoured their public pronouncements for evidence of infidel materialism.

By contrast, methodological naturalism is simply the claim that supernatural elements have no place in strictly scientific discourse; in this weaker form, it was perfectly permissible (indeed, rather commonplace) for the naturalist to accept the reality and importance of the spiritual in matters of morality and ultimate human purpose, as long as it was acknowledged that such matters lay outside science. This separation into what Stephen Jay Gould referred to as nonoverlapping magisteria was Hooker's goal in his Norwich address, just as it was Magee's in his sermon; Huxley and Tyndall held similar views, but their relish for theological controversy made them less acceptable to those who applauded Hooker.[96] Methodological naturalism allowed potentially controversial (and irresolvable) religious matters to be private; by excluding them from science, one could avoid forming any judgment as to the propriety of anyone's beliefs on nonscientific topics. As a result, as Barton illustrates in her chapter in this volume, practical scientific cooperation was possible in the absence of religious consensus.

The rise of scientific naturalism may therefore perhaps be best understood within the wider context of that most characteristic of Victorian debates—over who was a gentleman rather than over who was a scientist. In the middle decades of the century, a gentleman had to be a Christian; in Sir Archibald Edmonstone's *The Christian Gentleman's Daily Walk* (1840), true gentlemanliness and Christian behavior are more or less synonymous. David Castronovo argues that this generalized, spiritualized, democratic gentleman becomes so loosely defined as to be ubiquitous and meaningless. Dickens's numerous attempts to replace the traditional ideal of birth with a more spiritual one, which emphasized hard work and virtue, played an important part in this.[97] In *Our Mutual Friend* (1865), when Eugene Wrayburn's marriage to

Lizzie Hexam is being discussed, Twemlow asserts that he thinks Wrayburn a true gentleman, adding, "When I use the word gentleman, I use it in the sense in which the degree may be attained by any man." As Castronovo notes, by this point the word's sense has been diluted to such an extent that it is effectively meaningless.[98]

In Anthony Trollope's final Palliser novel, *The Duke's Children* (1880), the Duke of Omnium's daughter says that the man she wants to marry—who the duke considers beneath her—is at least a gentlemen, to which the duke replies:

> "So is my private secretary. There is not a clerk in one of our public offices who does not consider himself to be a gentleman. The curate of the parish is a gentleman, and the medical man who comes here from Bradstock. The word is too vague to carry with it any meaning that ought to be serviceable to you in thinking of such a matter."[99]

As middle-class men like Hooker struggled to become gentlemen, they destroyed the term's meaning in the process. There is an ironic, but entirely understandable, parallel with the changes wrought by the impact of the "right to private judgment" on public professions of faith. The Broad Church movement within the Anglican communion was one symptom of this; for some members of the other Church parties, the Broad Church was so broad as to be no longer Christian, since its advocates denied fundamental aspects of Christian teaching in favor of a purely rational or natural religion.[100] One effect of a private faith is that everyone can be presumed to be some sort of Christian, and when everyone is a Christian, no one is—as Newman recognized. The scientific naturalists were laying claim to gentlemanliness while trying to avoid clarifying the degree or precise nature of their faith; *agnosticism* was an attempt to bracket the whole question, but unlike Tyndall and Huxley, Hooker seems to have felt no need of the label. Hooker's desire to be a gentleman (even a Christian gentleman) was stronger than any desire to engage in controversy, which helps explain, for example, why the newspapers rallied to his defense over the Ayrton affair.

For historians, the problem with the methodological sense of naturalism is that almost everyone turns out to be a naturalist. However, far from being a weakness of this argument, this objection highlights its strength. The most diverse ranges of beliefs could happily coexist as long as the underlying metaphysics of their shared naturalism were simply left unexamined. As Mathew Stanley shows in his chapter in this volume, even some of the Darwinians' potential foes—such as Richard Owen, or some of the North British physi-

FIGURE 6.3. Emperor of Kew. Joseph Hooker celebrating his ninetieth birthday by surveying his domain in the company of his second wife, Lady Hyacinth. (Courtesy of the trustees of the Royal Botanic Gardens, Kew.)

cists—were united by a common belief in the uniformity of nature's laws that ensured they used common scientific practices. To which might be added the observation that common standards of gentlemanly behavior were every bit as important as common laboratory practices. Given a broad enough "church" of scientific naturalism, everyone could be considered a naturalist, and when everyone is a naturalist, no one is; the term loses much of its analytical bite when extended so far. However, why should historians be spared the fate of our actors, who found their favorite terms were also being emptied of meaning?

Notes

1. Joseph Dalton Hooker, *Presidential Address to the British Association for the Advancement of Science* (Norwich, UK: Fletcher and Son, Printers, 1868), 28–29.

2. Quoted in "The British Association for the Advancement of Science," *Leeds Mercury*, 21 August 1868, 3.

3. Quoted in "Epitome of Opinion in the Morning Journals," *Pall Mall Gazette*, 21 August 1868, 3.

4. "The British Association for the Advancement of Science," *John Bull*, 22 August 1868, 571.

5. "Science Perverted to Infidelity," *Nottinghamshire Guardian*, 28 August 1868, 6.

6. Bernard Lightman, *Evolutionary Naturalism in Victorian Britain: The "Darwinians" and Their Critics* (Aldershot, UK: Ashgate, 2009), 1:10–11; and Frank Miller Turner, *Between Science and Religion: The Reaction to Scientific Naturalism in Late Victorian England* (New Haven, CT: Yale University Press, 1974), chap. 2.

7. For example, "The Victorian Conflict between Science and Religion: A Professional Dimension," in Frank M. Turner, *Contesting Cultural Authority: Essays in Victorian Intellectual Life* (Cambridge: Cambridge University Press, 1993), 171–200.

8. Frank M. Turner, "The Late Victorian Conflict of Science and Religion," in *Science and Religion: New Historical Perspectives*, ed. Thomas Dixon, Geoffrey Cantor, and Stephen Pumfrey (Cambridge: Cambridge University Press, 2010), 88–89.

9. Turner, *Between Science and Religion*; Robert M. Young, *Darwin's Metaphor: Nature's Place in Victorian Culture* (Cambridge: Cambridge University Press, 1985); Bernard Lightman, *The Origins of Agnosticism: Victorian Unbelief and the Limits of Knowledge* (Baltimore: Johns Hopkins University Press, 1987); Turner, *Contesting Cultural Authority*; and Turner, "Late Victorian Conflict."

10. E.g., Lightman, *Origins of Agnosticism*; and Frank M. Turner, "The Secularization of the Social Vision of British Natural Theology," in Turner, *Contesting Cultural Authority*, 101–30.

11. My argument has been much influenced by Ruth Barton. See her "'Huxley, Lubbock, and Half a Dozen Others': Professionals and Gentlemen in the Formation of the X Club, 1851–1864," *Isis* 89 (1998): 414.

12. J. F. M. Clark, "'The Ants Were Duly Visited': Making Sense of John Lubbock, Scientific Naturalism and the Senses of Social Insects," *British Journal for the History of Science* 30 (1997): 156.

13. For Huxley, see Adrian Desmond, *Huxley: The Devil's Disciple* (London: Michael Joseph, 1994); and Adrian Desmond, *Huxley: Evolution's High Priest* (London: Michael Joseph, 1997). A rather different interpretation of Huxley's career and motives emerges in Paul White's *Thomas Huxley: Making the "Man of Science"* (Cambridge: Cambridge University Press, 2003).

14. See Herbert Schlossberg, *Conflict and Crisis in the Religious Life of Late Victorian England* (London: Transaction, 2009).

15. See David Castronovo, *The English Gentleman: Images and Ideals in Literature and Society* (New York: Ungar, 1987), 1–52.

16. Turner, *Contesting Cultural Authority*, 180–81.

17. Roy M. MacLeod, "The Ayrton Incident: A Commentary on the Relations between Science and Government in England, 1870–1873," in *Science and Values*, ed. Arnold Thackray and Everett Mendelsohn (New York: Humanities Press, 1974), 70.

18. Ray Desmond, *Kew: A History of the Royal Botanic Gardens* (London: Harvill Press, 1995), 225–37.

19. [Quoted in Maxwell Tylden Masters?], "Mr. Ayrton and Dr. Hooker," *Gardener's Chronicle*, no. 28 (1872): 939.

20. J. D. Hooker to James Hector, 14 December 1871: John Yaldwyn and Juliet Hobbs, eds., *My Dear Hector: Letters from Joseph Dalton Hooker to James Hector, 1862–1893* (Wellington: Museum of New Zealand Te Papa Tongarewa, 1998), 144–45. Ayrton, like his father, did practice law in Bombay, but had been born at Kew. His mother was the daughter of a British lieutenant colonel; W. P. Courtney, "Ayrton, Acton Smee (1816–1886)," rev. H. C. G. Matthew, *Oxford Dictionary of National Biography*, Oxford University Press, 2004, accessed 5 July 2011, http://www.oxforddnb.com/view/article/947.

21. J. D. Hooker to C. Darwin, 20 October 1871: DAR 103: 87–92.

22. "Mr. Ayrton Again," *Saturday Review* 34 (1872): 169.

23. "House of Commons, Thursday, Aug. 8," *Times*, 9 August 1872, 3. Emphasis added.

24. [Maxwell Tylden Masters?], "Editorial (July 13)," *Gardener's Chronicle*, no. 28 (1872): 933.

25. The position at Kew initially paid five hundred pounds a year, rather more than any clerk earned in 1840, but probably slightly less than William Hooker's previous position in Glasgow had brought him. Richard H. Drayton, *Nature's Government: Science, Imperial Britain and the "Improvement" of the World* (New Haven, CT: Yale University Press, 2000), 168–69.

26. [Masters?], "Editorial (July 13)," 933.

27. J. D. Hooker to W. J. Hooker, 11 April 1849, Kew, India Letters: 158.

28. Drayton, *Nature's Government*, 197–99.

29. Quoted in Mea Allan, *The Hookers of Kew, 1785–1911* (London: Michael Joseph, 1967), 217.

30. Drayton, *Nature's Government*, 197–201.

31. J. D. Hooker to W. J. Hooker, 18 May 1843: Leonard Huxley, *Life and Letters of Joseph Dalton Hooker*, 2 vols. (London: John Murray, 1918), 1:166.

32. Darwin Correspondence Project Database, letter no. 3395, accessed 27 June 2011, http://www.darwinproject.ac.uk/entry-3395.

33. Ibid.

34. Darwin Correspondence Project Database, letter no. 3411, accessed 27 June 2011, http://www.darwinproject.ac.uk/entry-3411. *Kenspeckle* is a Scots dialect word meaning easily recognizable or conspicuous (*OED*).

35. Hooker, quoted in W. B. Turrill, *Joseph Dalton Hooker: Botanist, Explorer and Administrator* (London: Scientific Book Club, 1963), 197; C. Darwin to J. D. Hooker, 25 and 26 January 1862: Frederick Burkhardt et al., eds., *The Correspondence of Charles Darwin*, 20 vols. (Cambridge: Cambridge University Press, 1997), 10:48; and Adrian Desmond, "Redefining the X Axis: 'Professionals,' 'Amateurs' and the Making of Mid-Victorian Biology—A Progress Report," *Journal of the History of Biology* 34 (2001): 7.

36. Charles James Fox Bunbury, *The Life of Sir Charles J. F. Bunbury*, 2 vols. (London: John Murray, 1906), 2:226.

37. Hooker to Darwin, 30 August 1868: Burkhardt et al., *Correspondence of Charles Darwin*, 16(2):703.

38. See David Brown, "Equipoise and the Myth of an Open Elite: New Men of Wealth and the Purchase of Land in the Equipoise Decades, 1850–69," in *An Age of Equipoise? Reassessing Mid-Victorian Britain*, ed. Martin Hewitt (Aldershot, UK: Ashgate, 2000), 139.

39. Thomas Hay Sweet Escott, *England: Its People, Polity and Pursuits* (London: Cassell, 1881), 280.

40. Quoted in Robin Gilmour, *The Idea of the Gentleman in the Victorian Novel* (London: G. Allen and Unwin, 1981), 2–5.

41. Barton, "'Huxley, Lubbock, and Half a Dozen Others.'" See also Ruth Barton, "'An Influential Set of Chaps': The X-Club and Royal Society Politics, 1864–85," *British Journal for the History of Science* 23 (1990): 53–81.

42. See Clark, "Ants Were Duly Visited," 154.

43. Ibid., 154–55.

44. Barton, "'Huxley, Lubbock, and Half a Dozen Others,'" 426–27. See also James Elwick's chapter in this volume.

45. Noel G. Annan, "The Intellectual Aristocracy," in *Studies in Social History: A Tribute to G. M. Trevelyan*, ed. J. H. Plumb (London: Longmans, 1955), 243–87.

46. J. D. Hooker to C. Darwin, [11 April 1857]: Burkhardt et al., *Correspondence of Charles Darwin*, 6:369. I discuss the importance of the term *philosophical* at length in Jim Endersby, *Imperial Nature: Joseph Hooker and the Practices of Victorian Science* (Chicago: University of Chicago Press, 2008).

47. "Kew Gardens. Forenoon Opening, 1866–1889," Kew Archives (1875), f. 144. J. D. Hooker, memo, 29 October 1877, "Admission of the public, 1853–1925," Kew. Quoted in Desmond, *Kew*, 236.

48. Hooker, *Presidential Address*, 28–29.

49. Hooker to Darwin, 30 August 1868: Burkhardt et al., *Correspondence of Charles Darwin*, 16(2):703. Emphasis in original.

50. Castronovo, *English Gentleman*, 19–30, 46–65.

51. For examples of Hooker's church attendance, see Huxley, *Life and Letters of Joseph Dalton Hooker*, 1:33–34, 1:106, 1:68. For his wish to be buried at St. Anne's, see ibid., 2:480.

52. J. D. Hooker to T. H. Huxley, 4 January 1861, Huxley, *Life and Letters of Joseph Dalton Hooker*, 2:59.

53. J. D. Hooker to C. R. Darwin, 6 October 1865, quoted in Turner, *Contesting Cultural Authority*, 97.

54. J. D. Hooker to T. H. Huxley, 4 January 1861, Huxley, *Life and Letters of Joseph Dalton Hooker*, 2:59.

55. Solomon's aria from George Frideric Handel's oratorio *Solomon* (1749).

56. *Pall Mall Gazette*, 3 September 1868, issue 1112, 8. The item was reprinted in *The Bury and Norwich Post, and Suffolk Herald*, 8 September 1868, issue 4498, 7 (BL-NCBN).

57. J. D. Hooker to C. R. Darwin, 5 September 1868: Burkhardt et al., *Correspondence of Charles Darwin*, 16(2):718.

58. William Connor Magee, "The Christian Theory of the Origin of the Christian Life," in *The Gospel and the Age: Sermons on Special Occasions*, ed. William Connor Magee (London: William Isbister, 1884), 155.

59. John Lubbock, *The Pleasures of Life*, pt. 1, 3rd ed. (London, 1887), 161–62, quoted in Clark, "Ants Were Duly Visited," 156.

60. Magee, "Christian Theory," 160.

61. Ibid.

62. Ibid., 161–63.

63. The argument for such a separation was, of course, original to neither Magee nor Hooker,

but widely discussed at the time. See Lightman, *Origins of Agnosticism*, 1–28; and Schlossberg, *Conflict and Crisis*, 44–45.

64. Magee, "Christian Theory," 159, 70.

65. John Cotter Macdonnell, *The Life and Correspondence of William Connor Magee: Archbishop of York, Bishop of Peterborough*, 2 vols. (London: Isbister, 1896), 1:49, 1:188. J. C. Macdonnell, "Magee, William Connor (1821–1891)," rev. Ian Machin, *Oxford Dictionary of National Biography*, Oxford University Press, 2004, online ed., May 2006, accessed 19 May 2011, http://www.oxforddnb.com/view/article/17779.

66. Magee, "Christian Theory," 170–71.

67. Ibid., 171–72. Magee's quote is from 1 Corinthians 15:51.

68. Many of the X Club shared the belief that science and religion were in harmony. See Lightman, *Origins of Agnosticism*, chap. 5.

69. Hooker, *Presidential Address*, 26.

70. Palgrave (1824–97) is, understandably, best remembered as an anthologist, responsible for the much-reprinted *Golden Treasury of the Best Songs and Lyrical Poems in the English Language* (1861), usually referred to as "Palgrave's Golden Treasury."

71. F. T. Palgrave, "The Reign of Law." First published in *Macmillan's Magazine*, March 1867, quoted in Hooker, *Presidential Address*, 28–29. The poem was revised quite substantially when it appeared in book form in Francis Turner Palgrave, *Amenophis and Other Poems* (1892).

72. Hooker borrowed Argyll's book from Darwin ("which I want to understand if I can") while he was preparing his address, but did not discuss it in detail, merely referring favorably to Alfred Russel Wallace's critical review of it ("Creation by Law," *Journal of Science*, October 1867); J. D. Hooker to C. R. Darwin, 7 April 1868: Burkhardt et al., *Correspondence of Charles Darwin*, 16(1):384.

73. Hooker, *Presidential Address*, 27–28.

74. C. R. Darwin to J. D. Hooker, 23 August 1868: Burkhardt et al., *Correspondence of Charles Darwin*, 16(2):691.

75. "The Parliament of Science," *Liverpool Mercury*, 21 August 1868, 6.

76. Alvar Ellegård, *The Readership of the Periodical Press in Mid-Victorian Britain* (Göteborg: Göteborg Universitets Årsskrift, 1957), 12.

77. "The British Association," *Guardian*, 2 September 1868, 977.

78. *Manchester Guardian*, 24 August 1868, 2 (BL-NCBN). The *Pall Mall Gazette* expressed similar views: "Dr Hooker on Religion and Science," 22 August 1868, vol. 8, issue 1102, 593 (BL-NCBN).

79. C. R. Darwin to J. D. Hooker, 23 August 1868: Burkhardt et al., *Correspondence of Charles Darwin*, 16(2):691.

80. J. D. Hooker to C. R. Darwin, 30 August 1868: ibid., 16(2):702.

81. "Dr Hooker on the Evidences," *Spectator*, 22 August 1868, 986.

82. Lightman, *Origins of Agnosticism*.

83. I discuss this in relation to Darwin's faith in Jim Endersby, "Editor's Introduction" to Charles Darwin, *On the Origin of Species by Means of Natural Selection*, ed. Jim Endersby (Cambridge: Cambridge University Press, 2009).

84. Schlossberg, *Conflict and Crisis*, 1–8.

85. J. D. Hooker to John Lubbock, 29 February 1861: Huxley, *Life and Letters of Joseph Dalton Hooker*, 2:54–55.

86. Barton, "'Huxley, Lubbock, and Half a Dozen Others,'" 434–37.

87. Endersby, *Imperial Nature*, 218–19, 54–75.

88. J. D. Hooker to B. H. Hodgson, 19 April 1863: Huxley, *Life and Letters of Joseph Dalton Hooker*, 2:57–58. Emphasis in original.

89. J. D. Hooker to J. D. la Touche, 24 December 1893: ibid., 2:337.

90. John Henry Newman, "Discourses on the Scope and Nature of University Education" (1852), published as *The Idea of a University* (1873). Quoted in Gilmour, *Idea of the Gentleman*, 91.

91. Castronovo, *English Gentleman*, 40.

92. Quoted in Castronovo, *English Gentleman*, 63.

93. My thinking on this point has been greatly influenced by Gowan Dawson's arguments in *Darwin, Literature and Victorian Respectability* (Cambridge: Cambridge University Press, 2007).

94. I am indebted to Katherine Anderson and others for helping me to clarify this distinction during two very productive days of discussion at a conference of the contributors to this volume, York University, 6–7 May 2011.

95. See, for example, Ruth Barton, "John Tyndall, Pantheist: A Rereading of the Belfast Address," *Osiris*, 2nd ser., 3 (1987): 111–34.

96. Stephen Jay Gould, "Nonoverlapping Magisteria," *Natural History Review* 106 (1997): 16–22.

97. Castronovo, *English Gentleman*, 49–50.

98. Ibid., 51–52.

99. Anthony Trollope, *The Duke's Children* (1880; repr., Harmondsworth, UK: Penguin Books, 1995), 54; see also Gilmour, *Idea of the Gentleman*, 13–14.

100. Schlossberg, *Conflict and Crisis*, 7–8.

Broader Alliances

7

Sunday Lecture Societies: Naturalistic Scientists, Unitarians, and Secularists Unite against Sabbatarian Legislation

RUTH BARTON

The Sunday Lecture Society (SLS), founded in 1869, and its predecessor and competitor, Sunday Evenings for the People, initiated in 1866, have almost been lost to history. They appear in the history of scientific naturalism as sites for T. H. Huxley performances and as organizations of interest to the X Club. J. Vernon Jensen first noticed the X Club's interest in Sunday lectures.[1] Adrian Desmond describes the excitement of the audience when Huxley lectured "On the Desirableness of Improving Natural Knowledge" at St. Martin's Hall in 1866 under the auspices of Sunday Evenings for the People. James Moore has reprinted an auditor's account of a later Huxley lecture, "On the Origins of the English People" (1870), sponsored by the Sunday Lecture Society, and noticed the association of other X Club members with the SLS. More recently, David Knight found a collection of pamphlets that identified Huxley as a president of the SLS. Knight characterized the SLS as directed against "ignorance and superstition."[2] Sunday lecture–promoting organizations are also mentioned in accounts of Sabbatarianism. The fullest information is provided by John Wigley, who identifies the founders and objectives of five Sunday societies, both metropolitan and provincial, and mentions a few others, but all in less than ten pages.[3] Thus, the information available has been sketchy, enabling little interpretation. I have been able to go much further in determining the aims of the SLS and identifying those involved because James Moore generously drew my attention to a large collection of SLS pamphlets in the British Library.

Although I am analyzing the SLS because the leading representatives of scientific naturalism were involved, my argument emphasizes the other religious and ideological groups with whom they were allied in promoting Sunday lectures. The activities and objectives of the Sunday lecture societies take us beyond Huxley, beyond the X Club, and beyond scientific naturalism, and in doing so, I will argue, provide some possible explanations for the apparent success and effectiveness of Huxley, of the X Club, and of scientific naturalism. In Sunday lecture societies, scientific men, literary men, and reforming lawyers joined to promote improving lectures as an alternative to Sunday sermons. Unitarians, secularists, and advocates of naturalistic science were particularly conspicuous in the leadership of the Sunday societies. The objective that brought the supporters together was the secularization of the state, in particular, the removal of state support for a widely held religious vision of how Sundays should be spent.

The first section of this chapter identifies the early efforts of Sunday Evenings for the People; the second section analyzes the Sunday Lecture Society and its modest achievements over its first twenty years. A brief discussion of other efforts to promote improving Sunday activities follows. The third section moves from local detail to the grander narrative of the secularization of the British state in the nineteenth century. I interpret the alliance of diverse groups in promoting Sunday lectures and opposing Sabbatarian legislation in terms of secularization. In conclusion, the local detail leads to questions as to who were the "scientific naturalists," and the grander narrative leads to suggestions about how to assess the achievements of the "scientific naturalists" in comparison to the achievements of competing interpreters of science, such as the North British circle of physicists and engineers around William Thomson identified by Crosbie Smith, and the devout popularizers whose contributions Bernard Lightman has contrasted with those of Huxley, Herbert Spencer, and their friends.[4]

Sunday Evenings for the People and the National Sunday League

The first series of Sunday Evenings for the People, begun in 1866 by the National Sunday League (NSL), lasted only four weeks before being closed down under threat of legal action. The NSL had prepared for opposition by enlisting a large and respectable group of eminent supporters. The flyer for the first lecture (fig. 7.1) announced the names of fifty-four distinguished men who approved the plan to offer "discourses on science and the wonders of nature," especially to those who "do not attend places of worship." The first lecturer was Huxley, "On the Desirableness of Improving Natural Knowledge," the other lecturers and topics were listed, and the flyer asserted the social and even religious benefits of the lecture program, but I start my analysis with the list of names.

Victorian organizations enlisted worthy gentlemen (and ladies if the cause

SUNDAY EVENINGS FOR THE PEOPLE

WITH THE APPROVAL OF

Sir J. BOWRING, LL.D., F.R.S.
Prof. BEESLY, University College.
THOS. HORLOCK BASTARD, Esq.
FRANCESCO BERGER, Esq.
SIR JAMES CLARK.
W. B. CARPENTER, M.D., F.R.S.
M. D. CONWAY, Esq.
W. S. COOKSON, Esq.
L. F. CLARK, Esq.
ERAS. DARWIN, Esq., F.R.S.
CHAS. DICKENS, Esq.
W. H. DOMVILLE, Esq.
JOHN DILLON, Esq.
THOS. HENRY FARRER, Esq.
Prof. FRANKLAND, F.R.S.
F. J. FURNIVALL, Esq., M.A.
HENRY F. FARBROTHER, Esq.
Dr. J. E. GRAY, F.R.S., &c.

THOMAS GRAHAM, Esq., F.R.S.
W. B. HODGSON, Esq., LL.D.
JAMES HEYWOOD, Esq., F.R.S.
Prof. T. H. HUXLEY, F.R.S., &c.
Prof. A. W. HOFMANN, Ph.D. F.R.S.
FREDERICK HARRISON, Esq., M.A.
GAVIN HARDIE, Esq.
Prof. THOS. H. KEY, F.R.S.
Rev. THOS. KIRKMAN, M.A., F.R.S.
Sir CHAS. LYELL, Bart., F.R.S.
Sir J. LUBBOCK, Bart., F.R.S.
A. H. LAYARD, Esq., M.P.
J. BAXTER LANGLEY, Esq., M.R.C.S.
R. B. LITCHFIELD, Esq., B.A.
VERNON LUSHINGTON, Esq., B.C.L.
GODFREY LUSHINGTON, Esq.
JOHN STUART MILL, Esq., M.P.
Prof. HENRY MORLEY.

Rev. Prof. J. MARTINEAU.
Prof. RICHARD OWEN, F.R.S., &c.
WM. SCHOLEFIELD, Esq., M.P.
H. J. SLACK, Esq., F.G.S.
WM. SHAEN, Esq., M.A.
HERBERT SPENCER, Esq.
SPENCER SHELLEY, Esq.
P. A. TAYLOR, Esq., M.P.
Prof. J. TYNDALL, LL.D., F.R.S.
Rev. Prof. J. J. TAYLER.
Rev. CHAS. VOYSEY, M.A.
Sir JOSHUA WALMSLEY.
Prof. WILLIAMSON, Ph.D., F.R.S.
E. P. WOLSTENHOLME, Esq.
W. H. WILLS, Esq.
HENSLEIGH WEDGWOOD, Esq.
Sir J. G. WILKINSON, D.C.L., F.R.S.
ERASMUS WILSON, Esq., F.R.S., &c.

· The Sunday, as a day of rest and leisure, when the thoughts of men are released from the engrossing labour of mere existence, is the time most fitted for the exercise of the reflective faculties: and the Winter Sunday evenings would be so employed, if opportunities were afforded, by large numbers of those who at present do not attend places of worship, who would listen to discourses on science and the wonders of the universe. thus producing in their minds a reverence and love of the Deity, and raising up an opposing principle to intemperance and immorality.

A SERIES OF DISCOURSES IN

ST. MARTIN'S HALL, LONG ACRE,

WILL BE COMMENCED ON

SUNDAY EVENING, JANUARY 7th, 1866,

BY

PROFESSOR HUXLEY, F.R.S.,

ON
leneth improof
"THE DESIRABILITY OF PROMOTING NATURAL KNOWLEDGE."

To be followed by (Jan. 14) Sir J. BOWRING, LL.D., F.R.S. : "Religious Progress outside the Christian Pale, among Buddhists, Brahmins, Parsees, Mahomedans," &c. ; (Jan. 21st) W. B. CARPENTER, Esq., M.D., F.R.S.. ·· The Antiquity of Man ;" (Jan. 28th) W. B. HODGSON, Esq., LL.D., "Many members, but one body :" JAMES HEYWOOD, Esq., F.R.S., Rev. J. MARTINEAU. Prof. OWEN, F.R.S., CHAS. DICKENS, Esq., and many other gentlemen who have nobly offered their services.

To the many kind friends who have voluntarily tendered pecuniary help the Committee offer their thanks. and. at the same time, their assurance that to make the Sunday Evenings for the People " self-supporting " will be their study. A portion of the Hall will be free. so that the poorest may not be excluded : other parts will be reserved for holders of tickets.

Sacred Music will precede and follow each Discourse. and thus a social. moral. and really beneficent purpose will be served.

R. M. MORRELL, Hon. Sec..
108 Great Portland Street, W.

FIGURE 7.1. Huxley's edited copy of the flyer for the first Sunday lecture. Only the first four lectures were given in 1866 before the lectures were suspended under threat of legal proceedings. (Reproduced by courtesy of the College Archives, Imperial College London.)

was philanthropic or moral) as patrons, presidents, and vice presidents as a form of assurance that the organization was worthy of public support. Depending upon the cause, titled aristocrats, bishops and other higher clergy, army and navy officers, members of Parliament (MPs), or individuals of high public reputation were appropriate assurance.[5] The supporters of Sunday Evenings for the People had relatively modest social positions. The fifty-four names include a sprinkling of baronets, knights, and MPs, eighteen FRSs (fellows of the Royal Society), and a few clergymen.[6] To some extent the presence of a few eminent, widely respected individuals with intellectual and cultural standing compensated for the modest social standing of the listed supporters: the names of Charles Dickens, *Sir* Charles Lyell, John Stuart Mill, *MP*, *Professor* Richard Owen, and *Sir* James Clark (physician to the queen) endowed the Sunday Lecture Society with credibility and respectability. Dickens and Mill already had credentials as opponents of Sabbatarianism. Thirty years earlier Dickens had attacked a bill that would have banned all popular entertainment on Sundays and had defended "the right of the poor to spend their one day of rest as they wished." Mill had written against Sabbatarian legislation in *On Liberty* (1859).[7] The supporters of these Sunday lectures were at the radical end of respectability. For example, Mill and Frederic Harrison, and many others listed, were members of the controversial Jamaica Committee, established in late 1865 to campaign for Governor Edward John Eyre's recall from Jamaica and trial for murder.[8]

Science is strongly represented in the list of supporters of Sunday Evenings for the People. Scientific men make up about one-third of the list. Eighteen supporters are identified as FRS, one as FGS (fellow of the Geological Society), and there are two medical men (J. Baxter Langley, in addition to Clark). It is apparent that the FRS was being used as a mark of respectability and intellectual achievement, or even of superior reasoning power, as in the claim of one contemporary journal editor that scientific men "reason with more accuracy than other men."[9] Some of the scientific names clearly represent naturalistic science: Edward Frankland (professor), T. H. Huxley (professor), Lubbock (Sir John, baronet), J. Tyndall (professor), and Herbert Spencer (merely esquire), all members of the X Club; W. B. Carpenter, physiologist, Unitarian, and registrar of the University of London, who had declined an invitation to be a member of the X Club in 1864; and Alexander Williamson (professor) of University College, chemist and positivist. Other scientific men on the list were more eminent, older, and either less radical or, perhaps, kept their opinions to themselves. The names of Lyell, Owen, Thomas Graham (master of the Mint, previously professor of chemistry at University College), and (Professor) August Hofmann (of the Royal College of Chemistry) gave breadth and standing to the scientific part of the list. The FRSs were not merely the party that J. D. Hooker once described as "the young progressionists in Science."[10] One notable name, Sir John Herschel's, is missing and this is probably significant. The ever-cautious Herschel may well have judged the project too provocative. Also, at least one of the ascriptions is inaccurate—Erasmus Darwin, the wealthy, freethinking, but circumspect elder brother of Charles, was never an FRS.

Other smaller groupings—Unitarians, Christian socialists, and positivists—can be identified on the list of supporters: four gentlemen have the
title "reverend," but James Martineau and J. J. Tayler were Unitarian, so,
although they were socially respectable, they did not represent theological
respectability. (Five years previously W. B. Carpenter warned Lubbock that
his name might not be much use on a memorial as Unitarians had recently
been described as "direct natural enemies of the Church of Christ.")[11] That
leaves Charles Voysey and Thomas Kirkman, but Voysey was almost as much
of a theological liability as the Unitarians. He had been removed from one
London parish in the early 1860s for preaching against eternal punishment,
but was fortunate that the liberal bishop of London, A. C. Tait, had recommended him to another parish, albeit a poor East End one.[12] That leaves one,
Kirkman, clergyman and mathematician, a broad liberal in theology, being
a supporter of Bishop Colenso. But Kirkman was also a critic of naturalistic philosophies, and the source of the parody of Spencer's definition of
evolution as "a change from a nohowish untalkaboutable all-alikeness, to a
somehowish and in-general-talkaboutable not-all-alikeness, by continuous
somethingelsifications and sticktogetherations."[13] In addition to Martineau
and Tayler, Unitarians on the list include three of the scientific men—Carpenter, Heywood, and Slack—together with Hodgson; Peter Taylor, a radical
MP representing Leicester; and William Shaen, a lawyer and University of
London activist (Heywood, Shaen, Slack, and Hodgson are identified further below).[14] Representatives of Anglicanism include five men who had been
associated with the Christian socialists: T. H. Farrer; F. J. Furnivall; R. B.
Litchfield (who later married Henrietta Darwin); and the Lushington twins,
Vernon and Godfrey, both of whom were moving toward positivism in the
1860s. Other identifiable positivists are Edward Beesly, professor of history at
University College; Frederic Harrison, a lawyer; and Alexander Williamson.
The numbers are small but may indicate an association between the positivist
movement and support for Sunday lectures.[15]

The organization behind the Sunday Evenings for the People was the National Sunday League, which had been established in 1855 to campaign for
the opening of museums, art galleries, libraries, and gardens on Sunday af-
ternoons. It also promoted Sunday excursions, band performances in parks,
and, beginning in the mid-1860s, lectures. The league was associated with the
Owenite socialist and secularist movements.[16]

The flyer for the first Sunday Evening did not name the NSL but the advertisement was signed by "R. M. Morrell, Secretary." Morrell (b. 1823), a
secularist, was founder and honorary secretary of the NSL. His name appeared on all the league's advertisements and he was prominent in its pub-

lic activities for many decades. Morrell was a jeweler, a skilled workman, as
were most of the working committee members of the NSL.[17] More eminent
persons took the roles of president and vice president. Sir Joshua Walmsley
(1794–1871), radical Liverpool politician and proprietor of the liberal (London) *Daily News*, was president. Henry J. Slack (1818–96), the vice president,
was a Unitarian; a microscopist who wrote for, and edited, popular science
magazines; and a liberal. Walmsley campaigned, for example, for the bal-
lot. Slack's causes included the abolition of slavery, the higher education of
women, and the condemnation of Governor Eyre.[18] Both Slack and Walmsley
publicly supported Sunday Evenings for the People (fig. 1). Slack, identified
as FGS, is the only representative of science on the list without an FRS, which
indicates that he had a claim to inclusion beyond his scientific reputation.
The flyer lists neither Morrell nor any other working members of the NSL's
committee as supporters. To persuade the public of the respectability of the
lecture plans, names of people with social standing were required. Similarly,
although secularists were opponents of Sabbatarianism, none are identifiable
on this first list; the names of known secularists were unlikely to confer respectability on any cause.[19]

    One contemporary critic was not persuaded that the fifty-four names
proved the respectability of the lectures. When the *Standard*, a conservative
daily newspaper, reported the lectures, it satirized "professors," denigrated
the honorary LL.D degree, and reminded readers of the controversial reputation of the first knight on the list:

> There were plenty of "patrons," mostly of the class of persons who call them-
> selves professors—a sort of people who have grown rather rapidly of late, . . .
> very glib of utterance, very clear about their own omniscience, and ready to
> lecture at a moment's warning *de omnius rebus et quibusdam abiis* [on all top-
> ics with many diversions]; they even caught a live knight, though rather *passé*,
> and a good deal the worse for wear; there were LL.D's by the dozen.[20]

Sir John Bowring, the seventy-three-year-old ex-governor of Hong Kong, was
the worse-for-wear knight. The critic did not mention the other two knights
(Clark and Walmsley) and two baronets (Lubbock and Lyell) listed, nor did
he attack the standing of FRSs, who were more conspicuous than LL.Ds on
the list of supporters.

    The promoters claimed that they did not intend to draw hearers away
from sermons; rather, according to their flyer, the lectures were for those who
did not attend "places of worship." Unitarians, Christian socialists, and positivists believed in the intellectual capacities of the people and the reforming
power of education. The Christian socialists had set up both a Working Men's

College and a Women's College in London in the 1850s.[21] The flyer expressed the promoters' enormous optimism about the unfulfilled desires of nonattendees at church and chapel and the transforming potential of the proposed lectures. On Sundays (glossed as a time for "rest and leisure," which was not how Sabbatarians would have put it), when "the thoughts of men are released from the engrossing labour of mere existence," many men will be drawn to "the exercise of the reflective faculties," and "if opportunities were afforded . . . large numbers . . . would listen to discourses on science and the wonders of nature." Such topics would stimulate "a reverence and love of the Deity" and raise up "an opposing principle to intemperance and immorality" (fig. 1). Like other proponents of Sunday lectures, the NSL implied that, by providing an alternative to the public house, lectures would diminish drunkenness and depravity among the working classes.

The NSL sought lecturers "of the highest eminence in science and literature."[22] The opening list of eminent lecturers (fig. 1) indicates the Unitarian associations of the movement. First was "Professor Huxley, F.R.S." on "The Desirableness of Improving Natural Knowledge." That Huxley, who was not Unitarian, was chosen to lead the series suggests that he already had a reputation as an outstanding lecturer. The following three scheduled lecturers were Unitarians: "Sir J. Bowring, LL.D., F.R.S." on "Religious Progress outside the Christian Pale, among Buddhists, Brahmins, Parsees, Mahomedans"; "W. B. Carpenter, Esq., M.D., F.R.S." on "The Antiquity of Man"; and "W. B. Hodgson, LL.D." on "Many Members, but One Body." Bowring was a devout Unitarian who had been a close associate of Jeremy Bentham in the 1820s. As an MP in the 1840s, he stood for many reforming and humanitarian causes: "abolition of corn duties, a more humane application of poor relief, the extension of popular education, . . . abolition of flogging in the army, suppression of the opium trade, and the worldwide abolition of slavery."[23] Although his reputation had been tainted by various financial scandals, Whig governments often used him as a representative in negotiations with foreign governments. As governor of Hong Kong, he had introduced many utilitarian-type reforms—for example, in public health—but his handling of conflicts with China in the mid-1850s, during his period as plenipotentiary and chief superintendent of trade in the Far East, was widely condemned. Carpenter was a well-known Unitarian; he had written a very popular introduction to physiology, and through his position at the University of London was associated with many utilitarian and secularizing reformers. Hodgson was a Unitarian, phrenologist, political economist, and a supporter of such radical causes as Church disestablishment[24] and higher education for women. Lectures could be expected in the future from James Heywood (an-

other Unitarian, an FRS who had been active in Manchester scientific societies, an advocate of women's suffrage and the higher education of women, and
a radical MP of the 1850s who had been a leading advocate of parliamentary
intervention to reform Oxford and Cambridge),[25] James Martineau (Unitarian theologian), Professor Richard Owen (the eminent naturalist was highly
regarded as a lecturer), and Charles Dickens. Thus, of the eight advertised
lecturers, five were Unitarians and three of these were politically active in
liberal or radical causes. The dominance of Unitarians on the list of lecturers
suggests that they were influential among the organizers of the lecture series.
The dominance of representatives of science on the list of supporters and
their showing alongside the Unitarians on the lecture plan suggest that the
organizers considered that science (rather than literature or art or political
economy) would be particularly appropriate in raising human thoughts to
the Creator and that *FRS* was a valuable label.

Owen's position raises questions: either he felt out of place in this assem-
blage or, more likely, he was not as straightforwardly conservative as he is
often portrayed—which supports Nicolaas Rupke's conclusion that some of
the antagonism between Owen and the Hooker-Huxley circle was over institutional politics. Forty years later Morrell gratefully remembered Owen's
"sympathy and assistance," which indicates that his support was more than
just a name on a list.[26] Dickens, a friend of Owen, was pleased that he was
lecturing and wrote to Morrell, offering to follow Owen and Carpenter with
a reading from his *Christmas Carol.*[27]

The announced topics do not match the announced aim of discussing
the wonders of nature in ways that would produce "reverence and love of the
Deity." Rather, the lecture topics were highly controversial. According to the
Standard's satirical report, Bowring "told the world all about Brahma and Indian Pantheism—'conceptions' which, he [was] quite clear, 'are a noble advance on the familiar picturings of the Book of Genesis'; and . . . bestowed a
good deal of compassion on this latter document and the believers in it."[28] The
Standard brushed aside the content of Carpenter's lecture with the comment
that "we are all sufficiently informed" on pre-Adamite man. Lyell, although
a more sympathetic observer, described Carpenter as "very aggressive." Sir
Charles told Walter White, the assistant secretary to the Royal Society, that
Carpenter "attacked some Psalms as immoral and claimed that 9/10 of clergy
teach what they don't believe." But, although Carpenter had criticized Anglican clergy, he had defended Christian belief and had urged his audience that
"revelations of science were not in opposition to Christianity."[29] Hodgson's
lecture was the only one designed to produce reverence for the Deity. His
scriptural text, "There are many members but one body," was a peg on which

to hang a lecture on the human body. Hodgson concluded that the "wondrous working" of the body was "highly calculated to inspire the mind with a deep reverence for the Divine Maker."[30]

The inaugural lecture, though, said the satirist with heavy irony, was "the great one." According to Thomas Hirst, who attended with the Huxley family, it was a great success. The hall, which held two thousand, was crammed and another two thousand were turned away. The seats at the back, which were free, filled first; others were priced at sixpence, one shilling, or, for a seat in the stalls, two shillings and sixpence.[31] Huxley entered to the music of Haydn's *Creation*, and further excerpts became a concert at the end. The audience "behaved admirably," noted Hirst, exhibiting a typical middle-class worry about the disruptive potential of the lower orders.[32] "On the Desirableness of Improving Natural Knowledge" was one of Huxley's most skillful pieces of rhetoric. Huxley presented an image of natural knowledge as a kind but firm mother providing all that her children needed. He managed to claim all the achievements of technology for science — ocean liners, steam engines, improved public health — but then claimed that the moral and intellectual benefits of science were even greater and the former mere "toys." He twisted the Reformation principle of "justification by faith" into "believing what you are told to believe," and claimed for science a higher morality based on "justification by verification."[33] The crowd, the stirring music, and the "noble utterances" delivered, said Hirst, "in an earnest and impressive tone," made the lecture an awesome occasion. It would probably — or so Hirst thought — make "a new epoch" for both science and Sabbatarianism.[34]

Although the *Standard*'s report is recognizably about the same lecture, the judgments are entirely different from Hirst's. The critic shared Huxley's command of witty innuendo:

> The object of it seems to be to tell us that if there is a God, which is at least doubtful, nature is the entity in question, and physical science is its prophet; that the world has been groping about in darkness through philosophies, theologies and what not for centuries, and is only at last discovering that "scepticism is the highest of duties, blind faith the one unpardonable sin"; . . . and that when he has once fairly learnt to "break in pieces the idols built up of books (Bibles, for instance) and traditions and fine-spun ecclesiastical cobwebs," he will be able to "cherish the noblest and most human of man's emotions by worship, for the most part of the silent sort, at the altar of the Unknown and Unknowable"—a very scientific prospect.[35]

The ultimate irony was the music. "What on earth were they thinking of to select their music from the *Creation*? The thing is a myth, as the orators have been proving until everybody is tired of hearing their clatter." The *Standard* proposed a more suitable hymn:

> For the great God Pan is alive again,
> He lives and reigns once more.
> With deep intuition and mystic rite
> We worship the Absolute-Infinite;
> The great Nothing-Something, the Being-Thought,
> That mouldeth the mass of Chaotic Nought,
> Whose beginning unended and end unbegun
> Is the One that is All and the All that is One.

The original of this satire was written by the High Church Anglican Henry Mansel and directed at German idealist philosophy, but the *Standard*'s version was subtly edited to apply to Huxley and Spencer.[36]

Hirst was mistaken. There was no new epoch for science and Sabbatarianism because the Sabbatarians fought back effectively. The Lord's Day Observance Society (LDOS) told the directors of St. Martin's Hall that, under the 1781 Lord's Day Act, they were conducting a "disorderly house" because admission was being charged for entertainment on a Sunday. If those who owned the premises were successfully prosecuted, their penalty would be two hundred pounds per day. The NSL, which considered that it was neither providing entertainment nor interpreting "texts of Scripture" as defined by the act, agreed to test the law; thus, the LDOS took a case against representatives of the NSL.[37] Meanwhile, after making some slight changes to the organization of the lectures, the NSL promoted another series of lectures in 1867. This time the paid seats were described, following Anglican custom, as "sittings." But when the promoters of the lectures and the proprietors of the hall were again threatened with prosecution, the lectures were abandoned.[38] On losing the test case, supporters of the NSL tried unsuccessfully to get the act repealed.[39]

Thus, the first two attempts at Sunday lectures collapsed under the threat of prosecution. The sacred music that the Sunday Evenings for the People advertised (as serving "a social, moral and really beneficent purpose") became the focus of the accusation that entertainment was being provided and the original movement split over music. The NSL took protective action against the threat of prosecution by setting itself up as a religious body, the "Recreative Religionists." Religious music accompanied its lectures and attendees were charged only for reserved seats; hence, the LDOS prosecution against the Recreative Religionists in 1868 was unsuccessful.[40] Carpenter and Huxley publicly criticized the name as "a sham" and an "unworthy evasion."[41] To their allies in the Sunday League this may have seemed like treachery. The Sabbatarians made the most of it. According to the Working Men's Lord's Day Rest Association, the league had won "on a point of law," but it claimed the high moral ground for the Sabbatarians and cited Professors Huxley and Carpenter in support.[42] As is well known, Huxley used religious metaphors very freely for himself and his activities. He happily described his lectures as "lay sermons," as in his first compilation of essays, entitled *Lay Sermons, Essays and Addresses* (1870). He frequently called the enemies of science Amalekites, identifying them with the enemies of the people of Israel in the Old Testament and thereby classing himself and his friends as God's chosen people. His religious metaphors were often used ironically, as an in-joke with his friends. Their reaction to the recreative-religionist ploy shows that Huxley and Carpenter were not willing to present science as a form of religion even for the strategic purpose of providing improving entertainment on Sundays.

Sunday Evenings for the People continued under the auspices of the Recreative Religionists, although they later renamed themselves less provocatively as the National Sunday League of Protestant Dissenters.[43] The dissenters from the Recreative Religionists began to plan for a more honest organization to promote Sunday lectures.

The Sunday Lecture Society: Reforming Lawyers and the X Club

While accusing the Recreative Religionists of unworthy evasion, Carpenter and Huxley were active in setting up a new Sunday lecture organization. Their Sunday Lecture Society would not pretend to be a religious group, but not taking the protection of religion meant that the SLS could not attract hearers to lectures by offering music, even sacred music, because that would be entertainment. The promoters of the new Sunday evening lectures proceeded cautiously: they sought supporters, took legal advice, and raised money. In January 1869 Huxley reported at an X Club meeting that the question of Sunday evening lectures had been revived, but "independently" of the secularists' Sunday League.[44] It was not until November 1869 that the promoters held a meeting, not a public meeting, but a meeting to which they invited people thought to be sympathetic to the aims of the proposed society. Huxley chaired this preliminary meeting.[45]

It was resolved that a "Sunday Lecture Society" be formed, its chief objective being "to provide for the delivery on Sundays in the Metropolis, and to encourage the delivery elsewhere, of Lectures on Science—physical, intellectual, and moral,—History, Literature and Art; especially in their bearing upon the improvement and social well-being of man."[46] The aims were grand: a metropolitan society with national ambitions; the subject matter encompassed all science and added history, literature, and art; the social goals—"the improvement and social well-being of man"—sound unchallengeable to modern ears, but Victorians would have noticed the omission of any spiritual aims. Nevertheless, these were carefully chosen words. The lectures would not be primarily entertainment, not even rational entertainment. Rather, they claimed to be directed to moral and social improvement. The Sunday Lecture Society was close to claiming for itself the concern for "the health and well-being of the nation in its broadest sense" that Francis Galton suggested would mark a new "scientific priesthood."[47]

The committee elected to manage the affairs of the Sunday Lecture Society was dominated by lawyers. This was appropriate in view of the legal problems likely to arise from giving lectures on Sundays but also indicates a depth of concern for social and religious reform among lawyers. There were initially nine committee members, with a tenth added in 1870. Of the eight whose occupations I have been able to identify, six were lawyers, one a political economist, and one a "retired merchant." Of the five for whom I have been able to find religious affiliation, three were Unitarians, one Church of England (a Christian socialist of the 1850s), and one described himself as agnostic.[48] (For details, see table 7.1, which includes an additional, Anglican, committee member added in 1877.)

William Henry Domville, who established the SLS and who was the chief activist for decades, is identifiable only through newspaper reports of his will. He was the son of Sir William Domville, who had published against Sabbatarianism in the 1850s, and his wife, daughter (Louisa), brother (Sir James), and sisters were also supporters of the Sunday Lecture Society.[49] Domville was a wealthy solicitor and self-described "Agnostic." He gave generously to the society, not claiming repayment for expenses incurred on society business, and, from his estate, valued at eighty-four thousand pounds in 1898, bequeathing one thousand pounds to the Sunday Lecture Society.[50]

Respectable names would, it was hoped, attract further respectable support, encouraging those who did not want to be in the vanguard and reassuring those who deferred to authority. But the Sunday Lecture Society also advertised the number of its supporters rather than selecting only the most respectable and wellborn. One hundred and thirty people had promised their support before the November meeting; the first published list of members and donors, produced just after the meeting, contains over 250 names. In late 1871, at the second annual meeting, there were nearly 350 names.[51] The subscribers, donors, and supporters were a broad group: scientific men, medical men, lawyers, clergymen, a few musicians, writers and publishers, and a few

TABLE 7.1. Sunday Lecture Society, committee members, 1869–mid-1880s

Name	Occupation	Religious affiliation	Causes supported
J[ohn] G[ordon] Crawford			• Member of Royal Institution, 1883 • Helped raise money for a memorial to the Jewish community leader, Sir Moses Montefiore
William Henry Domville (1819–98), hon. treas. 1869–	• Solicitor	Agnostic	• Member of Voysey Defence Committee
Richard Chester Fisher (1840–)	• Barrister (admitted 1866)		• Married daughter of Richard Cobden, radical MP • Committed to free trade • SPCA • Antivivisection
Allen Dowdeswell Graham (1837–1905)	• Curate • Income "from dividends" (1881)	Church of England (resigned clerical career due to "scruples of conscience")	• Active in Charity Organisation Society • Established the Invalid Children's Aid Association
William B. Hodgson (1815–80)	• Political economist • Professor of commercial law, Edinburgh (1871–)	Unitarian	• Church disestablishment • Higher education for women • Pro-Confederacy in American Civil War • Antivivisection
William A. Hunter (1844–98)	• Barrister (admitted 1867) • Professor of Roman law (1870–78) and jurisprudence (1878–82), University College London		• Women's suffrage • Higher education for women • Friend of Charles Bradlaugh • MP 1885–95, for Irish home rule
Mark Eagles Marsden (1811–90), hon. sec. 1871–	• "Retired merchant" (1881)	Treasurer at South Place Chapel (Unitarian/Ethical Society)	• Association for the Repeal of Taxes on Knowledge • Zulu rights • Association for Abolishing the State Regulation of Vice • Friend of Charles Bradlaugh

(continued)

TABLE 7.1. (continued)

Name	Occupation	Religious affiliation	Causes supported
H. M. Phillips			
William Shaen (1821–87)	• Solicitor	Unitarian	• University of London activist • Italian unity and democracy • Abolition of slavery • Supporter of George Holyoake • Women's education and women's suffrage • Legal defense of Colenso • Solicitor to the Jamaica Committee
John Shortt (1840–1932), hon. sec. 1869–71	• Barrister (admitted 1866) • Judge 1894		
John Westlake (1828–1913)	• Barrister (admitted 1854) • QC 1874 • Whewell Professor of International Law, Cambridge (1888–1908)	Church of England (Christian socialist)	• Women's suffrage • Supporter of Colenso, *Essays and Reviews*, and Church reform in 1860s • Legal defense of Kossuth • National Association for the Promotion of Social Science • MP 1885–86, against Irish home rule

Sources: As in note 48.

Note: All those listed were founding committee members and served until at least 1885, with the following four exceptions: Hunter was appointed to the committee in 1870; Hodgson resigned when he moved to Edinburgh in 1871; Phillips joined in 1872, replacing Hodgson; and Graham was not appointed until 1877.

military men. Eminent scientific men were on the list: Charles Darwin donated one pound; Sir Charles Lyell donated two pounds and committed himself to an annual subscription of one pound. One of the clergy was Rev. J. D.
La Touche, whose reforming inclinations are clear in his long correspondence
with J. D. Hooker;[52] John Stuart Mill was again a supporter. The aging radical
Robert Grant, professor of comparative anatomy at University College, was
a member. Poor and nearing eighty, he paid his annual subscription, and for
five years "he scarcely missed a single lecture." Other listed supporters on the
edge of respectability were William Pare, an old Owenite who had been active
in the Harmony Hall project, and Robert Chapman, the radical publisher.[53]
More surprising, John Ruskin was on the list, suggesting that the SLS was
seen as a forum for those who wanted to address "the people." Some supporters preferred not to fully identify themselves: for example, an "F.R.S." and a
"Lord H." (one of only two peers) were on the first 1869 subscriber list.

The female members included the cultured and learned elite (for example,
Lady Lyell) and new independent women, such as Frances Power Cobbe. Sara
Hennell of Coventry, a longtime friend of Marian Evans, was a supporter.
Women were not only subscribers and donors; from the second season they
appeared regularly on the lecture program. The involvement of women is
consistent with both the Unitarian and Christian socialist interest in Sunday
lectures. Both groups, unlike the leading representatives of naturalistic science, supported the higher education and political participation of women.

Although there were no scientific men on the committee that managed
the society's affairs, scientific men and scientific topics were conspicuous in
the program, especially in the early years. The majority of the early lecturers
were associated with either medicine or science. Carpenter was prominent.
He opened the first series of lectures in 1870 with two lectures on the deep sea,
and took the opening position again in the 1870–71 season with two lectures
on the microscope. He was the most regular contributor to the society's lecture program in the 1870s. Huxley, who had taken such a prominent position
at the founding meeting, did relatively little thereafter. He gave one lecture
in early 1871, "On the Founding Fathers of the English People." (Admittedly,
he was busy on a host of committees and projects, including the Devonshire
Royal Commission, and at the end of 1872 he became biological secretary of
the Royal Society.) None of the other X Club members gave lectures. The
most prolific and eminent of Carpenter's early supporters in the lecturing
schedule was Thomas Spencer Cobbold (1828–86). He was Britain's leading
parasitologist and a lecturer at Middlesex Hospital Medical School, but chose
to advertise himself as Swiney Lecturer on Geology at the British Museum.
The majority of lecturers on scientific topics were either less eminent in sci-

ence, dependent on popular writing for an income (for example, Richard Proctor), or younger. From 1871 the brilliant young mathematician W. K. Clifford (1845–79), was a regular lecturer.

In spite of its caution, the SLS faced many difficulties. Shopkeepers who displayed the bills were accused of "aiding infidels" or being atheists themselves, and others declined to display the society's advertising, fearing loss of custom. Potential lecturers did not want to lecture on Sundays, and, in spite of the lengthy list of supporters, gentlemen of social and intellectual eminence were unwilling to take prominent patron roles. The committee was unable to find "four or five eminent men" willing to fill the positions of president and vice presidents, and thereby confer respectability on the society, and abandoned the attempt.[54]

When, at last, after nine years in operation, the London Sunday Lecture Society found a president and vice presidents, the range of representation was narrow and the list was short on traditional titles and honors. Naturalistic science dominated. Carpenter, just retired from the University of London, took the presidency.[55] Of the ten vice presidents, five were members of the X Club: Frankland, Huxley, Spencer, William Spottiswoode (identified as "P.R.S."), and Tyndall. Alexander Bain, professor of logic and English at the University of Aberdeen and a founder of the journal *Mind*, was, like Carpenter, interested in the relations of mind and body and, like Spencer, a representative of the moral and social sciences. Most astoundingly, Charles Darwin was on the list of vice presidents. Someone—there is no record of whom—persuaded the reclusive and single-minded Darwin to lend his name to the cause.[56] In addition, there were three lawyers and men of affairs among the vice presidents. James Heywood had been listed as a potential lecturer in 1866. He was a Unitarian and a radical MP, one of those prepared to intervene, in what many thought to be private property, to reform Oxford and Cambridge. James Booth was a Unitarian from Liverpool who had been secretary to the Board of Trade from 1850 to 1865.[57] The tenth vice president, the only one who was outside both science and Unitarianism, was Sir Arthur Hobhouse, knighted two years previously when he retired as law member of the Council of the Viceroy of India. When Carpenter died suddenly at the end of 1885, Huxley replaced him. Huxley had declined to take the SLS presidency in 1884 because he was president of the Royal Society, and he felt that as president he should not take controversial positions on which the society was divided.[58] However, neither Huxley nor his friend Spottiswoode showed any such caution about appearing on the list of SLS *vice* presidents as "P.R.S." Fortunately for the SLS, Huxley retired from the Royal Society presidency at the end of 1885 so felt free to take the SLS presidency.

After experimenting with different times, prices, and formats, from the 1871–72 season the committee settled upon a regular pattern of three series of eight lectures through the winter (November to May). They were held in the late afternoon, which suited the middle-class subscribers rather better than the working people whom the society had initially hoped to attract. The first annual report admitted the different interests of "working classes" and "members and subscribers." One reason for choosing the afternoon time was that the hall cost more to hire for evenings.[59] This contrasted with the Sunday Evenings for the People, which were evening events. Sunday Lecture Society prices were one penny, sixpence, and, for reserved seats, one shilling, with good discounts for series tickets. Prices, time (4:00 p.m.), and the central London location (St. George's Hall, Langham Place) were unchanged from 1871 until 1888, when the committee proposed moving the lectures to the East End.[60]

Literary, artistic, and classical topics appeared alongside scientific and medical topics in the lecture program. The deep sea, Michelangelo, fossils and their meaning, the prevention of infectious diseases, and "The Ideas of the Ancient Greeks Respecting Death and Immortality" were a few of the topics offered in the first year. Attendance initially grew in an encouraging manner: an average of 300 in the first year, over 450 in the second year. But average attendance then fell back to about 400 and remained at that level for many years.[61] By the 1880s many lectures were illustrated by oxyhydrogen lantern and these may have attracted a larger audience but the numbers are not reported for specific lectures. Over time physical science topics occupied a lesser proportion of the program and comparative religion became more common. The Greeks on death and immortality, in the first series, was only indirectly relativizing; the "Sacred Poetry of the Hebrews, Hindoos, and Other Oriental Races" in the tenth series directly so. Questions about mind and brain were recurrent topics. Carpenter lectured on "The Doctrine of Human Automatism"; Henry Maudsley on "The Physiology of Mind," "Mind and Body," and "The Physical Basis of Will," all in the first decade.

My impression is that topics became more provocative over time. Clifford, for example, was taking on the North British circle. In 1871 his title unfolded to identify his target: "The History of the Sun; an explanation of Laplace's Nebular Hypothesis, and of recent controversies in regard to the time which can be allowed for the evolution of Life." (William Thomson and his engineering associate Fleeming Jenkin had argued that the sun was not old enough to allow time for Darwin's slow process of transmutation by natural selection.) In 1874 Clifford attacked the arguments of Maxwell and Thomson for the universe having a beginning in "The First and Last Catastrophe; a criticism on some recent speculations about the duration of the Universe." It was surely provocation when, in the 1878–79 season, Richard Proctor gave a lecture entitled "Sabbath Superstitions: The Human and Astronomical Origin of the Week."

Many lecturers are not well known to modern historians of scientific naturalism. One of the most regular lecturers was George Gustavus Zerffi, a lecturer in art at South Kensington National Art Training School (the art equivalent of the College of Science). Zerffi had been an agent employed by the Austrian government to watch Hungarians and was a committed member of the secularist movement.[62] He lectured on Mexican art (1872), on "Natural Phenomena and Their Influence on Different Religious Systems" (1873), and from 1874 gave two lectures in most years. As a secularist, Zerffi was deeply committed to Sunday lectures. He gave numerous lectures for the National Sunday League—for example, "Preadamites, or Prejudice and Science"—as well as for the Sunday Lecture Society.[63]

The roles of the president and vice presidents were largely honorific. There is no evidence that the X Club vice presidents had continuing input to the affairs of the society—for example, there is no extant correspondence in the Huxley or Tyndall papers. John Morley, a promoter of the Tyneside SLS, wanted Huxley and Tyndall to aid the London SLS by giving lectures in the late 1870s, but they did not follow his advice.[64] They were not regular attendees, although Adrian Desmond records one occasion on which Huxley, Robert Grant, and Clifford attended together.[65] Nor do the vice presidents seem to have exercised control over the program. Lectures on socialism were becoming regular features by the late 1880s. Sidney Webb's 1892 lecture on "The Progress of Socialism" would surely have presented a viewpoint unacceptable to Huxley, the president, and highly unacceptable to Spencer, a vice president.

The Sunday Lecture Society was not the only society promoting Sunday lectures. "Sunday Evenings for the People" (under the auspices of the Society of Rational Religionists) had been neither given up nor closed down; moreover, other groups were initiating similar programs, with and without music. On 20 March 1870, when the Sunday Lecture Society offered a lecture at 4:30 p.m. in St. George's Hall, Langham Place, two "Sunday Evenings for the People" were also advertised: 7:00 p.m. at South Place Chapel (a Unitarian congregation that had converted to an Ethical Society) and 7:30 p.m. at the same St. George's Hall, where the "Church of Progress" offered J. Kay Applebee on "Religious Lessons from Shakespeare: *The Merchant of Venice*," followed by Weber's *Mass in G*.[66] There was some cooperation between these groups. Moncure D. Conway, of South Place Chapel, suggested the formation of an "Association of Liberal Thinkers" in collaboration with the Sunday Lecture Society.[67] Other Sunday groups were founded through the 1870s. F. J. Furnivall, a Christian socialist, established a Sunday "Shakspere Society" in 1873 and, from the agnostic side, Leslie Stephen was active in founding a walking group, the "Sunday Tramps," in 1879.[68]

The National Sunday League was considerably more successful than the Sunday Lecture Society. The league reported twenty-four thousand visitors at twenty-six lectures, almost a thousand people per lecture, in 1872, more than twice as many as the SLS. Certainly the music contributed to their greater popularity; but also, they met at 7:00 p.m., which the Sunday Lecture Society recognized was more convenient for working people. The league sponsored a range of other activities. It continued to campaign for the opening of museums and libraries on Sundays, and set up local "Band Committees" to promote Sunday concerts in parks. In summers, when the lecture season had finished, it ran weekly railway excursions that also attracted an average of about one thousand people.[69] These more recreational activities probably helped to give the NSL a broader following than the SLS.

The movements to promote both lectures and band concerts spread to the rest of the country, although to what extent provincial movements modeled themselves on London or were independent local initiatives is unclear. Wigley identifies Sunday lecture societies in Bradford, Salford, and Leicester, by 1875, as "branches" of the London SLS, but the annual reports of the London SLS made no such claim and even announced in 1874 that no Sunday lecture societies had yet been formed in the country.[70] It was some years before the London SLS reported the establishment of provincial societies. The SLS reported in 1881 that "Sunday Evening Meetings for the People," including "very successful" concerts of sacred music, had been held in Birmingham and in nearby Smethwick, and that a Sunday Lecture Society, similar to their own, had recently been inaugurated in Birmingham.[71]

The dominance of scientific men among the vice presidents of the Sunday Lecture Society suggests that they had achieved cultural leadership. In announcing its first president and vice presidents, the Sunday Lecture Society must have felt that the names of eminent scientific men gave some level of respectability to the controversial society. But comparison of the SLS list with the patronage of a later Sunday society suggests that the Sunday Lecture Society was making the best of difficult circumstances.

In 1875 a new "Sunday Society" took up the aims of the National Sunday League and began to campaign for the opening of museums, art galleries,

libraries, and gardens on Sundays. The supporters of the Sunday Society included more representatives of the Church of England and of elite society
than either the NSL or the SLS—suggesting that the cause was more respectable in the 1870s than in the 1850s, when the NSL was founded, and that anti-
Sabbatarianism received greater support when Sunday lectures were not the
chief cause. The clergy supporting the Sunday Society were more eminent
and ranged across more denominations than those who supported the Sunday Evenings for the People; also they were prepared to take more prominent
roles than those who supported the SLS. In addition to the Unitarians and
liberal Anglicans who could be expected among the supporters of a Sunday
society, Wigley identifies Baptist and Congregational clergy, and the Roman
Catholic bishop of Salford. Dean Stanley of Westminster chaired a public
meeting. On the platform, supporting Stanley in the stand against Sabbatarianism, were many members of the SLS, including Professors Huxley
and Tyndall, Mr. James Heywood, and Mr. W. H. Domville. Apologies for
absence were read from particularly eminent persons: the bishop of Exeter,
the Marquis of Ripon and the Marquis of Huntly, a series of MPs, and the
wealthy philanthropist Lady Burdett Coutts.[72] These were the kind of eminent supporters that the SLS had failed to obtain.

   With only one title (and that a lowly knighthood), no clergy, and no MPs,
the list of president and vice presidents that the SLS announced in 1879 was
a modest one, even allowing for the official standing of the president of the
Royal Society (Spottiswoode, from 1878 to 1883) and the personal eminence
of Darwin. Although the scientific names show that scientific men were recognized as cultural leaders, the SLS list would have been more authoritative if
the names had represented a greater variety of social authority. The difficulty
the SLS had in finding even its unrepresentative list shows how controversial
Sunday lectures were.

   It is significant that with the exception of the much younger Clifford, the
leading representatives of scientific naturalism were patrons, not activists in
the SLS. After his energetic participation in establishing the SLS in 1869–70,
Huxley was not active. Carpenter, the Unitarian, anchored the lecture series
for years, and when he retired from the University of London, was prepared
to take the presidency. His most active scientific supporters in lecturing were
Henry Slack and Clifford. The men who did the regular work were lawyers.
In the earlier Sunday Evenings for the People, it seems that the activists were
Unitarians and secularists. I therefore want to step back from considering
what the representatives of scientific naturalism achieved and consider instead the larger movements in which they were participants and what those
movements achieved.

Grander Narratives: Anti-Sabbatarianism and the
Secularization of the British State

Opposition to Sabbatarian legislation was one of many campaigns conducted in the Victorian period against legislation that privileged particular religious beliefs and practices. At the beginning of the nineteenth century, the British state was Anglican. By the time Victoria became queen in 1837, the repeal of the Test and Corporation Acts (1828), the Act of Emancipation (1829), and the Municipal Government Act (1835) gave civil rights to Nonconformists and Roman Catholics. They could now become magistrates or officers in the military, and stand in local and parliamentary elections.[73] Although some describe the legislative changes of this period as a revolution, the principle of religious equality was worked out only slowly in political practice, often through bitter controversy. Administrative procedures and long-established institutions were not quickly changed. In 1837 the registration of births, deaths, and marriages was removed from the local parish administration to a new office of the state, but in small towns and villages, where the only graveyard was the churchyard, burial rites were under Church control until 1880. Parliamentary legislation in the 1850s forced Oxford and Cambridge to remove the religious tests that had prevented non-Anglicans from studying at Oxford or graduating from Cambridge. Another step toward religious equality in the 1850s was the Jewish Emancipation Act (1858), whereby the state first extended civil rights to those outside the Christian tradition.

In the 1860s, when Sunday Evenings for the People began, Anglicans still had many privileges and advantages over their countrymen. Enormous resentment was felt over the Church rate, which required all rate payers in a parish to contribute to the upkeep of the local church.[74] This grievance was removed when the rate was made voluntary in 1868. Education provoked many disputes. The Elementary Education Act of 1870 seemed to non-Anglicans to give more support to Anglican schools than was warranted by the Anglican share of the population. One outstanding grievance, although felt only by those with the education and money to aspire to Oxford and Cambridge, was that the vast majority of college fellowships, which supported young graduates in their early years of scholarly or professional life, were available only to Anglicans. Another disability, impacting on those who were openly secular or atheist, was that the oath required in order to take a seat in Parliament or to be heard as a witness in court cases had to be sworn on the Bible. Thus, Anglican privilege was not an issue of the long-distant past for the scientific naturalists, agnostics, Unitarians, and secularists of the SLS and the NSL.

Not surprisingly, the activists of the SLS were committed to these causes.

James Heywood had supported parliamentary intervention to reform Oxford and Cambridge—because they would not reform themselves. He had passed the Mathematical Tripos at Cambridge in the early 1830s but was not awarded the BA until 1857, *after* Cambridge had been reformed by legislation in 1856.[75] William Shaen, the Unitarian solicitor who was a member of the SLS committee, had served his articles under a leading radical who acted for many of those who refused to pay the Church rates.[76] The youthful Huxley used the Nonconformist arguments against the Church rate in discussion with his Anglican parents.[77] Sometimes the far liberal end of the Church of England was involved in these campaigns. On the strength of his reputation as a reformer, the promising young scholar A. P. Stanley, later dean of Westminster and vice president of the Sunday Society, had been appointed secretary of the Royal Commission that inquired into Oxford and Cambridge in the 1850s and provided the information that enabled Parliament to legislate for the reform of the universities. John Westlake, of the SLS committee, had been involved in a Church reform movement of the 1860s and had discussed with John Lubbock how the doctrinal formularies of the Church of England might be broadened, although to no effect.[78]

Among the many challenges to Anglican orthodoxy, Sunday lectures were a particularly radical proposal. Many who were willing to support the opening of museums and libraries on Sundays were not willing to support Sunday lectures, which were in more direct opposition to Sunday sermons. As already noted, although Dean Stanley was prepared to chair a meeting of the Sunday Society, his name was not to be found among the supporters of the Sunday Lecture Society. Sunday lectures divided Huxley from leading Nonconformist industrialists whom Desmond has identified as supporters of his campaigns for science education and as personal friends. The Newcastle iron and arms manufacturer Sir William Armstrong declined to take the presidency of a proposed Tyneside SLS in 1883. At age seventy-three, he was "too old to accept new presidentships," but also, Armstrong told Huxley, he "inclined to the opinion that Sunday should be more devoted to rest & recreation than to education"—which latter seemed to be the purpose of the proposed scientific lectures. Moreover, the lack of "influential names" associated with the society (he was assessing the listed supporters) led Armstrong to fear it would be a focus for "religious contentions" with which he did not wish to be involved.[79]

The removal of Sabbatarian legislation was one of the later steps in this slow secularization of the state. Sabbatarian legislation was not *Church*-privileging legislation, as Sabbatarianism was an issue that united many Anglicans and Nonconformists, but Sabbatarian legislation expressed a particular theological vision of how Sundays should be spent. Public opinion had greatly changed since the 1850s, but when legislative change at last came, it was stimulated by an outrageous charge brought by the LDOS against the Leeds Sunday Lecture Society and its lecturer. The audience had laughed. This demonstrated that the Leeds SLS was providing entertainment on Sundays for money, said the LDOS, and it proceeded to press charges. The LDOS action provoked the formation of a National Federation of Sunday Societies, committed to the removal of all "vexatious restrictions" on the opening of museums, art galleries, and libraries; the delivery of lectures; and the performance of music on Sundays.[80] Hobhouse, Lord Hobhouse since 1885, introduced a bill in the House of Lords to modify the act of 1781.[81] The outcome, announced in letters to newspapers in May 1896 by the secretary of the National Federation of Sunday Societies, was that "next Sunday afternoon . . . all the great National Museums and Art Galleries in London will be open to the public."[82]

The success of the anti-Sabbatarian movement brought the decline of Sunday lectures. As Edward Royle, historian of Victorian irreligion, emphasizes, irreligion borrowed the forms of institutional religion and gained its energy from having an opposition.[83] Like the secular and socialist churches and Sunday schools, Sunday lectures rapidly declined once the anti-Sabbatarians had won their cause. Scientific and literary lectures had always been a minority attraction, only four hundred attendees on average at the SLS events and about a thousand at Sunday Evenings for the People. In the increasingly free religious world of the early twentieth century, Sunday lectures had more competition, not only from church and chapel services, but from museums, libraries, bands playing in parks, and new recreations such as bicycling. Sunday lectures rapidly disappeared. There are only a few records of Sunday lecture societies between 1900 and 1914, and, so far as I can determine, the London SLS did not survive the war.

The account of secularization here is not an instance of the grand sociological theory that religion disappears as societies become modern. Rather, I follow the argument of Jeffrey Cox that secularization is a phenomenon of Western Europe. Cox's historiographical article has the evocative title "Provincializing Christendom: The Case of Great Britain." To provincialize Europe, he argues, is to discuss it without assuming that Europe is the model for the rest of the world—the exemplar of modernity that other states, in so far as they become modern, will come to be like. Europe is different because of Christendom: "Christendom defines a particularly European form of Christianity, one that is territorial, parochial, elitist, political, and closely allied with governments."[84] The problem with the grand sociological theory is that it has extrapolated the European situation to the rest of the world. As Royle puts it, a church that was dominant invited opposition, and not only from the atheists, republicans, and secularists, but from Unitarians, Quakers, Baptists, and other Protestant Nonconformists. Frank Turner has emphasized that Nonconformists sought a secular state as a means to achieve liberty of conscience and true religion.[85] The result, by the 1890s, was religious pluralism.[86] Anglicans, Unitarians, agnostics, secularists, and even the rowdy Salvation Army were equal before the law, although the Church remained established and, as David Bebbington points out, the higher reaches of political power were dominated by Anglicans.[87]

The Sunday Lecture Society and Scientific Naturalism: Some Conclusions and Questions

By paying attention to the lists of patrons, supporters, committee members, and lecturers of the SLS and related organizations, the analysis here moves beyond the leaders of scientific naturalism who have been the focus of most studies of the movement. It identifies other scientific men who shared the secularizing objectives of Huxley and his friends and also other significant groups who were allied with them in the Sunday lecture societies. Recognition of allies points to reasons for the success of Huxley and company that go beyond the power of their arguments and the effectiveness of their networking.

It is clear that in both the Sunday lecture societies and the broader anti-Sabbatarian movement, the scientific naturalists were but one group within broad coalitions. In the Sunday Evenings for the People, secularists from the NSL and Unitarians took the initiatives and did the work. They brought in many scientific men as supporters, but the planned first lecture series had no specific association with those commonly identified as "scientific naturalists." True, Huxley gave the first lecture, but he was to be followed by Owen. When the Sunday lecture movement split over how to meet the LDOS challenge, Huxley and Carpenter were among the projectors of the alternative Sunday Lecture Society. Initially, Huxley was very active; he chaired a public meeting and gave one of the early lectures, but then disappeared from view for a decade. In the SLS, Unitarians showed more commitment than the representatives of scientific naturalism; the lawyers rather than the scientific men did the organizational work. In 1879, when Spencer, Huxley, and their friends were named as vice presidents, they were valued patrons and figureheads, not activists. By the late 1870s Huxley and Tyndall were also associated with a broader anti-Sabbatarian movement, supported by liberal Anglicans and Nonconformists more generally (rather than only the Unitarians) and represented by the Sunday Society. Thus, secularists began the movement, Unitarians and advocates of naturalistic science cooperated enthusiastically, but legislative change was achieved in the 1890s only after liberal Anglicans and Nonconformists joined the cause. Each of these groups needed support from other groups, and all shared in the aura of success.

The repeal of Sabbatarian legislation was just one aspect of the secularization of the state in Victorian Britain. The secularization of the state that Nonconformists had worked so hard to achieve for religious ends was supported by advocates of naturalistic science as part of their campaign for a fully secular vision of human society. The preachers of naturalistic interpretations of nature and culture wanted more than the secularization of the state; they hoped—and expected—that popular appeal to the supernatural would be replaced by scientific explanations of natural phenomena and that the authority of religious leaders would decline. Although the secularization achieved by the 1890s was far from what they sought, the scientific naturalists appeared to be on the winning side, unlike the devout Anglican popularizers and the North British circle of physicists and engineers, for those who had previously suffered institutional discrimination were now legally equal and socially respectable. Unbelief in most of its varieties now had the same rights as varieties of Christian orthodoxy. A particularly remarkable example of rising respectability is that in the mid-1880s Spencer was invited to follow the extremely wealthy and well connected Hugh Lupus Grosvenor, first Duke of Westminster, as president of the Sunday Society.[88] Whether Huxley and the scientific naturalists, or the physicists and engineers around Thomson, or the devout popularizers of science had the largest audiences or made the most converts in the latter decades of the nineteenth century is an interesting, difficult-to-answer question that is irrelevant to this argument.[89] The legal position of scientific naturalism had advanced; hence, the promoters of naturalistic science appeared successful.

Finally, the Sunday Lecture Society raises questions about who counts as a "scientific naturalist." In the SLS the great figureheads of scientific naturalism—Huxley, Tyndall, Spencer, Clifford—were joined by Carpenter, Slack, Cobbold, and the more obscure Thomas Graham, William Domville, and G. G. Zerffi. Can theists such as Carpenter be counted as scientific naturalists? Should Zerffi and Domville, who were not scientific men but who seem to have accepted scientific knowledge as authoritative, be included? What about Proctor, the popularizer?[90] Does the category of *scientific naturalist* become amorphous, as Paul White argues in his chapter in this volume?

Although my analysis has paid only passing attention to the subject matter of the SLS lectures, two emphases were noted. Lectures on the human mind and on comparative religion were conspicuous in the SLS program. The SLS's lecturers agreed that science could be applied to human affairs, that mind and conscience did not put humans outside nature. Hence, my hypothesis for further investigation is that the touchstone issue, which divided the advocates of a fully naturalistic science and the lecturers of the SLS from scientific men more generally, was whether humans are included in nature.[91]

Acknowledgments

I thank Gowan Dawson, Bernie Lightman, and the audience at the York conference for their assistance in clarifying the argument and filling out the evidence presented here, and Sharon Haswell for discussions about Sabbatarianism. The University of Auckland supported this research through its granting of research and study leave in 2004.

Notes

1. J. Vernon Jensen, "The X Club: Fraternity of Victorian Scientists," *British Journal for the History of Science* 5 (1970): 70. In this chapter I use *scientific naturalism* but avoid *scientific naturalist.* Although, as the introduction to this volume shows, *scientific naturalist* had been introduced in polemical contexts, the primary Victorian referent of *naturalists* was people who collected and classified flora and fauna; some, the "philosophical" naturalists, rose to more systematic and theoretical analyses than others, but all considered their activities to be scientific. I use the anachronistic *scientist* in my chapter title to avoid the ambiguous "scientific naturalist."

2. Adrian Desmond, *Huxley: The Devil's Disciple* (London: Michael Joseph, 1994), 344–45; "C. M. Davies on Professor Huxley's Sunday Lecture, 1870," in *Religion in Victorian Britain,* vol. 3, *Sources,* ed. James Moore (Manchester, UK: Manchester University Press, 1988), 456–60; and David M. Knight, *Science and Spirituality: The Volatile Connection* (London: Routledge, 2003), 119.

3. John Wigley, *The Rise and Fall of the Victorian Sunday* (Manchester, UK: Manchester University Press, 1980), 102–4, 125–31. Brian Harrison, "Religion and Recreation in Nineteenth-Century Britain," *Past and Present* 38 (1967): 98–125, mentions three metropolitan societies (109–13).

4. Crosbie Smith, *The Science of Energy: A Cultural History of Energy Physics in Victorian Britain* (London: Athlone Press, 1998); and Bernard Lightman, *Victorian Popularizers of Science: Designing Nature for New Audiences* (Chicago: University of Chicago Press, 2007).

5. The "Seaside Convalescent Hospital" (*Times,* 8 January 1887, 13) is typical of money-raising organizations in the high social status of its patrons, but atypical in the high proportion of female patrons.

6. Both baronets and knights had the title "Sir" and ranked below the aristocracy. Barons, who had the title "Lord," belonged to the lowest order of aristocracy and, with higher ranks (such as viscounts, earls, and dukes), sat in the House of Lords. "Esquire" was an ascription given to those considered to be gentlemen or taken by men who considered themselves gentle-

men. According to Jim Endersby, whom I thank for advice, there were no formal criteria for the ascription "Esquire" at this time.

7. Dickens: Ian Bradley, "The English Sunday," *History Today* 22 (1972): 362; and Mill: Harrison, "Religion and Recreation," 123.

8. Bernard Semmel, *Jamaican Blood and Victorian Conscience: The Governor Eyre Controversy* (1962; repr., Westport, CT: Greenwood Press, 1976). Eleven of the supporters of the Sunday Evenings for the People are among the supporters of the Jamaica Committee identified by Semmel. At least two (Dickens and Tyndall), however, were on Eyre's side.

9. "Science, Politics and Religion," *Quarterly Journal of Science* 2 (1865): 189.

10. Carpenter declined the X Club invitation: Ruth Barton, "'Huxley, Lubbock and Half a Dozen Others': Professionals and Gentlemen in the Formation of the X Club," *Isis* 89 (1998): 412n3. Graham later supported the SLS, which, combined with his position at the secular University College, suggests that he was a circumspect supporter of secularizing and naturalistic projects. On the "young progressionists" being unrepresentative of science: Hooker to Lubbock, 29 February 1861, Royal Botanic Gardens, Kew, JDH/2/3/10.240 – 41.

11. Carpenter to Lubbock, 27 February 1861, British Library, Avebury Add. MSS 49639. 32 – 33.

12. Five years later, in a celebrated, controversial case, Voysey was deprived of his living. He then set up independently, holding theistic services in London at St. George's Hall, Langham Place—where the SLS also held its lectures.

13. Thomas Kirkman, *Philosophy without Assumptions* (London: Longmans, Green, 1876), 292.

14. R. K. Webb distinguishes two schools of Unitarians in the Victorian period. See his "The Faith of Nineteenth-Century Unitarians: A Curious Incident," in *Victorian Faith in Crisis*, ed. Richard Helmstadter and Bernard Lightman, 126 – 49 (Houndmills, UK: Macmillan, 1990). The distinction is difficult to apply here due to lack of information, but also, supporters of Sunday lectures seem to have represented both schools. Webb identifies Martineau and Carpenter as belonging to the "new" school, but Hodgson expressed the evidential theology of the old school in his Sunday lecture (see below).

15. To identify names on lists, I use the *Oxford Dictionary of National Biography*, ed. H. C. G. Matthew and Brian Harrison, 61 vols. (Oxford: Oxford University Press in association with the British Academy, 2004), cited as *ODNB*; the *Dictionary of Nineteenth-Century British Scientists*, ed. Bernard Lightman, 4 vols. (Bristol, UK: Thoemmes Continuum, 2004), cited as *DNBS*; and Frederick Boase, *Modern English Biography*, 6 vols. (1892–1921; repr., London: F. Cass, 1965). The religious affiliations of lesser figures are often not mentioned; hence, there may be more positivists and Unitarians than I have been able to identify.

16. On the NSL: Edward Royle, *Victorian Infidels: The Origins of the British Secularist Movement, 1791–1866* (Manchester, UK: Manchester University Press, 1974), 258 – 61; Wigley, *Victorian Sunday*, 102 – 4; Harrison, "Religion and Recreation," 109–11; and advertisement, *Times*, 24 June 1865, 8.

17. Morrell's contribution to the NSL was honored on his eightieth birthday ("National Sunday League," *Times*, 28 September 1903, 13). For other *Times* reports that identify members of the NSL, see "Court Circular," 7 July 1858, 9, and "National Federation of Sunday Societies," 2 July 1895, 8.

18. On Slack: Thomas Seccombe, rev. Peter Osborne, "Slack, Henry James (1818–1896)," *ODNB*; Ruth Barton, "Just before *Nature*: The Purposes of Science and the Purposes of Popular-

ization in Some English Popular Science Journals of the 1860s," *Annals of Science* 55 (1998): 8. On Walmsley: C. W. Sutton, rev. Matthew Lee, "Walmsley, Sir Joshua (1794–1871)," *ODNB*.

19. Edward Royle, *Radicals, Secularists and Republicans: Radical Freethought in Britain, 1866–1915* (Manchester, UK: Manchester University Press, 1980), 297.

20. "The *Standard* on the Suppressed Lectures," *English Leader*, 31 March 1866, 155.

21. On Christian socialism: Edward Norman, *The Victorian Christian Socialists* (Cambridge: Cambridge University Press, 1987).

22. "National Sunday League," *Times*, 4 April 1867, 7.

23. Gerald Stone, "Bowring, Sir John (1792–1872)," *ODNB*.

24. The "Church" was the Church of England, often described as "by law established." Being established by law gave it a different legal status to every other religious sect or denomination. The extreme radicals among Church reformers wanted to disestablish the Church, making it merely one among other religious bodies. Disestablishment, like many other changes to Church organization, required an act of Parliament.

25. M. C. Curthoys, "Hodgson, William Ballantyne (1815–1880)," and "Heywood, James (1810–1897)," *ODNB*.

26. Morrell emphasized the assistance of Owen, Dickens, and Lyell in his eightieth birthday speech, reported in "National Sunday League," *Times*, 28 September 1903, 13. Nicolaas A. Rupke, *Richard Owen: Victorian Naturalist* (New Haven, CT: Yale University Press, 1994).

27. Dickens to Morrell, 9 November 1865, in *The Letters of Charles Dickens*, ed. Madeline House, Graham Story, and Kathleen Tillotson, 12 vols. (Oxford: Clarendon, 1965–2002), 11:107.

28. "*Standard* on the Suppressed Lectures," 155–56.

29. *Journals of Walter White*, ed. William White (London: Chapman and Hall, 1898), 196–97 (26 January 1866); and "The Sunday Movement," *Caledonian Mercury*, 24 January 1866, 3.

30. "S[us]pension of the 'Sunday Evenings for the People,'" *Reynold's Newspaper*, 4 February 1866, 8.

31. Note on money: there were twelve pence (d) in a shilling (s) and twenty shillings in a pound (£). For comparison a cheap, small book might cost one shilling, and daily newspapers in London cost one penny or threepence.

32. For Huxley's lecture, see *Natural Knowledge in Social Context: The Journals of Thomas Archer Hirst*, ed. William H. Brock and Roy M. MacLeod (London: Mansell, 1980), 7 January 1866; and Desmond, *Huxley*, 344–45.

33. The lecture was published almost immediately in the *Fortnightly Review* 3 (1866): 626–37, but with the revised title "On the Advisableness of Improving Natural Knowledge." Huxley assessed it as important, as shown by his placing it first in *Lay Sermons, Addresses and Reviews* (1870; repr., London: Macmillan, 1887, 1–16), his first collection of essays, and immediately after his "Autobiography" in the first volume of his *Collected Essays* (*Method and Results*, 1893; repr., London: Macmillan, 1904), 18–41. James Paradis analyzes the lecture (although without reference to its being a Sunday lecture) and sums it up as "the most succinctly representative of his vision." See his *T. H. Huxley: Man's Place in Nature* (Lincoln: University of Nebraska Press, 1978), 76–81.

34. *Natural Knowledge in Social Context*, 7 January 1866.

35. Only a few textual changes can be identified: Huxley changed "Desirableness" in the title to "Advisableness" between the publication of the flyer and the *Fortnightly* version; edited out

"and Unknowable" between the 1870 and 1893 versions. The interpolation "Bibles, for instance," is not in the first published version, so is probably the satirist's emphasis.

36. H. L. Mansel, *Scenes from an Unfinished Drama, Entitled Phrontisterion, or, Oxford in the 19th Century*, 4th ed. (Oxford: J. Vincent, 1852), 15–16. The cited section is sometimes printed separately as "Hymn to the Infinite by Full Chorus of Cloudy Professors."

37. For a particularly full account, see "S[us]pension of the 'Sunday Evenings for the People,'" 8. The issue was reported throughout the country; for example, "Stoppage of the 'Sunday Evenings for the People,'" *Glasgow Herald*, 31 January 1866, 6.

38. "Police," *Times*, 8 March 1867, 9; "The Sunday Evenings for the People at St. Martin's Hall," *Daily News*, 8 March 1867, 3; and "National Sunday League," *Times*, 4 April 1867, 7.

39. "House of Commons," *Times*, 20 June 1867, 8.

40. Wigley, *Victorian Sunday*, 125.

41. Quoted in "Lord's Day Rest Association," *Times*, 29 June 1869, 5.

42. Ibid.

43. See Mark Judge, "The London County Council and Sunday Observance," letter to the editor, *Times*, 28 November 1898, 7, for a history of the Recreative Religionists.

44. "X Club Notebooks," 7 January 1869, Royal Institution, Tyndall Papers.

45. See the pamphlets in *Miscellaneous Proceedings of the Sunday Lecture Society* (hereafter *Proceedings SLS*), held in the British Library at 4355.df.17. The printed letter of invitation to the preliminary meeting is bound as #1a.

46. Report of the Preliminary Meeting (*Proceedings SLS*, #1b), item 1.

47. Francis Galton, *English Men of Science* (1874), as quoted in Frank Turner, "Rainfall, Plague and the Prince of Wales: A Chapter in the Conflict of Religion and Science," *Journal of British Studies* 13 (1974): 65.

48. Biographical sources checked: *ODNB; DNBS*; Boase, *Modern English Biography; Dictionary of Legal Biography*, ed. A. B. Schofield (Chichester, UK: Barry Rose, 1998); Joseph Foster, *Men-at-the-Bar: A Biographical Handlist of the Members of Various Inns of Court* (1885). Some individuals are identifiable only through newspaper reports. For Crawford, see "The Metropolitan Board of Works," *Times*, 24 February 1883, 10; "Royal Institution of Great Britain," *Times*, 4 July 1883, 11; "Sir Moses Montefiore Memorial," *Times*, 21 December 1883, 4; Domville: "Wills," *Times*, 11 February 1898, 10; Graham: "Obituary," *Times*, 17 July 1905, 6; Marsden: "Injustice in Zululand," *Times*, 17 April 1890, 13; "The Late Mr. Mark Marsden," *Daily News*, 19 November 1890, 7. I have also checked the 1881 census records, but, even when the addresses that the SLS gave for committee members are used, some remain unidentifiable.

49. On Domville's father, see "Domville, Sir William," Boase, *Biography*, vol. 5; and Wigley, *Victorian Sunday*, 103, 125. Supporters from Domville's family can be identified by matching legatees of his will ("Wills and Bequests," *Morning Post*, 11 February 1898, 3) with names on an 1869 "List of Members, Donors, and Subscribers" (which is appended to the report of the Preliminary Meeting, *Proceedings SLS*, #1b).

50. "Wills and Bequests," 3; and "Wills," *Times*, 11 February 1898, 10. The "List of Members, Donors, and Subscribers" (updated to April 1871, and published with the report of the Second AGM, *Proceedings SLS*, #24) records that Domville donated five pounds, seven shillings and fourpence, spent in cab hire, to the society.

51. See the 1869 and 1871 lists of members referenced in the previous two notes.

52. There are numerous letters between Hooker and La Touche in Royal Botanic Gardens, Kew, JDH/2/3/10 and JDH/2/1/12.

53. Grant (1793–1874) had been a Lamarckian and a medical reformer in the 1830s and 1840s. When Grant's death was reported, his consistent support was emphasized ("Report," Fifth AGM, *Proceedings SLS*, #68, 7). Pare: Royle, *Victorian Infidels*, 314.

54. Problems were mentioned discreetly at the First General Meeting (*Proceedings SLS*, #16, 6–7). See also Domville's address to the society in 1888 (*Proceedings SLS*, #353). J. A. Froude explained to Huxley why he turned down an invitation to lecture (23 December [1869, 1870, or 1871], Imperial College, Huxley Papers [hereafter HP], 16.289–90).

55. The list was announced at the Tenth AGM (*Proceedings SLS*, #186, 8).

56. I thank Paul White for checking Darwin's correspondence for any reference to the Sunday Lecture Society.

57. M. C. Curthoys, "Booth, James (1796/7–1880)," *ODNB*.

58. Carpenter: Margaret Deacon, "Carpenter, William Benjamin (1813–1885)," *DNBS*. Huxley: Moore, *Sources*, 456n12. Huxley twice explained his position to Michael Foster; see Leonard Huxley, *Life and Letters of Thomas Henry Huxley*, 2 vols. (London: Macmillan, 1900), 2:56, 2:174.

59. Meeting time was discussed at the First General Meeting (*Proceedings SLS*, #16, 5). The format was fixed from the Second AGM ("Report," *SLS Proceedings*, #24, 5).

60. Domville announced the proposal in March 1888 (*Proceedings SLS*, #353).

61. Information on numbers, prices, times, and lecture topics can be followed through the twenty-year collection of annual reports in the British Library.

62. G. Le G. Norgate, rev. Anne Pimlott Baker, "Zerffi, George Gustavus (1820–1892)," *ODNB*.

63. "National Sunday League," *Times*, 11 January 1879, 8; and "Report," Tenth AGM (*Proceedings SLS*, #186), 6–7.

64. Morley to Huxley, 26 January 1879, HP 23.55–56.

65. Adrian Desmond, "Huxley, Thomas Henry (1825–1895)," *ODNB*.

66. Advertisements, *Times*, 19 March 1870, 13. Both Sunday lecture groups used the same hall.

67. Conway to Huxley, 13 November 1878, HP 12.298; and John Morley to Huxley, 26 January 1879, HP 23.55–56.

68. Harrison, "Religion and Recreation," 112; and Wigley, *Victorian Sunday*, 126.

69. Report of the NSL's annual meeting, "Sunday," *Times*, 24 April 1872, 9.

70. "Report," Fifth AGM (*Proceedings SLS*, #68), 6.

71. "Report," Twelfth AGM (*Proceedings SLS*, #240), 7.

72. Wigley, *Victorian Sunday*, 126–27; and "The Sunday Society," *Times*, 14 May 1877, 8.

73. This section draws on Owen Chadwick, *The Victorian Church*, 2nd ed., 2 vols. (London: Adam & Charles Black, 1970–72); G. I. T. Machin, *Politics and the Churches in Great Britain, 1832 to 1868* (Oxford: Clarendon, 1977); David Bebbington, *The Nonconformist Conscience: Chapel and Politics, 1870–1914* (London: Allen & Unwin, 1982); J. C. D. Clark, *English Society, 1660–1832* (1985; 2nd ed., Cambridge University Press, 2000); and Frank O'Gorman's review of Clark. "The Recent Historiography of the Hanoverian Regime," *Historical Journal* 29 (1986): 1005–20.

74. Taxes based on propery ownership or occupation were called rates and levied locally, traditionally by each parish. The income tax, by contrast, was a national tax.

75. Curthoys, "Heywood," *ODNB*.

76. Judy Slinn, "Shaen, William (1821–1887)," *ODNB*. Also see Hodgson and Hunter in table 7.1.

77. As Adrian Desmond has emphasized, this is one of the ways in which Huxley's thinking developed in interaction with Coventry Unitarians (*Huxley*, 8–10). Both Desmond and Bernard Lightman emphasize Huxley's opposition to Anglican privilege and his alliances with Dissent: see Desmond's second volume, *Huxley: Evolution's High Priest* (London: Michael Joseph, 1997); and Lightman, "Interpreting Agnosticism as a Nonconformist Sect: T. H. Huxley's 'New Reformation,'" in *Evolutionary Naturalism in Victorian Britain: The "Darwinians" and Their Critics* (Farnham, UK: Ashgate Variorum, 2009), article 4.

78. Ruth Barton, *The X Club: Power and Authority in Victorian Science*, forthcoming, chap. 3.

79. Armstrong to Huxley, 24 November 1883, HP 10.123–24. Armstrong did not identify the specific SLS. I propose Tyneside because Armstrong lived near Newcastle and a Tyneside SLS was established in 1884 ("Report," Sixteenth AGM, *Proceedings SLS*, #317, 7).

80. "Reid v. Wilson and Ward," *Times*, 30 June 1894, 17; "Bristol Sunday Society," *Bristol Mercury*, 28 November 1894, 3; and "National Federation of Sunday Societies," *Times*, 4 February 1895, 14.

81. The extent to which public opinion had changed is illustrated by the unsympathetic questioning that LDOS representatives faced from the Lords' Select Committee: "Lord's Day Observance," *Times*, 28 May 1895, 12; 31 May 1895, 12; and 21 June 1895, 13.

82. H. Mills, "Sunday Opening of Museums," *Standard*, 13 May 1896, 5.

83. Royle, *Radicals, Secularists and Republicans*, 224, 328.

84. Jeffrey Cox, "Provincializing Christendom: The Case of Great Britain," *Church History* 75 (2006): 129. The sociologists Peter Berger, Grace Davie, and Effie Fokas present similar arguments in *Religious America, Secular Europe?* (Aldershot, UK: Ashgate, 2008).

85. Frank M. Turner, "The Religious and the Secular in Victorian Britain," in *Contesting Cultural Authority: Essays in Victorian Intellectual Life* (Cambridge: Cambridge University Press, 1993), 31–32.

86. Jeffrey Cox, *The English Churches in a Secular Society: Lambeth, 1870–1930* (Oxford: Oxford University Press, 1982), chap. 8, for pluralism. Some say religion was "privatized"—for example, Hugh McLeod, *Religion and Society in England, 1850–1914* (Houndmills, UK: Macmillan, 1996), 224; but in the 1890s the Nonconformists developed new Free Church organizations to express their common identity, and in 1902 education again became a center of political controversy between Church and Nonconformity. See Bebbington, *Nonconformist Conscience*.

87. Bebbington, *Nonconformist Conscience*, 153–54.

88. David Duncan, *The Life and Letters of Herbert Spencer* (London: Methuen, 1908), 247.

89. There is evidence that Huxley was widely popular. His standing among working-class self-improvers in the late Victorian and Edwardian eras is indicated by many examples in Jonathan Rose, *The Intellectual Life of the British Working Classes* (New Haven, CT: Yale University Press, 2001).

90. On Proctor, see Bernard Lightman, "Science, Scientists and the Public: The Contested Meaning of Science in Victorian Britain," article 1 in *Evolutionary Naturalism in Victorian Britain*, 27–30; and Helge Kragh, *Entropic Creation: Religious Contexts of Thermodynamics and Cosmology* (Aldershot, UK: Ashgate, 2008), 109.

91. My suggestions here have been sharpened by discussion with Paul White and Matt Stanley.

8

The Conduct of Belief: Agnosticism, the Metaphysical Society, and the Formation of Intellectual Communities

PAUL WHITE

One of the prevailing assumptions about *scientific naturalism* is that it defines an intellectual community. In the work of Robert Young and Frank Turner, *scientific naturalism* is used to describe both a broad "movement" closely linked with secularization and professionalization, and an ideology for a group of reformers keen to establish the autonomy of science from aristocratic structures of patronage, from Anglican institutions of learning, and from theology.[1] Scholars have rarely agreed, however, about precisely whom to identify with "scientific naturalism" or about the actual coherence of such a group. Bernard Lightman has recently suggested that the boundaries might be expanded or that the criteria of scientific naturalism might be loosened in order to accommodate persons who, on the grounds of vocation, religious commitment, or philosophical orientation, might otherwise be excluded. He has also made the important point that this group, however defined, was by no means uniformly successful in its bid for cultural authority.[2]

The expression *scientific naturalism* had very limited currency in the Victorian period. If we look at one of the rare instances, Thomas Huxley's 1892 prologue to *Essays upon Some Controverted Questions*, we can see how the terms functioned as a polemical construct. In the essay, Huxley presented scientific naturalism as the heir of Renaissance humanism. He positioned it in opposition not to supernaturalism, religion, or even theology per se, but specifically to the Protestant principle of "private judgment" and "inner light," a principle invoked by Reformation writers against the Catholic Church, thereby replacing one form of servitude (to the pope) with another (to the Bible). According to Huxley, scientific naturalism was indebted to the critical and inquiring spirit of several Catholic philosophers, Erasmus and Descartes, and he even located it, in the form of modern historical and literary criticism of scripture, in the very "pale of the Anglican establishment."[3] From a practical standpoint, the expression framed Huxley's later essays on religion and politics, and located this body of scientific work in proximity to other forms of learned culture (religious, political, literary). It did so in a period of eroding consensus over elite forms of cultural authority (including science), and challenges to the hegemony of traditional elites from a variety of quarters.

There are significant risks, then, in using *scientific naturalism* as a banner for a generation of reformers who sought to liberate science and education from the Anglican Church. It presents Victorian intellectual communities as fundamentally oppositional, often taking at face value distinctions and divisions that were drawn in the context of polemical debate. It obscures the importance of groups that were formed in the period across the boundaries of vocation, religious denomination, and class affiliation, often in alliance to challenges from more radical or excluded groups. It suggests that the cultural authority of science was a result of autonomy from, rather than participation in, a broadly Christian culture. Finally, it implies an unduly fixed model of intellectual identity and community, while in practice boundaries were often permeable and overlapping, and communities were formed and re-formed in particular settings for particular purposes, sometimes enduring, but often not. Polemics can be very effective in creating groups of a certain kind, especially idealized and polarized groups. They can be useful too in mobilizing actual communities, in creating solidarity. The important work of Ruth Barton on the "X" Club has shown the advantages of focusing on local communities, who composed them, how they came together, what they accomplished, and why they dissolved.[4] In this chapter, I will examine a somewhat larger and more heterogeneous body, the Metaphysical Society, in an effort to clarify our thinking about more abstract or imagined communities like "scientific naturalists," which may not have existed at all.

The Metaphysical Society was first proposed in 1868 at a gathering of the architect and editor of the *Contemporary Review* James Knowles; the poet laureate, Alfred, Lord Tennyson; and the Reverend Charles Pritchard, Savilian Professor of Astronomy at Oxford.[5] The original aim was to assemble a group of clergymen to build a united front against the encroaching forces of science and skepticism. This plan was aborted, however, when some of the clergymen themselves, such as the Unitarian James Martineau, insisted that "both sides" be represented. The formation of the society was then planned at a party held at Knowles's house, with members selected in order to promote diversity of belief. Some likened the society to a scientific body, others to a broad church. Thus, R. H. Hutton described how clerical, literary, and political leaders met with men of science over dinner to discuss religious questions "after the manner and with the freedom of an ordinary scientific society."[6] The bishop of Peterborough, William Magee, suggested that science was ranked on a par with other religious denominations: "We only wanted a Jew and a Mahometan to make our Religious Museum complete."[7] In composition and format, the society more closely resembled some of the gentlemanly philosophical societies and university debating clubs, of which many of the members had experience. It was a mixed community that defined itself in part through oppositions, embodying divisions of perspective or belief between learned groups.

It was at Knowles's party that Thomas Huxley first coined the word *agnostic*. He would not elaborate on its meaning until 1889, but the term evidently gained some currency within the society. James Martineau, who had been present at the organizational party, wrote to Knowles shortly thereafter, "For the equal chance of gaining and giving light I would gladly join in [discussion] with gnostics and agnostics alike; but a society of gnostics to put down agnostics I cannot approve and could not join."[8] Lightman has argued that the agnosticism preached by aggressive secularists like John Tyndall, William Clifford, and Leslie Stephen (all members of the society), developed from a tradition within Protestant theology regarding the limits of knowledge.[9] The term could signify an absence of religious belief, a general uncertainty, but also a more dynamic orientation toward belief as a practice: an active questioning, an openness toward the beliefs of others or toward evidence contrary to one's own beliefs, a process of conscientious doubt and inquiry. Such practices were integral both to scientific method and to the religious life of many liberal Anglicans and some Dissenters, like the Unitarian Martineau. One of the questions we might ask is whether the Metaphysical Society debates were designed to resolve differences of belief, or to subsume them within a larger process. If the aim was not consensus, then why was the community formed and what held it together? Could the practice of heated debate itself help to build a community that was not based on shared beliefs, but on one's conduct toward belief?

Evidence and Intuition

In a reminiscence of the society, Hutton reconstructed one of its monthly soirées. The topic of the evening was "Can experience prove the uniformity of nature?" But discussion gradually devolved upon the question of which of the learned members used the expression *I believe* with more earnestness and solemnity. The Roman Catholic philosopher William Ward argued that the moral nature of man dictated belief in God. The zoologist Thomas Huxley stressed that men of science humbly required firmer grounds for belief than the testimony of personal feelings. The Catholic priest and scholar John Dalgairns replied that theologians founded their creeds not on a "working hypothesis," but on a "higher intuition than any inductive law can engender." Assorted reflections on human and miraculous agency ensued, with comments by the art critic John Ruskin and the political economist Walter Bagehot.[10] The scope of the papers presented over the eleven-year history of the society confirms that the subject of belief was central to its discussions. Members debated particular principles and doctrines, but they were equally attentive to the process of belief, its psychology and physiology, its habits and manner.[11]

By contemporary accounts, one of the most memorable papers delivered to the society was William Clifford's "The Ethics of Belief."[12] A fellow of Trinity College Cambridge, and professor of mathematics at University College London, Clifford had gained a reputation as a leading spokesman on the physical sciences and scientific culture through participation in the Cambridge debating society, the Apostles, and later through periodical writing and public addresses on such topics as the history of the sun, the evidence for an ether, and the relation between science and poetry.[13] In his paper, Clifford outlined the "duty of inquiry" through a series of hypothetical examples, beginning with a shipowner who, though doubting the seaworthiness of his vessel, dismissed ungenerous suspicions about the honesty of builders and contractors, and trusted to Providence to deliver his crew and cargo safely. In one scenario, his ship went down, he collected his insurance money, and "told no tales"; in another, the ship proved to be sound, and completed its voyage safely. Yet in either case, Clifford stated, the man was guilty of harboring a conviction without sufficient evidence: "He had acquired his belief not by honestly earning it in patient investigation, but by stifling his doubts." The correctness of belief, according to Clifford, was not based on its truth or falsehood, nor on its sincerity, but on how it was acquired. The most conscientious of convictions could be founded on prejudice and passion. Clifford passed swiftly from evidence to ethics, arguing that no beliefs were ever strictly private; rather, they formed part of the stock of conceptions of a society, accumulated over the course of its history: "an heirloom, which every succeeding generation inherits as a precious deposit and a sacred trust." This common property could either be enriched and perfected, or debased, such was the "awful responsibility" of each believer.[14]

The evidential high ground taken by Clifford was challenged several months later by St. George Mivart, a zoologist who had gained notoriety for his criticisms of Darwin's *The Descent of Man* (1871), and for his efforts to reconcile evolutionary theory with Roman Catholicism.[15] Mivart remarked on the "great warmth" with which some of his colleagues regarded truth, as if it were "necessarily a good." Rather than pursuing truth with such unquestioning faith, however, should we not ask "Why is truth desirable?" Should all facts be brought before a dying man, even the illegitimacy of his children? In order for truth to be unequivocally good, one would have to believe in a beneficent Providence, the very assumption that condemned Clifford's shipowner to infamy. Finally, Mivart remarked that various peoples throughout history seem to have been strengthened by the acceptance of falsehoods, citing Herbert Spencer's claim that the survival of a nation may require that it not abandon its faith too early.[16]

The terms of debate initiated by Clifford and Mivart were taken up in later discussions. Speaking from his experience as a jurist, James Fitzjames Stephen defended the position that truth should be regarded with reverence and awe. To lie or use falsehood as a shortcut to some desirable end was in fact criminal: a temptation to which "every cheat, and thief, and forger, and coiner, gives way." Stephen also attacked the logic of Mivart's contention that the value of truth was only assured by belief in God. Was not knowledge about the cure for cholera beneficial to both believers and nonbelievers? Stephen passed on to beliefs closer to Mivart's heart: the Roman Catholic dictum that the wicked shall suffer the eternal flames of hell. Surely this was not objectively desirable, and yet it would seem an important thing to know, *if* it were true.[17] Reverence for truth was also defended by Stephen's younger brother Leslie, who joined the society in 1876, having distinguished himself as a critic, editor, and historian of thought.[18] He argued that beliefs in matters of fact rested on a "convergence of innumerable probabilities," a variation of the claim often made by empirical philosophers that established scientific theories, such as Newton's laws, though never absolutely certain, were nevertheless repeatedly tested and confirmed by the evidence of the senses. He underscored the duty of "candour" and the moral necessity of keeping the mind open and vigilant against the "intellectual contagion which leads us to accept beliefs current around us without examination."[19]

Leslie Stephen's arguments were not aimed directly at Mivart, but at another Roman Catholic, John Henry Newman. Though not a member of the society, Newman and his 1870 work *An Essay on the Grammar of Assent* served as the basis of discussion for a number of papers. He had acquired a controversial reputation through a series of subtle apologetics for ecclesiastical authority and a public defense of his sincerity on matters of belief against charges of deceit made by the liberal Anglican Charles Kingsley.[20] Newman's

Essay addressed some of the fundamental assumptions behind the empiricist philosophy promoted by many society members, including Clifford and the brothers Stephen. He argued that empiricism—namely, the doctrine of belief as conditional upon inference, and as proportional to evidence, a doctrine often traced to John Locke—was actually contrary to experience. Newman maintained that according to "psychological facts," belief was never provisional or proportional: "No one can hold conditionally what by the same act he holds to be true."[21] Instances in which assent was allegedly mingled with doubt were in fact cases in which assent was not really given. Newman also claimed, contrary to the ethics of Clifford and others, that belief was always highly personal. In effect, Newman reinstated the individual circumstances and good faith of Clifford's shipowner as legitimate foundations of belief. The faculty of belief, or "illative sense," as Newman called it, was seated in the mind of every individual. It was "an acquired habit . . . formed and matured in practice and experience," and variable according to each person's history, talents, and profession.[22]

If Newman was a target of some society members, aspects of his own empiricist critique were taken up by others in the monthly meetings. Thus, R. H. Hutton answered Leslie Stephen with the claim that evidence was by no means the strongest or most ethical ground of belief: "The less a child or a man weighs the arguments against yielding to impure thoughts, the sounder his conviction . . . will be, and the better he will be." Hutton likened sound beliefs to the possession of instinct, and offered the example of a "sceptical chicken" that might resist the inclination to chip at its shell because it lacked empirical evidence that it was likely to succeed.[23] Such arguments were not advanced exclusively by clergymen. The moral philosopher Henry Sidgwick declared empirical philosophy incoherent on the grounds that it was unable to distinguish real from apparent knowledge without introducing assumptions that could not be proved by experience, such as the trustworthiness of memory and the reliability of testimony. What is meant by "the verification of experience?" Sidgwick asked, only that "assumptions are accompanied by anticipations of feelings or perceptions which are continually found to resemble or agree with . . . the more vivid feelings or perceptions which constitute the main stream of consciousness."[24] Thus, Sidgwick, a Cambridge reformer and self-confessed convert to natural science, advanced a position on evidence and intuition that was not very different from Newman's illative sense.

Studies of the Victorian debates about belief have tended to arrange the participants on two sides: the "freethinkers" or "rationalists" on one hand, and the various Christian denominations on the other. The divide reproduces the traditional conflict between science and religion in the form of competing philosophies and methodologies: empiricism and intuitionism.[25] The Metaphysical Society discussions suggest, however, that the boundaries between ostensibly opposed schools of thought could shift or dissolve in the course of critical engagement, exposing considerable common ground and contested territory claimed by all parties. To appreciate this dynamic aspect of the society, we need to consider the beliefs of its members not as a set of fixed positions, but as a social process, something made and remade in the act of debate. Arguments advanced for an evidential model of belief often enrolled assumptions, imperatives, or principles whose foundations were moral or logical, not empirical. Intuitionism, in turn, was defended by recourse to the "facts" and "experience" of belief. Newman's *Essay* was presented as an empirical study, utilizing the observational methods of science to examine how belief was actually formed. Yet the results of his study indicated that belief was never a simple process of weighing evidence, but always rested on predispositions acquired over the course of one's upbringing and through the influence of traditions and other social sanctions, as well as the medium of the senses. Appeals to evidence and inner determinants, to intellectual duty and conscientious inquiry, were thus made on nearly all sides in different ways.

The Emotion of Conviction and the Will to Believe

The original name of the group had been the Metaphysical and Psychological Society. The interests of the members in the sciences of mind were indicated in a resolution passed at the first meeting:

> To collect, arrange, and diffuse Knowledge . . . of mental and moral phenomena . . .
> To collect trustworthy observations upon . . . the relations of brain and mind . . . the faculties of the lower animals, &c.[26]

Running through much of the debate on belief were attempts to define its nature and relationship to other mental faculties or functions. Traditional introspective methods of psychology were discussed alongside new researches in experimental physiology and pathology. One of the first papers presented to the society was titled "On the Emotion of Conviction," delivered by the editor of the *Economist*, Walter Bagehot. Bagehot remarked that while the intellectual aspect and evidential grounds of belief had been elaborately discussed, the emotional aspect had scarcely been mentioned. To illustrate the role that emotion played in the formation of belief, Bagehot described his own strange obsession with a parliamentary election in a London borough that he had contested years ago and lost by a mere seven votes cast at the last hour: "If I call up the image of the nomination day, with all the people's hands outstretched, and all their excited faces . . . the old feeling almost comes back upon me, and for a moment I believe that I shall be a Member for Bridgewater."[27] Referring then to a series of religious enthusiasts from John Knox to Ignatius Loyola, Bagehot declared that men for whom belief "seems to burn [like a hot flash] across the brain" have been the great movers of history, for better or worse. "When the subject is thoroughly examined," he concluded, "conviction will be proved to be one of the intensest of human emotions, and one most closely connected with the bodily state."[28] That emotion should be so neglected in studies of belief Bagehot attributed to the authorship of psychological treatises by calm, quiet, and careful minds for whom emotion rarely arose. He went on to complain of the works of John Stuart Mill and Alexander Bain, which failed to comprehend the "animal nature of belief" or the relation of conviction to "active disposition."[29]

Questions about the nature of belief and its relationship to emotional and bodily processes held important implications for how beliefs were formed and regulated. One of the members of the society most expert on these topics was William Carpenter, a Unitarian and author of influential works on comparative and mental physiology, as well as a series of papers on animal and human automatism. It was on the latter subject that he addressed the society in 1875, describing the human body as a complex of mechanical apparatuses, nervous, muscular, circulatory, and digestive. Physiologists had no doubt, he added, that there were also mechanisms of thought and feeling, of intellect and imagination. A vast proportion of mental activity was purely reflexive, arising spontaneously from the physical structure of the brain. Might the formation of beliefs, whether from sensory experience, instincts, or intuitions, be an automatic process, the mind passively impressed by the external world, or by its own internal drives and dispositions? According to Carpenter, the cerebral mechanism could itself be mastered through the exercise of will. Indivisible and irreducible to matter, the will was the only break in the continuous chain of automatic functions stretching from the lowest impulses to the highest mental activity.[30] For a fuller explication, Carpenter referred the members of the society to his *Principles of Mental Physiology* (1874). Here, Carpenter described how the process of belief was ultimately subject to the will, and he declared freedom of the will to be among the most fundamental and incontrovertible of beliefs. Belief was thus a dynamic process, crucial to self-development and the exercise of moral character: "a direct result of our recognition of a self-determining power within ourselves [and] the foundation . . . to the Conduct of our lives."[31]

Some of Carpenter's points had been raised in an earlier exchange in the society. Building on his expertise in invertebrates, Thomas Huxley had presented an alternative account of automatism in his paper "Has the Frog a Soul?" Huxley recounted the eighteenth-century debate between Haller and Whytt on the locus of the soul, and described recent experiments in which acid was applied to the thigh of a frog whose head had been severed just above the medulla oblongata. The headless frog tried to wipe off the irritant with the dorsal surface of its foot. When its foot was cut off, it tried with its leg, and eventually, after a pause, with the foot of the other leg. It then remained motionless for hours until a new stimulus was given. Huxley proposed a further experiment in which the decapitated head and trunk were sent several hundred miles in opposite directions. Each part, he maintained, would continue to perform equally purposive, though quite distinct, actions, adapting means to ends in a manner usually deemed volitional (but in the case of the headless frog, evidently not).[32] In the ensuing discussion, the purely automatic nature of the frog was queried by one member of the society, who remarked that frogs recently caught ate no flies because they were frightened. Huxley replied, "If you shake a watch, you disturb the regularity of its machinery." He then exclaimed, "I don't want your soul and your vital force to explain a living organism!"[33]

Huxley did not draw out the implications of his argument for human freedom. But his challenge to the doctrine of the immateriality of the soul bore upon the constitution of belief and its role in the moral fabric of man. This challenge was taken up at a later meeting by Archbishop Henry Manning, who argued that automatic functions of the body and the brain were guided by a moral agent—the will—which acted on the brain through the operation of attention. By fixing the mind or the sense organs on particular ideas or objects, the will could determine the course of action in a definite direction. The reflex action of the nervo-muscular apparatus, and the automatic and instinctive action of the cerebrum, were in fact instruments of the will. They could be regulated by a succession of thoughts and emotions, themselves directed by selective attention. What indicated that the exercise of attention was an act of will, rather than purely automatic, was the presence of effort. Attention was the result of a struggle in the mind, and was thus preeminently a moral function, mobilized in the overcoming of temptation, or in altering a course of habit. "The control of the will over thought runs through the whole moral culture and discipline of man," Manning wrote. "The replenishment of the mind, or brain . . . with thought and knowledge . . . is to a great extent all through life a voluntary act."[34]

Despite substantial differences and disagreements, the discussion papers, comments, and auxiliary writings of the society agreed that belief was a vital operation, an ongoing process, part of one's governing constitution, directing one's actions and defining one's public character. Clifford's celebrated paper raised these points urgently. Even if a belief was so fixed that one could not think otherwise, still one had a choice with regard to the action suggested by it, and could not escape the duty of investigating it. If the will was lax, then a belief could take root in the mind and prepare the ground for others like it, further weakening one's power of self-control: it "gradually . . . lays a stealthy train in our inmost thoughts . . . and [leaves] its stamp upon our character for ever."[35] Many of the disputes within the society might even be traced to this shared investment in the conduct of belief. If belief was an act of will, then how was the will to adjudicate conviction? Clifford, Huxley, Leslie Stephen, even Carpenter, all maintained that this was the work of scientific method, itself the epitome of the disciplined will. The function of the will was to exercise restraint, to withhold consent until sufficient evidence was gathered through methodical observation, interrogation, and experiment. But if belief was volitional, then was it not also emotional? Recall Huxley's outburst when his views on animal automatism were challenged. Or Mivart's remark that one tended to believe what was most gratifying for oneself. Or Bagehot's confession that once he had become convinced that he should be elected a member for Bridgewater, his actions had followed suit, however irrational. He failed to run for seats in other districts where his victory was more likely.

Systematic observations on the emotion of conviction and the will to believe would later be made by William James in his *Principles of Psychology* (1890), and reiterated in a famous essay. In order for an object to have reality, according to James, it must not only appear before the senses. It must also be interesting and important: "We disbelieve facts and theories for which we have no use." Moving well beyond the sober introspection of Mill and Bain, James described the deepened sense of reality that characterized belief as something akin to drunkenness or nitrous oxide intoxication, under which "a man's very soul will sweat with conviction." James also noted that the edifice of natural science seemed built on contrary principles: "What patience and postponement, what choking down of preference, what submission to the icy laws of outer fact are wrought into its very stones and mortar."[36] He cited Clifford and Huxley on the immorality of subjecting belief to private pleasures or passions, adding, however, that this scientific dictum of waiting upon evidence was itself a passional preference. Men of science who claimed that belief in the existence of God or the immortality of the soul should be withheld did so because those beliefs were irrelevant to them. Recent psychiatric studies had indicated, furthermore, that doubt could assume pathological proportions. The journal *Mind* reported in 1876 that patients had been admitted for observation who suffered from a "questioning mania," an incessant and subtle inquiry into the grounds of all things, and "an impulse to secure, at any expense of time and trouble, an absolute accuracy in the most trivial things."[37] Such accounts effectively eroded the distinction between men of science and religious enthusiasts. Given the passionate nature of belief, James urged, "none of us ought to issue vetoes to the other . . . but delicately and profoundly to respect one another's mental freedom: then only shall we bring about the intellectual republic."[38]

Good Manners

The spectacle of such a diverse assembly conversing over dinner was described by the politician Arthur Russell to his brother: "It is a wonderful thing to see a cardinal discussing, on equal terms, with professed atheists gathered round a social table. . . . Frederic Harrison declared that he considered belief in miracles as the commencement of insanity, and His Eminence [Manning] replied that he considered an incapacity to believe in the supernatural as a commencement of ossification of the brain."[39] Such a private gathering, combining sociability with controversy, was unusual in gentlemanly society. Disputation was typically a more public affair, involving audiences, debating platforms, or print. At the Athenaeum Club, by contrast, one might find a broad and divergent community in a place of conversation, dining, and drinking, and for some members, serious work. But the Athenaeum was not a stage for confrontations. It was a large communal space with many rooms and quiet corners, where one could come and go as one pleased, meet with friends and associates, and avoid the rest. The Metaphysical Society threw its members together in one room, expected them to be friendly, and yet positively encouraged antagonism. The sparring was not always cordial, however, and had to be underwritten by a formal code. At the society's first meeting, it was agreed "that the members crediting each other with a pure quest of truth would confer together on terms of respectful fellowship, and never visit with reproach the most unreserved statement of reasoned belief and unbelief."[40]

The need for such rules was evident in the early meetings. The third session featured a paper by Huxley on the immortality of the soul. Several months later, the Roman Catholic J. D. Dalgairns spoke on the same subject. But before he began, he felt compelled to reiterate the society's code— namely, that "no element of moral reprobation must appear in the debates." The statement brought a remark from W. G. Ward: "While acquiescing in this condition as a general rule, I think it cannot be expected that Christian thinkers shall give no sign of the horror with which they would view the spread of such extreme opinions as those advocated by Mr. Huxley." A pause ensued, and then Huxley replied, "As Dr. Ward has spoken I must in fairness say that it will be very difficult for me to conceal my feeling as to the intellectual degradation which would come of the general acceptance of such views as Dr. Ward holds." According to one of the members present, J. A. Froude, no further discussion of the matter took place, and it was "brought home" to all "that if such a tone were admitted the Society could not last a day."[41]

Mutual respect, disagreement without reproach, and purity of purpose: that the society stood in need of such a code bears some examination. Its members were specially chosen from among the most respected and learned professions, and the format of dinner and discussion was a convention among gentlemen. Indeed, given the importance of gentlemanly manners in the conduct of scientific disputes and scholarly exchange in the Victorian period, we might expect that a sense of propriety and consideration would prevail within the society and secure respectful debate, no matter how heated. There was something that the society aspired to, however, that was not prescribed by traditional manners, and may in fact have been contrary to them. Other gentlemanly communication, in specialist societies and clubs or correspondence, might involve participants with diverse or opposing viewpoints. Information and ideas could be exchanged despite marked differences in religious, political, or philosophical belief. But such differences were seldom dwelled upon, and usually submerged. The society, however, brought such differences to the fore. It dictated a level of frankness and even "unreserve" in expression that was not in keeping with gentlemanly forbearance. Disputes might cut to the core of personal identity, threatening the nature and value of one's calling: what would happen to the nation if Roman Catholics had exclusive charge of education? or if everyone believed (or disbelieved) like Mr. Huxley?

Some of the members were close friends—for example, Huxley and Tyndall, Knowles and Tennyson, Hutton and Bagehot. Yet friendship was not what held the group together. Instead, the society seemed to have been bound in part by mutual antipathies. Some members had a history of open and hostile exchange. Huxley had written a scathing attack on Mivart's *Genesis of Species* (1871), which was itself an extended criticism of Darwin.[42] Notoriously shy of public controversy, Darwin had sought to address his disagreements with Mivart in traditional gentlemanly fashion, through correspondence and private conversation. For Darwin, having someone to his home, or visiting someone in London, opened a space of friendship where differences could be aired and public controversy avoided, or if not, then rendered amicable and supportive. Yet Mivart could receive Darwin warmly, send cordial letters full of respect and concerns for his health, and then write acrimonious reviews (sometimes anonymously), claiming that the latter were entirely impersonal, animated by the spirit of truth. Huxley for his part seemed unperturbed. He relished the opportunity to denounce one of Darwin's critics in print, yet he assured the elder naturalist that Mivart was "by no means a bad fellow"; his mind had simply been poisoned by "accursed Popery and fear for his soul."[43] If not in religious belief, then in manner of engagement, Huxley was in fact much closer to Mivart than to Darwin. They valued the kind of disputation that Darwin found overly one-sided and divisive. For Huxley and Mivart too, the spheres of public and private need not correspond or conflict. They could engage in caustic debate and dine together on the same evening.[44]

The Metaphysical Society seems to have been a space carved out specifically for such polemical exchange and intellectual grandstanding (not unlike the sphere of the periodical, of which more below). Contrary beliefs were actively solicited and contested. It was not gentlemanly tact and toleration that secured this "respectful fellowship," but rather a reconstruction of credibility around new discourses and practices of learned behavior, especially those of belief. Moral discourses about method, the virtues of inquiry and conscientious doubt, codes of honesty, transparency, and plain speaking governing the formulation of belief were all being worked out here in the course of debate. Huxley's agnostic creed, though ostensibly a doctrine of unbelief, was in effect a formula for conducting oneself with regard to belief that nicely fit the professed aims of the society—namely, the pursuit of truth with absolute devotion, close and reasoned debate, and unreserved communication.

What threatened the proceedings, as indicated in Froude's account, was a tone of denigration, an assumption that anyone or any group occupied the moral or intellectual high ground. We might compare the troubling exchange between Huxley and Ward with another several months later. Following a paper by Martineau titled "Is There Any Axiom of Causality?" Huxley expressed agreement with the author's claim that, although we have no conclusive physical evidence, we are yet obliged to assume the principle of causation, and to regard it as analogous to the power of will. But Huxley objected to Martineau's identification of this "will," however supreme and uniform, with a personality (that is, with a personal God). As Martineau reported to a correspondent, Huxley

> could only say that Personality, as he measured it, was not equal to the immensity of the product (the universe). And it was easy to see that, as a physiologist, he was accustomed to look at the human personal attributes as merely emerging from the simply vital phenomena . . . and that this Naturalist's esti-

mate of them rose up to damp and quench his inward reading of them from
the consciousness of intellectual light and moral freedom.[45]

It is unclear how much of this report is a summary of Huxley's comments,
and how much is Martineau's own interpolation ("it was easy to see . . .").
But in Martineau's rendering of the debate, Huxley's objections were significantly qualified. He spoke from his own perspective "as a physiologist," or as
a "Naturalist." His comments thus served both to mark the boundaries of his
own identity and to gesture toward a common (high) ground with the clergyman, a shared intellectual calling and moral intuition.

Huxley would later describe the society as a "remarkable confraternity
of antagonists . . . [representing] every variety of philosophical opinion . . .
with entire openness." He contrasted its meetings with others marred by
the "uncharitableness of sectarian bigotry," the "spirit of exclusiveness and
domination," and "restraints on the freedom of learning and teaching."[46]
Huxley might have been alluding to religious corporations, or universities,
or party-political organizations, or indeed to almost any form of social organization in Victorian Britain. If society members were able to achieve such
openness, it was only by creating a group that was itself enclosed—sealed
off from other strictures governing the expression of belief, other forms of
public scrutiny. Henry Sidgwick was one of several society members who had
withdrawn from a Cambridge fellowship on the grounds that he could not
subscribe to the Thirty-Nine Articles of the Anglican Church. He had just resigned his position at Trinity College in the year that the Metaphysical Society
was formed. In a pamphlet, *The Ethics of Conformity and Subscription* (1870),
he justified his decision on the grounds that ministers of religion and learning had an obligation of absolute sincerity.[47] Martineau wrote to Sidgwick
shortly thereafter. He was generally supportive of Sidgwick's decision, but he
warned that such frank confessions of doubt would forfeit the clergy's power
and trust. A Unitarian, Martineau admitted his own sense of guilt when recit-
ing prayers he did not believe, and yet he held that "the clergyman's instinct
must be to conceal his theological struggles in his normal functioning."[48] The
Metaphysical Society was filled with men whose official duties and social authority rested on a command of certainties and a display of conviction, not
on openness and methodical doubt.

The Society Serialized

Of the society's sixty-two members, ten were editors of leading periodicals.
Virtually all of the serials that together constituted the "higher journalism"

of Victorian Britain in the 1870s were represented: the *Spectator*, the *Dublin Review*, the *Economist*, *Fraser's Magazine*, the *Cornhill Magazine*, *Macmillan's Magazine*, the *Contemporary Review*, the *Fortnightly Review*, and the *Nineteenth Century*. Many of the papers presented at the meetings were later published in these journals. But the relationship between the society and the periodical was more dynamic. Sometimes journal articles were produced in response to society papers and discussions. Carpenter's 1871 essay "The Physiology of the Will" appeared in the *Contemporary Review* following the exchange of papers by Huxley and Ward on the soul, though neither of the latter was published. On the other hand, a society paper might be generated in response to a periodical piece. James Stephen's unpublished paper "On the Utility of Truth" addressed Mivart's essay "On Force, Energy, and Will," which had appeared several months earlier in the *Nineteenth Century*. Was the elite society, with its closed and largely unreported meetings, part of a larger system of communication, cultural production, and exchange? Or were the two realms, the society and the periodical, in important respects discontinuous, even opposed?

The editors within the society, including Knowles, Bagehot, Hutton, Froude, Leslie Stephen, and John Morley, explicitly conducted their journals as media for lively debate, reflecting the format of learned societies and clubs. They drew on a mid-Victorian model of the periodical as an organ of liberalism, in which free and reasoned discussion was considered an agent of social progress and reform. They also capitalized on a new development in periodical publishing, the signed article, and sought to attract authors who were recognized leaders of opinion in different spheres. The rise of the named author system was hailed by contemporaries as a triumph of freedom of expression: so liberal had the sphere of print become that writers no longer needed to hide behind the veil of anonymity.[49] But the trend could also be linked to other developments, such as the division of learned culture into specialist fields, the emergence of intellectual celebrity, and the commercialization of controversy skillfully stirred and managed by editors.[50] Many society members had already established reputations as contributing writers.

Knowles even tried to repackage the whole society in serial form. After quitting the *Contemporary Review*, he launched the *Nineteenth Century* in 1877 with specially created forums that mimicked Metaphysical Society discussions.[51] "A certain number of gentlemen," he wrote, "have consented to discuss from time to time, under this title, questions of interest and importance."[52] These "Modern Symposia" were organized around typical society topics, such as "The Influence on Morality of a Decline in Religious Belief." Papers were arranged in succession so as to emulate a live debate.

Knowles seems to have enforced a practice of serial authorship, determining the order of presentation in advance; for each writer, he announced, was allowed to read what preceded his own contribution, but not what followed. A similar format was implemented for some of the full-length articles as well. Gladstone's "On the Influence of Authority in Matters of Opinion" was responded to by J. F. Stephen, and Gladstone concluded with a "Rejoinder." Although the actual society was never mentioned, the formula for such a group was itself a periodical feature, and readers were invited to witness the spectacle of learned men in confrontation. In Knowles's "Symposia," the authors' names appeared at the head of each article in large capitals, rather than in the usual place, at the end. It was as if the Metaphysical Society had been converted into an intellectual parliament with a large gallery of onlookers and a stenographer to record the proceedings and issue them to the press.

It has been suggested that these periodical roundtables tended to accentuate the divisions between learned groups, and that they signaled the demise of the society, since members could do no more than publicly announce their disagreements without hope of resolution.[53] Such an interpretation misses what the society and serials like the *Nineteenth Century* were about in the first place. These debates arose within a community that was explicitly composed of different learned professions, between individuals who were taken as representative of widely divergent and perhaps irreconcilable perspectives. By engaging each other in debate, members affirmed the value and credentials of even those whose viewpoints were antithetical. By christening their collective body as one of openness, freedom, and diversity, they affirmed that their highly selective society defined the scope of legitimate professions, the range of respectable perspectives, and the philosophic tone in which disputes about belief should be conducted. Theologians, philosophers, "higher" journalists and critics, men of science and letters, constituted themselves as joint authorities of culture. It was not only the aforementioned Jew and "Mahometan" who were absent, but social radicals, industrialists, evangelicals, women, workers, and any number of others who might, according to different criteria of openness and diversity, have contributed substantially to debates about belief and the production of culture.

Journals like the *Nineteenth Century* promoted and extended the society's model of cultural authority by a diverse, yet exclusive elite. Liberalist principles of free discussion and equal representation notwithstanding, the Victorian periodical was the preeminent forum in which men (and exceptionally, women) of a certain type appeared together as educators and leaders of opinion. Learned elites not only displayed their reasoned convictions individually. The manner in which they held forth was exhibited collectively before a public perceived as fundamentally divided by religious beliefs, political convictions, and class feeling. The demise of the Metaphysical Society and the contestation of this model of authority were more evident in other publishing ventures, such as William Stead's *Review of Reviews*. As Gowan Dawson has shown, Stead popularized and undermined the elite periodical through a series of appropriations: seizing the role of higher journalism as an organ of public opinion and change, and capitalizing on the cult of intellectual celebrity, but in a manner that shifted power away from the contributing author.[54] Stead incorporated the work of writers that had appeared in other journals, in a condensed and extracted form, injecting a new "voice of the people" embodied by the editor as mediator. Stead used extracts from Huxley's "Agnosticism" article to encourage openness and a spirit of inquiry toward subjects of broad public interest, like spiritualism. Huxley clashed with Stead over the use of some of his articles, even accusing the publisher, as he would socialist land nationalizers and international copyright infringers, of (intellectual) theft.[55]

Stead's populism and redistribution of intellectual property were more in keeping with the new politics of mass democracy than with mid-Victorian liberalism. In their various capacities as authors and editors, professors and tutors, government ministers and MPs, royal commissioners on schools, preachers and presiders over cathedral chapters, the members of the Metaphysical Society were all engaged in the shaping of beliefs, and in the establishing of criteria of legitimate belief, *for others*. A recurrent theme in the society discussions on belief was popular credulity and ignorance—that is, public opinion conceived as a problem and a threat. Convinced that the hope of a future life, and the fear of eternal damnation, were still essential moral sanctions, James Stephen justified authority rather than freedom on matters of belief for the "mass of men."[56] In a letter to his brother in 1874, Arthur Russell described a "pleasant debate" at the society over a paper by Bagehot on "The Metaphysical Basis of Toleration": "By far the most interesting fact which came out was the entire agreement between Archbishop Manning and Professor Huxley. Both maintained that the duty of self-preservation fully justified the State in suppressing opinions which are not compatible with its safety, the State of course being the sole possible judge of what opinions are dangerous to its existence."[57] Which was more remarkable: that an ultramontane should agree with a scientific reformer with a reputation for bishop eating, or that members of a society dedicated to intellectual freedom should appeal to force rather than evidence to quell the ignorance and enthusiasm of the masses?

Conclusion

Debates about belief have traditionally been examined within the context of the Victorian "crisis of faith." Influential works by Walter Houghton, Gertrude Himmelfarb, and others characterized the Victorian age as a struggle between the firm will to believe and the earnest spirit of doubt.[58] This portrait has formed part of a larger picture of the period as one of cultural transition and gradual secularization, culminating in the eventual dominion of a scientific worldview.[59] "Scientific naturalism" emerged as an organizing category in the 1970s and 1980s as part of a social turn in Victorian intellectual history and in the history of science in particular. The social frameworks here were class struggle and the rise of professions. The traditional "conflict between science and religion" was thus recast in terms of industrial middle-class professionals and Dissenters, and conservative Anglican elites with their monopolies of privilege and power. Building on the work of Young, Turner, and John Greene, James Moore defined "scientific naturalism" as a rival worldview or theodicy for a group of "dissident intellectuals" who opposed the authority of the established Church.[60]

If we try to fit the Metaphysical Society into this picture, then we might regard it in Houghtonian fashion as reflecting "Victorian minds in crisis," struggling (and failing to resolve) the deep social and intellectual divides that characterized Britain during the second half of the nineteenth century. We might view it as a case of incipient professionalization, involving the division of knowledge and intellectual communities into discrete domains.[61] Or we might dismiss it as an idle social gathering, mere intellectual theater and posturing, without effect. In any case, *scientific naturalism* could function as a convenient watchword or shorthand for those members of the society who positioned themselves against established religious authorities and institutions, who opposed religious beliefs on evidential grounds, or proposed an alternative "religion of unbelief."

But we should not be too complaisant or too jaded about what the Metaphysical Society stood for or accomplished. We can use this community that cut across vocational and confessional boundaries to reexamine our construct of "scientific naturalism" and the frameworks (secularization, professionalization, and class conflict) that have structured our view of intellectual groups as diametrically opposed. *Agnosticism*, as Huxley tried to define it, was less a slogan for "scientific naturalists" than a bridge between intellectual groups, a means of drawing them together despite differences of belief. Agnosticism and the Metaphysical Society were conceived together in a period when the role of elite culture was challenged from a variety of quarters. Society meetings affirmed the value of the culture produced by its members against forms of authority and popular belief excluded from its body. In its monthly gatherings and ritualized debates, the members' personal identities and authority, as Catholic theologians, moral philosophers, or professors of zoology, were distributed within a larger collective body of eminent men of the day, custodians of belief, and ministers of culture, roles that were further elaborated and amplified in the periodical domain to which so many of the members actively contributed. An oppositional structure could thus serve as the basis for a broad community of elites—religious, literary, journalistic, scientific—sustained in part by an agnostic code of openness and inquiry, and also by a sense of exclusiveness and embattled authority over those who, it seemed, would do better to believe what they were told by their intellectual betters.

Notes

1. Robert Young, "The Impact of Darwin on Conventional Thought," in *Darwin's Metaphor: Nature's Place in Victorian Culture*, 1–22 (Cambridge: Cambridge University Press, 1985); Frank M. Turner, *Between Science and Religion: The Reaction to Scientific Naturalism in Late Victorian England* (New Haven, CT: Yale University Press, 1974); Frank Turner, "Victorian Scientific Naturalism and Thomas Carlyle," *Victorian Studies* 18 (1975): 325–43; and Frank Turner, "The Victorian Conflict between Science and Religion: A Professional Dimension," *Isis* 69 (1978): 356–76.

2. Bernard Lightman, *Evolutionary Naturalism in Victorian Britain: The "Darwinians" and Their Critics* (Farnham, UK: Ashgate, 2009), xii–xvi.

3. T. H. Huxley, prologue to *Essays upon Some Controverted Questions* (London: Macmillan, 1892), 11, 22.

4. Ruth Barton, "'An Influential Set of Chaps': The X-Club and Royal Society Politics, 1864–85," *British Journal for the History of Science* 23 (1990): 53–81; and Ruth Barton, "'Huxley, Lubbock, and Half a Dozen Others': Professionals and Gentlemen in the Formation of the X-Club, 1851–1864," *Isis* 89 (1998): 410–44.

5. Alan W. Brown, *The Metaphysical Society: Victorian Minds in Crisis, 1869–1880* (New York: Columbia University Press, 1947), 20–23.

6. R. H. Hutton, "The Metaphysical Society: A Reminiscence," *Nineteenth Century* 18 (1885): 177.

7. J. C. MacDonnell, *Life and Correspondence of William Connor Magee*, 2 vols. (London: Isbister, 1896), 1:284.

8. J. Drummond and C. B. Upton, eds., *The Life and Letters of James Martineau*, 2 vols. (London: James Nisbet, 1902), 2:368.

9. Bernard Lightman, *The Origins of Agnosticism: Victorian Unbelief and the Limits of Knowledge* (Baltimore: Johns Hopkins University Press, 1987), 146–76. On the early uses of the term, see also Bernard Lightman, "Huxley and Scientific Agnosticism: The Strange History of a Failed Rhetorical Strategy," *British Journal for the History of Science* 35 (2002): 271–89.

10. Hutton, "Metaphysical Society," 182–91.

11. For a list of papers, see Brown, *Metaphysical Society*, 318–39. A nearly complete set of papers is in the Bodleian Library, Oxford.

12. W. K. Clifford, "The Ethics of Belief" [11 April 1876], published in *Contemporary Review* 29 (1877): 289–309. Whenever possible, I have quoted passages that are identical in the original papers and the published versions.

13. F. Pollock, introduction to W. K. Clifford, *Lectures and Essays*, ed. L. Stephen and F. Pollock, 2 vols. (London: Macmillan, 1879).

14. Clifford, "Ethics of Belief," 289–90, 292. On the convergence of empirical and ethical imperatives in Clifford's essay, see Helen Small, "Science, Liberalism and the Ethics of Belief: The *Contemporary Review* in 1877," in *Science Serialized: Representations of the Sciences in Nineteenth-Century Periodicals*, ed. Geoffrey Cantor and Sally Shuttleworth, 239–57 (Cambridge, MA: MIT Press, 2004).

15. J. W. Gruber, *A Conscience in Conflict: The Life of St. George Jackson Mivart* (New York: Columbia University Press, 1960); and Turner, *Between Science and Religion.*

16. St. G. J. Mivart, "What Is the Good of Truth?" [13 June 1876], Bodleian Library, much expanded as "Force, Energy and Will," *Nineteenth Century* 3 (1878): 933–48. See 945–47.

17. J. F. Stephen, "On the Utility of Truth" [11 February 1879], published in James McCarthy, ed., *The Ethics of Belief Debate* (Atlanta: Scholar's Press, 1986), 41–46. Stephen had previously replied to Mivart in his paper "What Is a Lie?" [11 July 1876].

18. N. Annan, *Leslie Stephen: The Godless Victorian* (London: Weidenfeld and Nicolson, 1984).

19. L. Stephen, "Belief and Evidence," 12 June 1877, Bodleian Library.

20. J. H. Newman, *Apologia Pro Vita Sua: Being a Reply to a Pamphlet Entitled "What, Then, Does Dr. Newman Mean?"* (London: Longman, Green, Longman, Roberts, and Green, 1864); and J. H. Newman, *History of My Religious Opinions*, 2nd ed. (London: Longman, Green, Longman, Roberts, and Green, 1865). The controversy with Kingsley is discussed in Frank Turner, *John Henry Newman: The Challenge to Evangelical Religion* (New Haven, CT: Yale University Press, 2002). On Newman's theology, see especially J. H. Walgrave, *Newman the Theologian: The Nature of Belief and Doctrine as Exemplified in His Life and Works*, trans. A. V. Littledale (London: Sheed & Ward, 1960); and B. M. G. Reardon, *From Coleridge to Gore: A Century of Religious Thought in Britain* (London: Longman, 1971).

21. J. H. Newman, *An Essay in Aid of a Grammar of Assent* (London: Burns, Oates, 1870), 155, 374, 163.

22. Ibid., 342.

23. R. H. Hutton, "On the Relation of Evidence to Conviction" [13 November 1877], in McCarthy, *Ethics of Belief*, 113–18.

24. H. Sidgwick, "Incoherence of Empirical Philosophy" [14 January 1879], published in *Mind* 7 (1882): 543.

25. See Brown, *Metaphysical Society*, 108–47.

26. Minute Book of the Metaphysical Society, in Brown, *Metaphysical Society*, 26.

27. W. Bagehot, "The Emotion of Conviction" [13 December 1870], published in *Contemporary Review* 17 (1871): 33.

28. Ibid., 34.

29. Ibid., 35. See J. S. Mill, in James Mill, *Analysis of the Phenomena of the Human Mind*, 2 vols. (London: Baldwin and Cradock, 1829), 1:412–13; and Alexander Bain, *The Emotions and the Will* (London: Longmans, Green, 1859), 20–21, 505–38, esp. 536–37.

30. W. B. Carpenter, "On the Doctrine of Human Automatism" [17 November 1874], published in *Contemporary Review* 25 (1874): 397–416.

31. W. B. Carpenter, *Principles of Mental Physiology* (London: King, 1874), 226–27. On the efforts of Carpenter and others to reconcile the ethical status of the will with new deterministic and materialistic models of mind, see Lorraine Daston, "British Responses to Psychophysiology, 1860–1900," *Isis* 69 (1978): 192–208.

32. T. H. Huxley, "Has a Frog a Soul; and of What Nature Is That Soul, Supposing It to Exist?" [8 November 1870], in A. Russell, *Papers Read at the Meetings of the Metaphysical Society* (London: privately printed, 1896), 82–83.

33. Ibid.

34. H. Manning, "The Relation of the Will to Thought" [11 January 1871], published in *Contemporary Review* 16 (1871): 475. See also W. B. Carpenter, "The Physiology of the Will," *Contemporary Review* 17 (1871): 192–217.

35. Clifford, "Ethics of Belief."

36. W. James, *The Principles of Psychology*, 2 vols. (London: Macmillan, 1890); and W. James, "The Will to Believe," in *The Will to Believe and Other Essays in Popular Philosophy*, 1–31 (London: Longmans, Green, 1897).

37. "The Questioning Mania," *Mind* 1 (1876): 413.

38. James, "Will to Believe," 30.

39. Russell, *Papers Read at the Meetings of the Metaphysical Society*, 84–85.

40. Drummond and Upton, *Life and Letters of James Martineau*, 2:368–69.

41. J. D. Dalgairns, "On the Theory of the Soul" [16 March 1870], published in *Contemporary Review* 16 (1870): 16–43; and W. G. Ward, *William George Ward and the Catholic Revival* (London: Longmans, Green, 1912), 309–10.

42. T. H. Huxley, "Mr. Darwin's Critics," *Contemporary Review* 18 (1871): 443–76; and St. G. J. Mivart, *On the Genesis of Species* (London: Macmillan, 1871).

43. T. H. Huxley and H. A. Huxley to Darwin, 20 September 1871, in *The Correspondence of Charles Darwin*, ed. F. Burkhardt et al., 20 vols. (Cambridge: Cambridge University Press, 1985–2012), 19:586; [St. G. J. Mivart], "Darwin's *Descent of Man*," *Quarterly Review* 131 (1871): 47–90; and St. G. J. Mivart, "Ape Resemblances to Man," *Nature*, 20 April 1871, 481.

44. On the challenge to gentlemanly relations posed by Mivart's harsh reviews, see Gowan Dawson, *Darwin, Literature and Victorian Respectability* (Cambridge: Cambridge University Press, 2007), 77–81.

45. J. Martineau, "Is There Any 'Axiom of Causality'?" [15 June 1870], published in *Contemporary Review* 14 (1870): 636–44; and J. Martineau to C. Wicksteed, 26 August 1870, in Drummond and Upton, *Life and Letters of James Martineau*, 2:374.

46. T. H. Huxley, "Agnosticism," *Nineteenth Century* 25 (1889): 169–94.

47. H. Sidgwick, *The Ethics of Conformity and Subscription* (London, 1870). On the wider context of debates on clerical subscription, see James Livingston, *The Ethics of Belief: An Essay on the Victorian Religious Conscience* (Tallahassee, FL: American Academy of Religion, 1974).

48. J. Martineau to H. Sidgwick, 18 March 1870, in Drummond and Upton, *Life and Letters of James Martineau*, 1:451–52.

49. Gowan Dawson, Richard Noakes, and Jonathan R. Topham, introduction to *Science in the Nineteenth-Century Periodical: Reading the Magazine of Nature*, ed. Geoffrey Cantor et al. (Cambridge: Cambridge University Press, 2004), 19–20.

50. Helen Small, "Liberal Editing in the *Fortnightly Review* and the *Nineteenth Century*," *Publishing History* 53 (2003): 75–96.

51. On the spectacle of liberalism created by Knowles's departure from the *Contemporary Review* following its takeover by evangelical patrons, see Small, "Science, Liberalism and the Ethics of Belief," 248–49.

52. [J. Knowles], *Nineteenth Century* 1 (1877): 331.

53. Brown, *Metaphysical Society*, 190.

54. Gowan Dawson, "The *Review of Reviews* and the New Journalism in Late-Victorian Britain," in *Science in the Nineteenth-Century Periodical*, 172–95.

55. Ibid., 180–81. On Huxley's debates over land nationalization and copyright, see Paul White, *Thomas Huxley: Making the "Man of Science"* (Cambridge: Cambridge University Press, 2003), 141–48.

56. J. F. Stephen, *Liberty, Equality, Fraternity* (London: Smith, Elder, 1873), 52–75.

57. W. Bagehot, "The Metaphysical Basis of Toleration" [10 February 1874], Arthur Russell to Odo Russell, 11 February 1874, in Russell, *Papers Read at the Meetings of the Metaphysical Society*, 83.

58. Walter Houghton, *The Victorian Frame of Mind, 1830–1870* (New Haven, CT: Yale University Press, 1957), 54–109, 218–62; and Gertrude Himmelfarb, *Victorian Minds* (London: Weidenfeld and Nicolson, 1968), 300–313.

59. For example, Basil Wiley, *More Nineteenth Century Studies: A Group of Honest Doubters* (London: Chatto and Windus, 1956); Owen Chadwick, *The Secularization of the European Mind in the Nineteenth Century* (Cambridge: Cambridge University Press, 1977); and H. Stuart Hughes, *Consciousness and Society: The Reorientation of European Social Thought, 1890–1930* (London: MacGibbon and Kee, 1959).

60. James Moore, "Theodicy and Society: The Crisis of the Intelligentsia," in *Victorian Faith in Crisis: Essays on Continuity and Change in Nineteenth-Century Religious Belief*, ed. Bernard Lightman and Richard Helmstadtler, 153–86 (Basingstoke, UK: Macmillan, 1990).

61. Stefan Collini, *Public Moralists: Political Thought and Intellectual Life in Britain, 1850–1930* (Oxford: Clarendon Press, 1991), 211–13.

9

Where Naturalism and Theism Met:
The Uniformity of Nature

MATTHEW STANLEY

In the waning of years of both his life and the nineteenth century, T. H. Huxley wrote of the earliest "thinking men" trying to understand the world around them. These primitive humans found that

> there is order amidst the confusion, and that many events take place according to unchanging rules. To this region of familiar steadiness and customary regularity they gave the name of Nature. But, at the same time, their infantile and untutored reason, little more, as yet, than the playfellow of the imagination, led them to believe that this tangible commonplace, orderly world of Nature was surrounded and interpenetrated by another intangible and mysterious world, no more bound by fixed rules than, as they fancied, were [their own] thoughts and passions.

This was the world of the supernatural, and Huxley asserted that from the earliest times of which we have any knowledge, "Naturalism and Supernaturalism have consciously or unconsciously, competed and struggled with one another."[1]

He told the story of human progress as the elimination of the supernatural from man's thoughts, and pointed to his century as the era when naturalism finally gained the upper hand: "The stream of tendency toward Naturalism . . . has, of late years, flowed so strongly that even the Churches have begun, I dare not say to drift, but, at any rate, to swing at their moorings." Naturalistic science was an enemy that had surrounded supernatural theology, and was daily driving it back.[2]

Huxley's narrative of the triumph of scientific naturalism in the second half of the nineteenth century explicitly gave credit to the uniformity of nature: the principle that events were always consistent, regular, and invariably governed by natural laws. The retreat of supernaturalism, he said, was identical with the growth of uniformity. But the growth of uniform thinking within Victorian science was not always seen as a threat by theists. Belief in uniformity often existed quite comfortably alongside belief in the supernatural, and theistic scientists made it an important part of their thinking. Contrary to Huxley's militaristic narrative, uniformity was not the dividing line between naturalistic and supernaturalistic thinking—rather, it was shared and embraced by both groups. Despite their bitter opposition, theists and naturalists not only accepted the principle in common, but each saw it as being closely tied to their worldview and argued for its incompatibility with their opponents' claims.

The importance of uniformity is brought into sharp relief through the use of an important aspect of recent historiography of science: the question of practice.[3] Huxley and the scientific naturalists needed to make the argument that the *way* they did science was different from the theists—that is, that their scientific *practice* was novel. And reliance on uniformity, they said, was what made them different. But, as we shall see, this fundamental aspect of their practice was not unique, and in fact it was the *lack* of difference that was key to the eventual victory of the scientific naturalists. Their strategies, strangely, relied on common ground with their enemies.

Uniformity

The term *scientific naturalism* was coined by Huxley in 1892, but its ideas, methods, and attitudes became widespread decades before.[4] In particular, the concept of uniformity that the scientific naturalists claimed as their own was well established before he and his allies began their ascendancy in Victorian science. The concept is straightforward: uniformity is the claim that the laws of nature are the same everywhere and everywhen in the universe, and that those laws do not break down or lapse anywhere in time or space. For example, uniformity demands that scientists conceive of the earth's surface being formed in distant ages by the same processes of matter and energy at work today—the crucial contribution of Charles Lyell and the uniformitarian geologists. This idea emerged from complex historical processes and comes in several different varieties, but by the mid-nineteenth century there was a core version of uniformity.[5]

A useful place to begin exploring the emergence of uniformity in the early nineteenth century is John Herschel's *Preliminary Discourse* (1830), a formative text for a generation or two of scientists. It helped articulate scientific orthodoxy before the scientific community split into theistic and naturalistic camps, and remained important for both. Herschel stressed that the regularity of natural phenomena was an essential part of the universe: it is "impressed on us from our earliest infancy . . . that events do not succeed one another at random, but with a certain degree of order, regularity, and connection."[6] Those phenomena that "happen uniformly and invariably" can then become the bases of laws of nature, which were both the proper focus of attention for the natural philosopher and the goal of his investigations.[7]

Herschel addressed directly the uniformity of these laws. Was it possible that laws were "subject to mutation" over time, or that they were restricted in their application? The clear answer was no: a natural law functions "uninterruptedly, for ages beyond all memory, [and gives] a strong expectation that it will continue to do so in the same manner; and thus our notion of an *order of nature* is originated and confirmed."[8] Herschel's position that the essence of science was the search for, and study of, universal, uniform laws was accepted by every scientist discussed here, whether theist or naturalist. Precisely what uniformity meant, and how one should think about it, was more complicated.

Naturalists

Scientific naturalism had its most important locus in the X Club.[9] This informal network (essentially a dining club) of young, ambitious scientists sought to professionalize their discipline and increase its social and cultural standing. A critical part of this effort was the exclusion of theology, the supernatural, and the clergy from science. Barton asserts, "They opposed all suggestions that there were supernatural interventions in the natural order and any attempts to constrain scientific investigation within theologically-determined boundaries."[10]

The triumphalist story of scientific naturalism—that science only became modern once it cast off the albatross of dogmatic theology—was precisely the story promoted by the X Club. Its leaders spent a great deal of time and energy discussing the foundations of science and explaining how those foundations excluded the supernatural. And the most important idea supporting that exclusion was uniformity.

Consider two of the most eloquent and aggressive spokesmen for this position, T. H. Huxley and John Tyndall. For Huxley, the purpose of science was to uncover the orderly system of the universe: "The fundamental axiom of scientific thought is that there is not, never has been, never will be, any disorder in nature." It would be impossible for science to admit "the occurrence of any event which was not the logical consequence of the immediately antecedent events, according to these definite, ascertained, or unascertained rules which we call the 'laws of nature.'"[11] The laws of nature admit no exceptions or interruptions.

This uniformity then allowed scientific discussion of the distant past, because one could extrapolate the past based on forces visible today (e.g., erosion).[12] Huxley argued that even violent, apparently nonuniform events such as volcanoes and earthquakes should be thought of as uniform phenomena:

> The working of a clock is a model of uniform action; good time-keeping means uniformity of action. But the striking of the clock is essentially a catastrophe; the hammer might be made to blow up a barrel of gunpowder, or turn on a deluge of water; and, by proper arrangement, the clock, instead of marking the hours, might strike at all sorts of irregular periods, never twice alike, in the intervals, force, or number of its blows. Nevertheless, all these irregular, and apparently lawless, catastrophes would be the result of an absolutely uniformitarian action.[13]

A natural law was universal and could not, by definition, be interrupted: "To speak of the violation, or the suspension, of a law of nature is an absurdity. All that the phrase can really mean is that, under certain circumstances the assertion contained in the law is not true; and the just conclusion is, not that the order of nature is interrupted, but that we have made a mistake in stating that order."[14] Any apparent disruption to the orderliness of nature was an indication of a human error, and nothing more.

But, of course, humans had throughout history claimed to see such disruptions, and attributed them to the intervention of a higher power. Huxley's position was that the progress of all of the sciences could be measured by the rejection of divine intervention in favor of natural causes.[15] What made a discipline scientific was its insistence on unbroken law, particularly in rejection of the supernatural. Huxley referred to the order of nature in almost every essay or lecture, and explicitly opposed it to the claims of theology.[16]

Huxley's friend and ally John Tyndall also spoke vigorously of the power of uniformity to banish an intervening God, particularly in his famous "Belfast Address" in 1874. A valid natural law, he said, "asserts itself everywhere in nature."[17] The advance of science was steadily crushing the mutability of nature demanded by religious believers:

> Now, as science demands the radical extirpation of caprice and the absolute reliance upon law in nature, there grew with the growth of scientific notions

a desire and determination to sweep from the field of theory this mob of gods and demons, and to place natural phenomena on a basis more congruent with themselves.[18]

Tyndall could hardly be clearer. The uniformity of natural laws left no room for divine action in science. A sterling example of the military metaphor of science and theology is seen in the martial language used:

> The impregnable position of science may be described in a few words. We claim, and we shall wrest, from theology the entire domain of cosmological theory. All schemes and systems which thus infringe upon the domain of science must, *in so far as they do this*, submit to its control, and relinquish all thought of controlling it. Acting otherwise proved disastrous in the past, and is simply fatuous to-day.[19]

Science, as a complete scheme of the universe, could have no interaction with theology other than accepting its surrender. Before the advent of science, Tyndall said the unlearned masses had no option other than filling the world with "witchcraft, and magic, and miracles, and special providences." The power of natural laws would simply squeeze the world until nothing else remained: "The law of gravitation crushes the simple worshipers of Ottery St. Mary, while singing their hymns, just as surely as if they were engaged in a midnight brawl."[20]

The subtext of these claims was that uniformity not only restricts theology from entering science, but that uniformity can only be justified in a world without divine intervention. How can scientists plan and conduct an experiment if they must worry that Jehovah will change the constants of nature? Uniformity can only be justified, then, if everyone agrees a priori that there can be no divine interventions.

Theists

While scientific naturalism made serious inroads into Victorian science, it did not seem to have much impact on theistic scientists themselves. The claims of Huxley and Tyndall that uniformity demanded a science completely free of divine action did not drive the theists to secularism. And yet these theistic scientists were in total agreement with the naturalists that uniformity was critical to the advance of science. How could they embrace the naturalistic methods but not the naturalistic conclusions?[21]

The answer is that the theists saw uniformity as *their* impregnable position, not Tyndall's. The consistency of natural laws over time and space was a sign pointing toward God, not warding him off. John Herschel, the great authority of Victorian science, wrote that natural laws had their origin in the "Divine Author of the universe" and that the uniformity of those laws came from "the constant exercise of his direct power in maintaining the system of nature, or the ultimate emanation of every energy which material agents exert from his immediate will, acting in conformity with his own laws."[22] The laws of nature were only stable because of God's constant and ubiquitous action. If matter and energy were left to their own devices, the universe would be a place of chaos. The orderliness of natural phenomena could only be explained, the theists argued, if God ensured that it was so. Victorian theists were following in a long tradition of allying God and the laws of nature that was widespread and productive. This was explicit in natural theology as well as implicit in the work of the scientists themselves. Natural laws were seen as instances of divine fiat, and they were constant because God is consistent in his actions.[23]

Despite the expectations of scientific naturalism, this link between uniformity and divine action remained widespread throughout the nineteenth century, and can be seen in the work of some of the most important scientists of the period. Consider William Thomson, better known as Lord Kelvin. Kelvin's debates with geologists and biologists over the age of the earth earned him a reputation as a defender of the old guard unwilling to adapt to modern scientific methods and perspectives. But Kelvin's position in these debates was actually built on precisely the same methodology as his opponents: uniformity.

It is common to think of attacks on geological estimates of the earth's age as relying on attacks on uniformity: radioactive dating is unreliable, or uniformitarian processes were disrupted by directly creative acts such as Noah's flood. It is easy to interpret Kelvin's arguments in this light, particularly with his statements such as "There cannot be uniformity."[24] However, this is misleading. He was actually attacking a particular kind of geological uniformitarianism that allowed for an extremely old earth, not uniformity in general, and his position was essentially that geologists were not being uniform *enough*. Huxley, and others, he said, assumed that geological forces should be thought of as being constant deep into the past, virtually forever. But this constancy of geology was made impossible by the constancy of physics: the second law of thermodynamics demanded that deep in the past, the surface of the earth would appear and behave quite differently from how it does today.

Far from attacking uniformity, Kelvin was vigorously defending uniformity. He believed strongly that natural laws were permanent and universal.[25] Indeed, he was actually attacked for this assumption by, among others, his own student John Perry.[26] Kelvin thought the whole scientific enterprise relied on the idea that currently observed natural laws could be extended throughout time: "The essence of science, as is well illustrated by astronomy and cosmical physics, consists in inferring antecedent conditions, and anticipating future evolutions, from phenomena which have actually come under observation."[27] He completely rejected the possibility of violations of natural laws, particularly supernatural intervention: "If a probable solution [to any scientific problem], consistent with the ordinary course of nature can be found, we must not invoke an abnormal act of creative power."[28]

Kelvin was a Latitudinarian Christian, and he was adamant that a uniform universe was still perfectly consonant with a religious worldview. Writing with his collaborator P. G. Tait, he said,

> We have the sober scientific certainty that heavens and earth shall "wax old as doth a garment" [Psalm 102:26]; and that this slow progress must gradually, by natural agencies which we see going on under fixed laws, bring about circumstances in which "the elements shall melt with fervent heat" [2 Peter 3:10].[29]

They made the case that a world running by "natural agencies" was an idea with strong scriptural support. The reason for this, of course, was that Kelvin saw natural laws as coming from God. Only his creative intelligence allowed for uniformity in the first place.[30]

Kelvin's fellow pillar of Victorian physics James Clerk Maxwell had similar feelings, although he was an evangelical. He clearly worked with the assumption of uniformity of natural laws in both time and space. He explained that a basic principle of science was that "place and time are not among the conditions which determine natural processes."[31] Maxwell did not welcome violations of uniformity. He rejected otherwise impressive hypotheses for the cause of gravity because they suggested interruptions of conservation of energy.[32]

Maxwell was a fairly conservative Victorian evangelical Christian, who took Scripture quite seriously. And as with Kelvin, Maxwell thought that the uniformity of the cosmic order had greater significance:

> I think that each individual man should do all he can to impress his own mind with the extent, the order, and the unity of the universe, and should carry these ideas with him as he reads such passages as the 1st Chap. of the Ep. to Colossians . . . , just as enlarged conceptions of the extent and unity of the world of life may be of service to us in reading Psalm viii.; Heb. ii. 6, etc.[33]

The unity of nature was a theological concept as well as a scientific one. The scriptural passages Maxwell referred to here emphasized God's role as creator of the natural world ("For by him were all things created, that are in heaven, and that are in earth, visible and invisible") and the awe that God designed his creation for man ("What is man, that thou art mindful of him?").[34] Thus, Maxwell was powerfully linking the unity and order of nature not just with divine creation itself, but also with the role of man in that creation. In the same letter he argued that we can see "wisdom and power" in the uniformity of natural laws just as effectively as in the beneficial adaptations of living creatures: "uniformity, accuracy, symmetry, consistency, and continuity of plan are as important attributes as the contrivance of the special utility of each individual thing."[35]

In addition to uniformity being another premise in the argument from design, it was a tool given to man by God to fulfill the commandment to subdue nature. Maxwell instructed his students that once they understood the constancy and universality of natural laws, they would

> begin to understand the position of man as the appointed lord over the works of Creation and to comprehend the fundamental principles on which his dominion depends which are these—To know, to submit to, and to fulfil, the laws which the Author of the Universe has appointed. Attend to these laws and keep them, you succeed, break them, you fail and can do nothing.[36]

Natural laws were a manifestation of the divine will, just like biblical commandments. The principles that ran the steam engine and created the wealth of industrial Britain were gifts from God. Conscientious people had an obligation to be mindful of the immutability of both natural and moral law.

Maxwell's best-known statements on the uniformity of nature appeared in his famous lecture "Molecules" at the British Association for the Advancement of Science in 1873. He argued that new techniques of spectroscopy showed that in the sun "there are molecules vibrating in as exact unison with the molecules of terrestrial hydrogen as two tuning-forks tuned to concert pitch, or two watches regulated to solar time. . . . Now this absolute equality in the magnitude of quantities, occurring in all parts of the universe, is worth our consideration."[37] Hydrogen in distant stars, or liberated from rocks buried since time immemorial, was identical to that in the Cavendish lab. This incredible uniformity among matter scattered through space and time indicated the hand of a divine manufacturer: again, uniformity could only be explained through God.

Religious defenses of uniformity were found in the life sciences as well. William Carpenter, the pioneering physiologist and liberal Unitarian, argued that "orderly uniformity" was the distinguishing feature of a law of nature, and that this uniformity revealed the law's divine origins:

It is thus that when we pass from the sphere of human government to that of the Divine, and speak of the universe as "governed" by the "laws" of a supreme Ruler, we mean that his power is exerted, not like that of an arbitrary potentate who changes his course of action as his own caprice or passion may direct, but like that of a benevolent sovereign whose rule is in uniform and orderly conformity with certain fixed principles, originally determined as conducive to the welfare and happiness of his people.[38]

Even the great weapon of the scientific naturalists, evolution, was fit comfortably into such schemes. Belief in theistic evolution—that is, evolution guided or supported in some way by God—was widespread among both scientists and Christians interested in science, and many commented on how Darwin's ideas had extended uniformity throughout the world of life. Frederick Temple, the future archbishop of Canterbury, stated, "Once more, the doctrine of Evolution restores to the science of Nature the unity which we should expect in the creation of God."[39] The Duke of Argyll, who did important work in both science and politics, argued that God's choice to create species via natural laws instead of direct fiat was no slight of his power: creation by process "is Creation still."[40]

In sum: far from uniformity being antithetical to religious thinking, many scientists and philosophers concluded that uniformity only makes sense in a theistic world. Without an ordering force (i.e., God), one would expect the universe to be a mishmash of chaotic events. The only guarantee for constancy of the laws of nature was the intent of the lawgiver. Temple and Argyll acknowledged that the uniformity assumption was critical for science ("on no other assumption can Science proceed at all"), that it was justified both by the results of science ("This idea is a product of that immense development of the physical sciences which is characteristic of our time") and by simple experience ("Millions on millions of observations concur in exhibiting this uniformity").[41] The theists did not reject empiricism, reasoning, testing, or theorizing. Rather, they said all of those things were dependent on God, and pointed to natural laws as showing his role in the universe.[42]

The Miracle Problem

Despite this, the scientific naturalists thought they had one attack for which there was no counter. Miracles, they said, are the essence of Christianity. And a miracle, it seemed, must be a violation of a natural law, and therefore a violation of uniformity, and thus cannot be consonant with science.[43] One must choose, then, between science and Christian theology. This was a vulnerable point emphasized repeatedly by Victorian scientific naturalists, many of whom were directly inspired by David Hume. Huxley's battles with Gladstone turned on precisely this issue, and in the infamous "prayer gauge" debate (an 1870s controversy over whether one could scientifically determine the efficacy of prayer), Tyndall was merciless in drawing this line.[44] He said that once science had demonstrated the uniformity of nature, "the age of miracles is past."[45] The only way out, he said, was to retort, "How do you know that a uniform experience will continue uniform? You tell me that the sun has risen for six thousand years: that is no proof that it will rise tomorrow; within the next twelve hours it may be puffed out by the Almighty."[46] He said someone attacking uniformity in this way, however, could barely function in the normal world, and had no reason to believe that Jack and the beanstalk was not a true story, since perhaps the natural laws governing bean growth had been suspended at some time. The Victorian naturalists felt they had hemmed the theists into an inescapable dilemma. To do science, the universe needed to be uniform. But uniformity forbade miracles, and without miracles, what was Christianity? Which, then, would the theists sacrifice: science or theology? As we have already seen, the theistic scientists refused to discard either, and had robust interpretations of uniformity in a religious framework. So it should be no surprise that they drew upon religious uniformity as a resource to explain miracles in an orderly world.

There was widespread agreement among theistic scientists that apparent violations of natural law were illusory. Many other Christians agreed—Temple declared, "There may be instances where this Order is apparently broken, but really maintained, because one physical law is absorbed in a higher."[47] That is, an event that appeared to be outside the laws of nature actually was lawful, but it simply obeyed a law of which humans were not yet aware. An analogy might be to consider someone who understood the law of gravity, but not that of buoyancy. A hot air balloon would appear to this person to be miraculous, but a better-informed observer would understand that no laws had been broken.

What, then, of the supernatural? Would not religious believers need violations of natural law to be assured of the existence of supernatural forces? One of the prices of this strategy was that, in an important sense, the category of the supernatural faded away (or was at least redefined). Argyll acknowledged that if something happened in our world, then uniformity demands that it be the result of natural law:

> The Reign of Law in Nature is, indeed, so far as we can observe it, universal. But the common idea of the Supernatural is that which is at variance with Natural Law, above it, or in violation of it. Nothing, however wonderful, which happens according to Natural Law, would be considered by any one as

Supernatural. The law in obedience to which a wonderful thing happens may not be known; but this would not give it a supernatural character, so long as we assuredly believe that it did happen according to *some* law.[48]

If scientists had total knowledge of all natural laws, then nothing would ever appear supernatural. What seems to be inexplicable is actually only temporarily obscured. The Duke of Argyll pointed out that the technological advances of the Victorian period allowed completely normal humans to achieve feats that earlier generations would have called supernatural (such as sending a message instantly across the Atlantic Ocean).[49] Perhaps, then, Jesus's healings in the Gospels simply relied on laws of medicine not yet understood.[50]

Writing earlier in the century, Baden Powell warned that asserting divine causation for apparent gaps in uniformity would be dangerous for religion, because "enlarged discovery shall disclose the connection and explanation of these appearances by regular laws, [and] their argument for a Deity will fall to the ground!" Reducing God to only a "confession of ignorance," as in the case of Newton's arguments for planetary stability, was bad science and bad religion. Some critics of this position claimed it restricted God's action, saying that a God who could not intervene in special circumstances was no God at all. But, again, it was uniformity, not interruptions of it, that truly showed us the nature of things: "To speak of apparent anomalies and interruptions as *special* indications of the Deity, is altogether a mistake. In truth, so far as the *anomalous* character of any phenomenon can affect the inference of presiding Intelligence at all, it would rather tend to *diminish* and detract from that evidence."[51]

How, then, does God watch over his creation and enact his plans, if not through interruptions of the natural order? As Argyll asserts, with natural laws: "There is nothing in Religion incompatible with the belief that all exercises of God's power, whether ordinary or extraordinary, are effected through the instrumentality of means—that is to say, by the instrumentality of natural laws brought out, as it were, and used for a Divine purpose."[52] God created laws as the means by which he exercises power in the world, like a craftsman who builds his own tools. The Deity could manipulate natural laws in a variety of ways without violating their essence, and could produce any of the fantastic events recorded in Scripture. However, how could uniform, unchanging laws produce singular events that appear to be obvious disruptions of nature? As early as the 1830s, Charles Babbage found a solution to this problem in a parlor trick performed with his calculating machine. The machine, which of course ran on fixed rules, could produce a steady, regular sequence of numbers only to make suddenly a great jump—thus demonstrating that what appear to be exceptional events could be easily generated by fixed laws.[53] This interpretation of miracles was widespread throughout the Victorian period and across sectarian lines. It was an efficacious and popular strategy for retaining God's action in a uniform world.

But if God only works through natural laws, in what sense can these events be miracles? Argyll argued that the marker of a miracle is not the presence of supernatural causes, but rather that it has its origin in divine intent. This view, he said, was perfectly harmonious with Scripture and allowed defense of all the essential events of Christianity.[54] Similarly, Frederick Temple argued that even if science were to someday give an explanation of all the miracles in the Bible, it would not in any way change their role in revelation. The miracle could be in their timing, or intent, or effect, rather than in their breach of uniformity.[55] This fit well with a traditional Protestant distinction between miracles, which required an objective witness to provide proof of supernaturalism, and special providence, which appeared to be normal events—except when viewed through the eyes of faith.[56] So this move would essentially eliminate the category of formal miracles and subsume all divine actions under special providence. Miracles in a uniform universe might no longer be particularly miraculous, but the critical issue could be resolved. A providential God, Temple declared, did not have to be incommensurable with uniformity, and therefore with science:

> Science will continue its progress, and as the thoughts of men become clearer it will be perpetually more plainly seen that nothing in Revelation really interferes with that progress. It will be seen that devout believers can observe, can cross-question nature, can look for uniformity and find it, with as keen an eye, with as active an imagination, with as sure a reasoning, as those who deny entirely all possibility of miracles and reject all Revelation on that account. The belief that God can work miracles and has worked them, has never yet obstructed the path of a single student of Science.[57]

The important religious function of miracles was retained, along with the power and potential of science. Some twenty-first-century theologians are proposing similar strategies as well.[58]

Scientific Practice across the Religious Boundary

I have shown how uniformity did not, as the Victorian scientific naturalists suggested, necessarily indicate the absence of religion in science. Religious scientists working at the time of the emergence of scientific naturalism, as well as Christian theologians, agreed completely with the uniformity of natural laws as a methodological precept. Indeed, they thought that precept only made sense in a world with an active Deity. Although theistic scientists saw it as a religious concept, and naturalistic scientists saw it as an areligious (or even antitheological) one, virtually no practicing scientists rejected it. This created a common space for the two groups to work side by side despite the fact that they disagreed on an enormous range of subjects (for an interesting example of such cooperation, see Ruth Barton's contribution to this volume). It is quite remarkable that, despite the jeremiads offered by each side against the other, the scientific community continued to function smoothly. No scientists suggested expelling James Clerk Maxwell from the BAAS; no scientists proposed evicting John Tyndall from the Royal Institution. This was possible because both groups in fact agreed on almost everything needed to actually *do* science, as the current example of the uniformity of nature illustrates. They might disagree bitterly about specific claims of fact or interpretation, but the activity of scientific research per se was surprisingly smooth.

There was a genuine shift within the scientific community marked by scientific naturalism, but it had little impact on scientific practice. In the 1820s, it would have been difficult to find a scientist at Cambridge University who was not religious. By the 1920s, the presence of a publically religious scientist at Cambridge was worthy of note.[59] But if Kelvin had time-traveled fifty years into the future, he would have been perfectly comfortable with the assumptions still driving science. The transition from theistic science to naturalistic science changed little in scientific practice, and, indeed, was made smoother by the essential similarities between the two positions because the actual methodologies did not have to be altered.

So if the scientific naturalists did not come to dominance because of their methodological superiority, how did they win? They did not have a unique way of thinking about science, and I do not think they convinced their opponents that their approach was superior.[60] This question requires more detailed research to give a thorough answer, but I would like to suggest a tentative possibility: the X Club and allies seized the means of production. Production, that is, of the next generation of scientists.

Why Did the Naturalists Win?

Members of the X Club worked very hard to place themselves in locales of scientific power.[61] One easy measure of their success was the staggering number of leadership positions they occupied in scientific societies.[62] But even beyond the personal achievements of its members, the group was able to have an enormous impact on the future of science by focusing on science education.

Huxley was the exemplar for this. He wrote and spoke a great deal about science education (even serving on the first London School Board), and thought strategically about how best to reform science to meet his naturalistic goals. A major part of his strategy was to shape the next generation of science teachers, so as to start a pipeline of like-thinking scientists. When the 1870 Elementary Education Act was passed, it created a huge demand for science teachers and, by implication, training for those teachers. Huxley pounced on this opportunity, writing textbooks and lab manuals, and running summer courses for these new teachers that would inculcate both teachers and students with a naturalistic, secular worldview.[63] He was not reticent to share his plans in colorful language: to one correspondent he described "a course of instruction in Biology which I am giving to Schoolmasters—with the view of converting them into scientific missionaries to convert the Christian Heathen of these islands to the true faith."[64] Many X Club members were also examiners for the Department of Science and Art, and thus could directly control what qualified someone to teach science.[65]

Two of the other chapters in this volume bring out the details of Huxley's work along these lines. Bernard Lightman documents how his energies on the Devonshire Commission were largely focused on elementary education and educators, trying to create an environment in which students would grow up with scientific naturalism. James Elwick's analysis shows how Huxley crafted examinations as a kind of scientific catechism for naturalism in which, for example, students were expected to discuss the human body as a machine. Huxley was clear that he intended the exams to repel "parsonic influence."[66]

This strategy worked on the university level as well. Huxley was deeply involved in the creation of biology professorships all over Britain in the 1870s and 1880s, and worked hard to influence who received those positions.[67] His goal was to place candidates who were ideologically sound (i.e., purely naturalistic) as well as scientifically talented. In this he was quite successful. In a short number of years Huxley had already managed to place his students, allies, and demonstrators at University College London, Edinburgh, Leeds, Johns Hopkins, New Zealand, and even Anglican institutions such as Oxford and Cambridge.[68] They were noted for bringing a naturalistic perspective with them and evangelizing for the new scientific outlook.[69] In addition to the effect of their own personalities, these protégés developed courses of study and training that had their roots with Huxley and became standard for generations to come.[70]

   This was not a magic bullet for Huxley and his allies, but theistic scientists and interpretations of science did become less and less common in British universities. With their allies in charge, the X Club could be assured that the next generation of science students was trained in a naturalistic perspective, which those students would then pass on to their students.[71] By the end of the century, Huxley's methods were well entrenched, and by the end of the Great War few could imagine it being otherwise.[72] This is not to say that the scientific naturalists corrupted science education. The professors they placed were talented and skilled, but their teaching methods were not value neutral either. Huxley designed his teaching to stand for what Adrian Desmond calls a "distinct ideological faction" that clearly marked off acceptable (naturalistic) from unacceptable (theistic) ways of thinking about science.[73] This faction was successful in propagating itself: as Melinda Baldwin shows in her chapter here, the generation of scientists following on the X Club saw themselves as the heirs of the scientific naturalists.

   A side effect of this is that once the scientific naturalists gained dominance in the scientific community, they were able to rewrite the history of their discipline to erase the long tradition of theistic science. Scientists frequently reimagine their past in order to support their vision for the future, and the wave of scientific naturalists at the end of the nineteenth century did so to establish a particular way of thinking about science; that is, that science as an enterprise only made sense in an areligious context.[74] Concepts like uniformity, which were both theistic and naturalistic in practice, became recast as *only* naturalistic.

   How could this be done? It might seem that overthrowing a centuries-old tradition such as theistic science would require a dramatic revolution, but in fact it was surprisingly smooth. This was because, as I have argued here, the positions of the theistic scientists and the scientific naturalists were actually quite similar in terms of basic concepts such as the uniformity of nature. So the practices and methods of theistic scientists could often be imported into naturalistic work with simple relabeling, or sometimes without comment at all. Huxley was particularly skilled at this. For example, he proclaimed that William Carpenter's work, particularly his textbooks, was the foundation of "rational" methods for thinking about living things.[75] Huxley argued that Carpenter's goal of explaining life in terms of laws (as opposed to vital forces) was the key step in removing religious and spiritual legacies from the life sciences. But this was quite different from the way Carpenter saw laws; of course he saw natural laws as manifestations of the Creator, and thus infused with religious significance. And it is not that Huxley was unaware of Carpenter's thoughts along these lines, as in his private correspondence he is quite hostile to Carpenter's spiritual interpretations.[76] But in his published work Huxley took a different approach. He simply stressed the points on which he and Carpenter agreed—natural laws—then elsewhere argued that natural laws were solely naturalistic. He did not have to persuade his colleagues that natural laws were important, because everyone already agreed on that. When theists read Huxley's discussions of laws in life, they could nod along happily, thinking of those laws in theistic terms. When scientific naturalists read about such laws, they too were happy, thinking of those laws in naturalistic terms. So in his published research Huxley could gain support from both camps, while his naturalistic interpretation of that research was passed on through his students and teaching.

Similarly, Bernard Lightman has shown elsewhere how Huxley co-opted literary strategies associated with natural theological writings to promote a naturalistic cosmology.[77] The "common object" strategy used by natural theologians to show God's contrivance in the world was clearly on display in Huxley's lectures and writings (such as "On a Piece of Chalk"). But he deftly used the same strategies to arrive at different conclusions, often subtly framed to paint a closed world of only natural phenomena with no room for God. Again, a theist could read many of Huxley's pieces, appreciate the science, and see the argument as familiar. Huxley even managed to do this in his teaching. He successfully made the fundamentals of his scientific methods palatable to theists, even to the point where his religious enemies such as St. George Jackson Mivart and J. W. Dawson sent their children to study under him.[78]

The overlap of theism and naturalism I have discussed here allowed a gradual transition over the course of a generation or two as the X Club's protégés more and more fully took over the universities. The strategies of the early scientific naturalists were highly effective at using their common ground with Victorian Christianity to create a space for propagating their values and aims. Older theistic scientists could read the work of younger naturalistic scientists and still see comforting continuities, and the younger naturalists could read classic work and see essential similarities. In this way the work of religious scientists such as Carpenter became secularized without requiring rejection of the work as unscientific.[79] Our modern understanding of the uniformity of natural laws as being purely naturalistic, then, is contingent and not inevitable, and a close historical examination of the issues shows that it can be, and was, a tool used both for and against religion. The scientific naturalists' triumph came from their ability to claim uniformity for themselves even as the theists felt that concept remained with them. Their victory in removing theism from the expectations and parlance of the scientific community had little to do with how science was done (despite their claims to the contrary) and much more to do with attempting to secure better access to professional positions, resources, and cultural authority.[80]

Acknowledgments

Some of the material in this chapter appeared previously in "The Uniformity of Natural Laws in Victorian Britain," *Zygon* 46 (2011): 536–60. The author is grateful for permission to reuse that material here.

Notes

1. Thomas Henry Huxley, prologue to *Essays upon Some Controverted Questions* (New York: D. Appleton, 1892), 3–4.

2. Ibid., 16, 22. Huxley noted that he did not think that science made supernaturalism per se impossible; rather, it simply disproved all existing forms of supernatural claims.

3. The literature on scientific practice is becoming large, but for a representative sampling, see Graeme Gooday, *The Morals of Measurement: Accuracy, Irony, and Trust in Late Victorian Electrical Practice* (Cambridge: Cambridge University Press, 2004); Sally Newcomb, *The World in a Crucible: Laboratory Practice and Geological Theory* (Boulder, CO: Geological Society of America, 2009); David Kaiser, ed., *Pedagogy and the Practice of Science: Historical and Contemporary Perspectives* (Cambridge, MA: MIT Press, 2005); and Jed Buchwald, ed., *Scientific Practice: Theories and Stories of Doing Physics* (Chicago: University of Chicago Press, 1995).

4. While it would be more accurate to speak of *natural philosophers* or *men of science* in this period rather than *scientists*, I will use the latter term for clarity. On the rise of scientific naturalism, and opposition to it, see Ronald L. Numbers, "Science without God: Natural Laws and Christian Beliefs," in *When Science and Christianity Meet*, ed. Ronald L. Numbers and David C. Lindberg (Chicago: University of Chicago Press, 2003); Frank Miller Turner, *Between Science and Religion: The Reaction to Scientific Naturalism in Late Victorian England* (New Haven, CT: Yale University Press, 1974); Robert M. Young, *Darwin's Metaphor: Nature's Place in Victorian Culture* (Cambridge: Cambridge University Press, 1985); Michael Ruse, "The Relationship between Science and Religion in Britain, 1830–1870," *Church History* 44 (1975): 1–18; and L. S. Jacyna, "Scientific Naturalism in Victorian Britain" (PhD diss., University of Edinburgh, 1980).

5. An excellent discussion of the history of uniformity and the subtleties in its meaning can be found in R. Hooykaas, *Principle of Uniformity in Geology, Biology and Theology* (Leiden, Netherlands: Brill, 1963). Hooykaas distinguishes between "actualism" (the attempt to explain past events by reference to causes now in operation) and "uniformitarianism" (the idea that these causes have always operated with the same intensity). This is an important distinction, but *uniformity* is the more commonly used term today, and I will rely on it here. See also Martin J. S. Rudwick, "Uniformity and Progression: Reflections on the Structure of Geological Theory in the Age of Lyell," in *Perspectives in the History of Science and Technology*, ed. Duane Roller (Norman: University of Oklahoma Press, 1971), 209–27.

6. John F. W. Herschel, *A Preliminary Discourse on the Study of Natural Philosophy* (London: Longman, Rees, Orme, Brown, and Green, and J. Taylor, 1830), 35.

7. Ibid., 119.

8. Ibid., 39–40, 35.

9. Ruth Barton, "'An Influential Set of Chaps': The X-Club and Royal Society Politics, 1864–85," *British Journal for the History of Science* 23 (1990): 53–81; Roy MacLeod, "A Victorian Scientific Network: The X-Club," *Notes and Records of the Royal Society* 24 (1969): 305–22; and J. Vernon Jensen, "The X Club: Fraternity of Victorian Scientists," *British Journal for the History of Science* 5 (1970–71): 63–72.

10. Barton, "'Influential Set of Chaps,'" 56.

11. Thomas Henry Huxley, "Scientific and Pseudo-scientific Realism," in *Collected Essays*, vol. 5 (London: Macmillan, 1901), 70.

12. Thomas Henry Huxley, "On the Persistent Types of Animal Life," in *Scientific Memoirs of Thomas Henry Huxley* (London: Macmillan, 1898–1902), 90.

13. Thomas Henry Huxley, "Geological Reform," in *Collected Essays*, vol. 8 (London: Macmillan, 1893–94), 324–25.

14. Thomas Henry Huxley, *Introductory Science Primer* (London: Macmillan, 1887), 14.

15. Thomas Henry Huxley, "The Darwinian Hypothesis," in *Collected Essays*, vol. 2 (London: Macmillan, 1893–94), 13.

16. Ruth Barton, "Evolution: The Whitworth Gun in Huxley's War for the Liberation of Science from Theology," in *The Wider Domain of Evolutionary Thought*, ed. D. Oldroyd (Dordrecht, Netherlands: Kluwer, 1983), 268.

17. John Tyndall, "Address," in *Report of the Forty-Fourth Meeting of the British Association for the Advancement of Science Held at Belfast in August 1874* (London: John Murray, 1875), lxxxviii.

18. Ibid., lxvii.

19. Ibid., xcv. A brief overview of the military metaphor can be found in David Lindberg and Ronald L. Numbers, eds., *God and Nature* (Berkeley: University of California Press, 1986), 1–18.

20. John Tyndall, *Fragments of Science for Unscientific People*, 4th ed. (New York: D. Appleton, 1875), 67, 49.

21. Numbers, "Science without God," 265, 281.

22. Herschel, *Preliminary Discourse on the Study of Natural Philosophy*, 37.

23. Peter Harrison, "The Development of the Concept of the Laws of Nature," in *Creation: Law and Probability*, ed. Fraser Watts, 13–36 (Aldershot, UK: Ashgate, 2008). The natural theological tradition used a number of different strategies to point to the existence of a Creator, including both this sort of lawgiver arguments and special adaptation arguments. See Matthew Stanley, "A Modern Natural Theology?," *Journal of Faith and Science Exchange* 3 (1999): 105–12; and Jonathan R. Topham, "Science, Natural Theology, and Evangelicalism in Early Nineteenth-Century Scotland," in *Evangelicals and Science in Historical Perspective*, ed. David Livingstone, 143–73 (Oxford: Oxford University Press, 1999).

24. Crosbie Smith and M. Norton Wise, *Energy and Empire: A Biographical Study of Lord Kelvin* (Cambridge: Cambridge University Press, 1989), 585. Hooykaas's distinction between actualism and uniformitarianism is helpful for understanding the issues in this controversy.

25. Joe Burchfield, *Lord Kelvin and the Age of the Earth* (New York: Science History Publications, 1975), 3, 28.

26. Smith and Wise, *Energy and Empire*, 544–48.

27. William Thomson, "Presidential Address," in *Popular Lectures and Addresses*, vol. 2 (London: Macmillan, 1891), 197.

28. Ibid., 200.

29. William Thomson and P. G. Tait, "Energy," *Good Words* 3 (1862): 606–7.

30. Burchfield, *Lord Kelvin*, 49; and Smith and Wise, *Energy and Empire*, 555.

31. James Clerk Maxwell, "On the Dynamical Evidence of the Molecular Constitution of Bodies," in *The Scientific Papers of James Clerk Maxwell*, vol. 2, ed. W. D. Niven (Cambridge: Cambridge University Press, 1890), 418. Hereafter *SP*.

32. James Clerk Maxwell, "Attraction," in *SP*, 2:491.

33. James Clerk Maxwell to Charles John Ellicott, bishop of Gloucester and Bristol, 22 November 1876, in *Scientific Letters and Papers of James Clerk Maxwell*, vol. 3, ed. P. M. Harman (Cambridge: Cambridge University Press, 1990), 418. Hereafter *SLP*.

34. Colossians 1:16, Hebrews 2:6, and Psalm 8, New International Version.

35. Maxwell to Ellicott, 22 November 1876, *SLP*, 3:417.

36. James Clerk Maxwell, "Introductory Lecture at Aberdeen," *SLP*, 1:543.

37. James Clerk Maxwell, "Address to the Mathematical and Physical Sections of the British Association," *SLP*, 2:224.

38. William Carpenter, "Nature and Law," in *Nature and Man: Essays Scientific and Philosophical* (London: Kegan Paul, Trench, 1888), 382, 367–68.

39. Frederick Temple, *The Relations between Religion and Science* (London: Macmillan, 1884), 121.

40. Duke of Argyll, *The Reign of Law*, 5th ed. (London: Strahan, 1867; repr., 1870), 29. Some modern views on theistic evolution can be found in Robert Pennock, ed., *Intelligent Design Creationism and Its Critics: Philosophical, Theological, and Scientific Perspectives* (Cambridge, MA: MIT Press, 2001), 471–536.

41. Temple, *Relations between Religion and Science*, 8; Argyll, *Reign of Law*, 3; and Temple, *Relations between Religion and Science*, 27.

42. This fusion of theism and apparently naturalistic reasoning is sometimes today called "religious naturalism." For more, see Willem Drees, *Religion, Science, and Naturalism* (Cambridge: Cambridge University Press, 1996); David Ray Griffin, *Religion and Scientific Naturalism: Overcoming the Conflicts* (Albany: SUNY Press, 2000); and J. A. Stone, "Varieties of Religious Naturalism," *Zygon* 38 (2003): 89–93.

43. For example, see Michael Ruse, *Can a Darwinian Be a Christian? The Relationship between Science and Religion* (Cambridge: Cambridge University Press, 2001), 95; and Michael Ruse, "Methodological Naturalism under Attack," *South African Journal of Philosophy* 24 (2005): 44.

44. Robert Bruce Mullin, "Science, Miracles, and the Prayer-Gauge Debate," in *When Science and Christianity Meet*, ed. Ronald L. Numbers and David C. Lindberg (Chicago: University of Chicago Press, 2003). Huxley wrote an entire book on Hume.

45. Tyndall, *Fragments of Science for Unscientific People*, 36.

46. Ibid., 409.

47. Temple, *Relations between Religion and Science*, 32–33.

48. Argyll, *Reign of Law*, 4.

49. Ibid., 12–13.

50. Temple, *Relations between Religion and Science*, 195.

51. Baden Powell, *Essays on the Spirit of Inductive Philosophy, the Unity of Worlds, and the Philosophy of Creation* (London: Longman, Brown, Green, and Longmans, 1855), 155.

52. Argyll, *Reign of Law*, 22. Also see Michael Ruse, "Relationship between Science and Religion in Britain," 509–10.

53. Ruse, "Relationship between Science and Religion in Britain," 510–11.

54. Argyll, *Reign of Law*, 17–30.

55. Temple, *Relations between Religion and Science*, 195–96.

56. Mullin, "Science, Miracles, and the Prayer-Gauge Debate," 205–6.

57. Temple, *Relations between Religion and Science*, 219–20.

58. Griffin, *Religion and Scientific Naturalism*, 38–40. The point that it is possible to retain the value of a miracle even after explaining it with natural laws is made in Ruse, *Can a Darwinian Be a Christian?*, 96.

59. Matthew Stanley, *Practical Mystic: Religion, Science, and A. S. Eddington* (Chicago: University of Chicago Press, 2007).

60. The common story that many Victorians were turned away from their faith by developments in science is addressed thoroughly in John Brooke's contribution to Ronald L. Numbers, ed., *Galileo Goes to Jail, and Other Myths in Science and Religion* (Cambridge, MA: Harvard University Press, 2009).

61. Barton, "'Influential Set of Chaps,'" 72.

62. Ibid., 59; and Roy MacLeod, "A Victorian Scientific Network: The X-Club," *Notes and Records of the Royal Society* 24 (1969): 310.

63. Adrian Desmond, "Redefining the X Axis: 'Professionals,' 'Amateurs' and the Making of Mid-Victorian Biology—A Progress Report," *Journal of the History of Biology* 34 (2001): 28. Desmond points out that many of these new teachers came from industrial, Dissenting regions of Britain and were thus particularly receptive to Huxley's outlook. Huxley's efforts toward science education are described in detail in Cyril Bibby, *T. H. Huxley: Scientist, Humanist, and Educator* (London: Watts, 1959), 123–93. For an overview of the development of science education in this period, see David Layton, "The Schooling of Science in England, 1854–1939," in *The Parliament of Science*, ed. Roy MacLeod (Northwood, UK: Science Reviews, 1981), 188–210.

64. Thomas Huxley to Anton Dohrn, 7 July 1871. Leonard Huxley, *Life and Letters of Thomas Henry Huxley*, 2 vols. (New York: D. Appleton, 1901), 1:389.

65. Graeme Gooday, "Nature in the Laboratory," *British Journal for the History of Science* 24 (1991): 334.

66. Thomas Huxley to Henry Cole, Henry Cole Diary 7, 1861, cited in Elizabeth Bonython and Anthony Burton, *The Great Exhibitor: The Life and Work of Henry Cole* (London: V&A Publications, 2003), 199.

67. For example, see L. Huxley, *Life and Letters of Thomas Henry Huxley*, 2:33.

68. Desmond, "Redefining," 33; and Gerald Geison, *Michael Foster and the Cambridge School of Physiology* (Princeton, NJ: Princeton University Press, 1978), 130–47.

69. Philip Pauly, "The Appearance of Academic Biology in Late Nineteenth-Century America," *Journal of the History of Biology* 17 (1984): 378.

70. Janet Howarth, "Science Education in Late-Victorian Oxford: A Curious Case of Failure?," *English Historical Review* 102 (1987): 339.

71. Geison and MacLeod point out the particular importance of Michael Foster's students in furthering Huxley's teaching system. Geison, *Foster*, 142–45; and Roy MacLeod, "The 'Naturals' and Victorian Cambridge: Reflections on the Anatomy of an Elite, 1851–1914," *Oxford Review of Education* 6 (1982): 185.

72. Bibby, *T. H. Huxley*, 111.

73. Desmond, "Redefining," 32.

74. On how scientists rewrite their discipline's history for current purposes, see Richard Staley, *Einstein's Generation* (Chicago: University of Chicago Press, 2009); Peter Galison, "Re-

reading the Past from the End of Physics," in *Functions and Uses of Disciplinary Histories*, ed. Loren R. Graham, Wolf Lepenies, and Peter Weingart (Dordrecht, Netherlands: Kluwer, 1983), 35–52; and Nigel Gilbert and Michael Mulkay, "Experiments Are the Key: Participants' Histories and Historians' Histories of Science," *Isis* 75 (1984): 105–25.

75. Carpenter, *Nature and Man*, 66–67.

76. Thomas Huxley to John Tyndall, 27 October 1876. Thomas Henry Huxley Papers, Imperial College London, 8.195.

77. Bernard Lightman, *Victorian Popularizers of Science: Designing Nature for New Audiences* (Chicago: University of Chicago Press, 2007), 372–77.

78. Desmond, "Redefining," 32.

79. Other examples of secularization or naturalization of previously theological concepts can be found in Frank M. Turner, "The Secularization of the Social Vision of British Natural Theology," in *Contesting Cultural Authority: Essays in Victorian Intellectual Life* (Cambridge: Cambridge University Press, 1993); James Moore, "Theodicy and Society: The Crisis of the Intelligentsia," in *Victorian Faith in Crisis*, ed. Richard Helmstadter and Bernard Lightman (Stanford, CA: Stanford University Press, 1990); Bernard Lightman, *The Origins of Agnosticism: Victorian Unbelief and the Limits of Knowledge* (Baltimore: Johns Hopkins University Press, 1987), 117–18; and Young, *Darwin's Metaphor*.

80. Frank M. Turner, "The Victorian Conflict between Science and Religion: A Professional Dimension," *Isis* 69 (1978): 356–76; and Numbers, "Science without God," 281.

New Generations

10

The Fate of Scientific Naturalism:
From Public Sphere to Professional Exclusivity

THEODORE M. PORTER

Frank Turner introduced his term of art *Victorian scientific naturalism* in the context of a study of late Victorian dissenters from it. In this way he consigned it to a specific epoch, brought to a close by the passing of a narrow and over-confident generation and their succession by subtler minds, men troubled by the loss of human meaning under arrogant scientism, hence less intolerant of faith and mystery. Turner adduced also the progress of knowledge in his tale of its decline, as if Rutherford's atom and Einstein's space-time must soften the asperity of scientifically naive naturalistic claims.[1] While the pertinence of relativity physics may be doubted, the heightened respectability of psychical research does indeed signify a momentary opening up of the ontology of science. What seems of much greater consequence for the fate of scientific naturalism, though, is the silencing of science as a voice of public critique.

This taming of science was gradual and fluctuating rather than sudden or monotonic, and it was and is far from absolute. Yet the waning of naturalism during the late Victorian period was a signal moment in the reshaping of public reason, one whose sources and significance extend to cultural and political as well as intellectual history. Turner, too, saw that the identity of science was at stake in the double reshaping of science through the formation and then dissolution of naturalism. He interpreted this, however, along the lines of a narrow and rather deterministic postwar sociology of the professions, assuming, in effect, that what scientists really want is the autonomy and the funding to go about their work, which may include providing expert advice, but that engagement with the public is, by and large, about public relations and a distraction from the serious business of research. Historians need rather to ask how, when, and to what extent scientists came to this view, and what alternatives there have been.

Naturalism and Professionalization

While it makes little sense to speak of the aims of *science*, which possesses nothing like unitary consciousness, scientists, being human, typically work with a purpose. An old-fashioned historiography, disinterred still from time to time by journalists and other scientific enthusiasts, supposes that true scientists are free spirits, unbounded by institutions and conventions, who accept no authority but truth. When social science was brought to bear on the problem, the persona of the scientist required to be socialized, at least in theory. As a corollary of the doctrine of modernization, much in fashion from the 1950s to the 1970s,[2] sociologists defined the progress of science according to a telos of professionalization and technical mastery. Through professions and disciplines, scientists are able to earn a living; to gain the facilities and funds required for their research; to enact standards of training, competence, and integrity; and to insulate themselves from external pressures, as of politics and religion. Their earnings are no charity but the wages of knowledge, since the complex societies that define "modernity" are, so the argument goes, functionally dependent on specialized expertise. It follows that professional standing, understood as public recognition for their vital role and the provision of means to fulfill it, is properly the great desideratum of scientists. Or sometimes, according to a more cynical historiography, professional organization gives ersatz experts the means to erect barriers to entry and possibly to hoodwink the public.

By 1980, modernization as a template for understanding economic development was giving way to dependency theory and other doctrines that were more skeptical of the march of progress through Western science and technology. In historical writing on science, though, the first sociological turn in the form of "externalism" was just then approaching its apogee.[3] During the next decade, an understanding that confined social interpretation to the institutions of science passed out of favor as the Edinburgh "strong programme" extended sociology into the very temple of scientific knowledge. This was a direct challenge to the old "internal" history of science as a record of intellectual progress. But it did not question the professionalization narrative, perhaps because the new social or "cultural" history of science involved often a turn to the microlevel, with a focus especially on laboratory practices. Such research might, like the best microhistory, cast a sharp beam of light on the bigger world, but often it did not. The inward-looking perspective provided little basis for a narrative that could address the larger historical role of science, and so left the teleology of professionalization largely untouched.

One of the more influential and enduring historical works on professionalization and its consequences was Turner's 1978 paper on Victorian naturalism. Proceeding in the apodictic mode of argument from definition, and invoking the Columbia sociologist Bernard Barber, he explained the establishing of a profession as a necessary phase in the institutional formation of science. "Within the group [scientists] must raise standards of competence, foster a common bond of purpose, and subject practitioners to the judgment of peers rather than of external social or intellectual authorities. Outside they must establish the independence of the would-be professional group, its right of self-definition, and its self-generating role in the social order." Under a tighter professional regime, many of the old practitioners must stand out as interlopers. Since science, especially natural history, in Britain had long been a favorite preserve of curates and rectors, whose duties within the Church of England left ample leisure for the observation and admiration of nature, the rising professionals faced a particular challenge in biology. It could only be overcome, Turner explained, by excluding these men of the cloth from the circle of true science. The influence of clergymen stood in the way of science becoming "self-defining in regard to its own function," an essential attribute of any profession.[4]

Turner's inspired move, somewhat reductionist in its implications, was to deploy this sociological narrative to account for a famous episode in intellectual history, the Victorian conflict between science and religion. In one breath, he supplied an ahistorical explanation of naturalism and filled out the analysis with historical particulars. Many of the leading advocates of the new scientific naturalism, such as Thomas Henry Huxley, came from backgrounds outside the establishment, and had accordingly been the more assiduous in acquiring credentials to back up their scientific claims. They seethed at ecclesiastical naturalists of lesser attainments earning tidy incomes for their commentaries on the divine order of nature while so many who pursued careers in science were obliged to accept lesser posts and even to survive on odd jobs. Some, including John Tyndall, had earned higher degrees in Germany, and many admired a university system that provided systematic training and support for research while making appointments on the basis of scientific merit. Above all, they resented the meddling of clerical dilettantes in serious questions of science such as Darwin's theory of species change.

Turner's interpretation has drawn continued attention and inspired various efforts at revision, though many historians continue to endorse his trajectory of professionalism and resistance. The most influential critique is really an extension. Adrian Desmond and James Moore interpreted the rise of scientific naturalism, with its professional ideals, in terms of social class, a bourgeois revolt against the political dominance of a conservative clerical order. Their biography of Charles Darwin situates the tranquil rural life of the great naturalist against a background of noisy urban protests by discontented artisans. Darwin himself, as he guarded for two decades the dangerous secret of his materialist-tending theory, paid the price in the form of deep social and personal anxiety, the source, they imply, of his disabling stomach ailments.[5] Desmond later portrayed Huxley, who stayed in London and did not shrink from battle, as spokesman for middle-class interests, challenging the sedate natural theology of smug bishops and complacent country parsons.[6] We have here the ascent of an intellectual class, an English *Bildungsbürgertum* of intellectuals, scholars, and men of science. Moore describes an alliance between the naturalists and modernizing religious writers who drew inspiration from the new German biblical criticism.[7] Darwin's *On the Origin of Species* (1859), after all, was published just months before the theologically audacious *Essays and Reviews*, written by progressive divines. The conflict here is not so much between natural science and religion or even theology, but between *Wissenschaft*, the whole domain of scholarly inquiry, and unreasoning orthodoxy.

If the naturalists were rebels, their rebellion was tempered by anxiety at the implications of their own challenge to the established order. Throughout, there were narrower and more personal interests at stake in the visionary campaign for a society based on knowledge and competence. As they gained social status and institutional power in a time of sharpened class tensions, they began to contemplate a campaign of reconciliation. The famous X Club, rather an aggressive outfit at its founding in 1864, soon moved toward respectability as, during the 1870s and 1880s, its members rose to the highest positions in British science. They also became noted public figures, leaders of culture. Desmond plots Huxley's trajectory from a modest background to his destination as "evolution's high priest." Moore turns the argument of changing patterns of class to the support of arguments like Turner's. As the naturalists gained control of the institutions of science, they became increasingly concerned to cultivate public and political support for science. In practice, this meant an alliance with governing elites, an aristocratic class that could support modern science on condition that it keep in its place and not pose a direct challenge to Christian faith, that anchor of an endangered social order. The men of science, now likewise fearful of working-class revolt, toned down their rhetoric. Huxley began to appear as the very model of English respectability. We find him in 1894 delivering the commendation after Salisbury's presidential address to the British Association for the Advancement of Science, a defense of evolution but rejection of Darwin's mechanism of natural selection in favor of something more purposeful.[8] Address and response, both irenic, were historically resonant, and quite opposed to the spirit of the last Oxford meeting of the BAAS in 1860, when Huxley had engaged in a fiery debate with Samuel Wilberforce about the descent of man, and more pointedly of grandmothers, from the apes.

By the late 1870s, the naturalists were becoming more conscious of a need to distance themselves from radical claims. Tyndall's "Belfast Address" at the British Association meetings in 1874, with its appeals to ancient atomism and its insistence that anyone who would make claims about nature must submit to the authority of science, was the high point of the naturalistic challenge to an established Church. Even then, he insisted that science revealed nature as meaningful. Soon afterward, Huxley and Tyndall entered on a campaign against spontaneous generation, whose whiff of materialism went too far for leaders of responsible science. That shift gives us a clearer view of their larger purposes. In their respectability, as in their former radicalism, the naturalists were thinking of science in the general culture, not merely the cultivation of professional autonomy and specialized disciplines.[9] Their opposition to bumptious scientific upstarts revealed a desire to restore the relative unity of elite culture of the early Victorian period—unified, now, against artisanal and working-class radicals who again seemed more threatening after the relative calm of the 1850s and 1860s.[10] From about 1873 to 1896, impressive economic growth was punctuated by crises leaving millions without work. The successful organization of trade unions was accompanied by a new labor politics including moments of acute disorder, as in the Hyde Park riots of 1886 and Bloody Sunday in 1887. Middle-class as well as working-class intellectuals wrote and lectured on the need for socialism, and on its inevitability. Also during the 1880s, the "woman's question" rose to political prominence. Some intellectuals, such as the effervescent young Karl Pearson, began thinking of science as the keystone of a new socialist order involving new structures of labor and the family. Not so the elite naturalists of Huxley's generation.

The Boundaries of Respectable Science

Scientific charlatanism had scarcely been possible before the period of British industrialization and the French Revolution, because scientific institutions did not have the capacity—nor, in the absence of popular unrest, was there sufficient incentive (apart from the profession of medicine)—to expel heterodoxy. In the new century, men of science were more often moved to attack the legitimacy of dissident claims about nature, not merely because the work of these autodidacts and mechanics appeared slipshod, but usually because it offended against morality and religion. Rarely had science faced so much pressure to be respectable as in the early nineteenth century. At this time, scientific insiders with the means to pursue their work full-time were more inclined than ever to defend Christian faith against radical doctrines of nature. This is when British "gentlemen of science" formed themselves into specialist societies devoted to geology, astronomy, mathematics, chemistry, and ethnology. There were hints of professional aspirations in the objections sometimes voiced against the scientific influence of aristocratic dabblers. Charles Babbage complained of the Royal Society in these terms in 1830, and the specialist societies as well as the British Association for the Advancement of Science, established in 1831, offered alternative venues for presenting, discussing, and judging scientific work. In subsequent decades, the Royal Society underwent a series of reforms to make membership more restrictive and to link it more closely to scientific achievement. None of these organizations, however, provided means to make a living from science, and a professional in this sense, who worked at science to earn his bread, suffered some loss of status within these communities of researchers for doing so.[11]

The gentlemen regarded working-class contributors, especially to field sciences such as botany and geology, with appreciative condescension. Natural history seemed a fine way for poor folk to spend leisure time, and their local knowledge enabled them to supply valuable observations and materials for collections.[12] If, however, they displayed too much ungentlemanly indifference to the hand of God in nature—if, for example, they veered too close to materialism—the indulgence of elite naturalists turned readily to condemnation. Professionalism, in the sense of gatekeeping, was here diametrically opposed to aggressive secularism. The leading early Victorian men of science worked hard to create institutions for persons like themselves, men for whom science was more vocation than avocation. They saw respectability in religious matters as allied to the pursuit of truth, and most were sincere in their Christianity. While few if any were biblical literalists, strict naturalism was another matter. Much more threatening to serious science than aristocratic idlers were radical materialists who denied human moral responsibility and excluded God from a causative role in nature.

The scientific gentlemen of the early nineteenth century deployed the filter of vocational commitment not to secure a secular profession, but to protect natural theology by keeping out radical phrenologists who reduced mind and spirit to organic matter, experimental enthusiasts who thought that life could be created from electricity or the forces of crystallization, medical materialists who endorsed Lamarckian transformations of species, and advocates of universal progress who practically dispensed with divine providence in their accounts of the creation and advancement of life. James Secord details the huge popular success of the *Vestiges of the Natural History of Creation* as well as the wide and sometimes bitter criticism by established men of science. This theory of the natural progress of the cosmos, the earth, life, and society was a bold work of synthesis that relied at every point on the writings of liminal as well as highly respectable figures in science. The author, who turned out to be Robert Chambers, could be recognized as a marginal man (or woman!) himself, even though his anonymity was not punctured until decades afterward.[13] A naturalist from the inner circle, even writing anonymously, who took a similar line could not have been dismissed so briskly, and the challenge to gentlemanly Christianity does not by itself account for the contumely that was sometimes heaped on *Vestiges*. It is striking, though, that very few scientific insiders in early Victorian Britain could have written a work of such comprehensive naturalism, leaving God so detached from the working out of his great plan of progressive creation. The near exception who proves the rule is Charles Darwin, already committed to the transmutation of species in 1844, who agonized for decades over the damage to his reputation, not least his scientific standing, were he to publish his theory without superabundant evidence.[14]

The naturalistic temper of post-Darwinian debates, this effort to establish science on a basis independent of religion, was not simply a product of Darwinism. Turner's argument about professionalizing science is corroborated by the case for Darwinism as only one element in a larger intellectual configuration. If Huxley was Darwin's bulldog, Tyndall barked (and bit) mainly for atomism and energy principles. James Joule, Hermann von Helmholtz, and William Thomson had given the first precise formulations of energy conservation in the language of physics, though others, such as the German Robert Mayer, were concerned rather with physiological issues. By the late 1850s, when physicists and chemists began vaunting energy conservation as the most fundamental of scientific laws, and when enthusiasts began advertising its metaphysical significance, the distinct status of life and mind was very much at issue. Tyndall supported his claim for the autonomy and sufficiency of naturalistic science in 1874 by invoking a new atomism along with the doctrine that energy could neither be created nor destroyed: not by man, and not by God. In a world where energy, including brain energy, came only from other forms of energy, there was perhaps no room for free will or for divine activity. No chinks should remain into which clerics or idealist philosophers of mind could insert theological or metaphysical explanations of worldly phenomena.[15]

But did this leave science to the scientists? Naturalistic ambitions were by no means confined by the boundaries of professional conquest. Evolution and energy doctrines flourished in the public sphere as elements in a *Kulturkampf* over the place of clergy in state, education, and intellectual life that was by no means limited to Bismarck's battles with Roman Catholicism. Other elements of the rising naturalism had still less to do with professional science. Henry Thomas Buckle, whose *History of Civilization in England* created as great a stir in 1857 as did Darwin's book two years later, was a strong political liberal who moved in the circle of John Stuart Mill. An education at home by his devoted mother made him a champion of instruction for women, but he did not look for intellectual progress through the institutionalization of knowledge. Buckle contended that a science of history presupposed the exclusion of human free will and divine activity, and he invoked the lawlike regularities

FIGURE 10.1. Henry Thomas Buckle, who became famous as a historian, first achieved note as a brilliantly idiosyncratic chess player. The first volume of his *History of Civilization in England* (1857) gave an account of progress in terms of the balanced expansion and diffusion of knowledge. From Alfred Henry Huth, *The Life and Writings of Henry Thomas Buckle*, 2nd ed., 2 vols. (London: Sampson, Low, Marston, Searle, and Rivington, 1880), vol. 2, frontispiece.

of statistics of birth and crime to demonstrate that human actions were free of such extraordinary influences. He appropriated his statistical principles from Adolphe Quetelet, who in turn had extracted them from administrative numbers.

These statistical "laws," having already entered the domain of public discussion, were not the achievements of specialized scholarship or science. While Buckle, who lived from an inherited fortune, was determined to establish a science of history, this had nothing to do with universities, disciplines, or professional societies. Indeed, he argued on the basis of comparative historical investigation that healthy intellectual progress depended on knowledge being neither too dispersed, as in the United States, nor too much concentrated, as in Germany. (England was neither too hot nor too cold, but just right.) He published and was debated in a field of political contestation. Professionalizing historians in England and elsewhere were almost unanimous in their opposition to his naturalistic account, and very few were drawn to his anticlericalism. Buckle's writing, as social science, implied the self-determination of social processes and conveyed his vehement support for liberal doctrines of legal equality and free trade.[16]

Buckle had much in common with Herbert Spencer, as prominent an evolutionist for most of the nineteenth century as Darwin. Spencer, raised up as a radical and a phrenologist, began his writing career as a subeditor for the strongly free-market *Economist* magazine, and later, with George Eliot and George Henry Lewes, became associated with the *Westminster Review* of the philosophic radicals. He earned a living from his writing, a string of high-profile reviews and ponderous best sellers. While he participated in the milieu of the X Club and was respected by Huxley, Tyndall, Francis Galton, Joseph Dalton Hooker, and other scientific naturalists, he did not seek a university position or participate in specialist scientific societies. Even more than Darwin, he inspired an efflorescence of public social science involving the biogenetic law of recapitulation and relying on analogies of the evolution of society with the embryological development of the organism. What he shared with his fellow men of science was not any professionalizing impulse but a desire to reach a large audience and to reshape the common culture. The Darwinian debates, so-called, were about biology and who was competent to practice it, but they were also about law, labor, morality, education, public health, race, colonialism, rights of women, social welfare, socialism, and the terrestrial power of churches. The professional organization of science and scholarship mattered for all of this, but points of ideological difference were indirectly, and often directly, at stake, and there was no movement to exclude the public from such debates. Quite the contrary.[17]

Public and Professional Science

If victory in the conflict between science and religion had depended on the successful outcome of a struggle for professionalization, the prospects during the reign of Victoria and Albert could not have appeared favorable. Although the elite of science, like the public generally, was divided over Darwinian evolution, the debates that counted did not take place in specialist scientific publications. These arguments were aired in the public sphere of popular lectures, newspaper discussions, and highbrow reviews such as the *Quarterly*, the *Edinburgh*, and the *Fortnightly*. In general, few scientific institutions in the era of high naturalism were closed off from elite and popular audiences. Rather, they were layered, with a few privileged insiders supported by a much more numerous base. The annual gathering of the BAAS was the most important event on the calendar, but attendees generally numbered in the thousands, exceeding by an order of magnitude the number of scientific professionals by even the most inclusive definition. The section for physics did not attract such crowds, but the leading British physicists, including William Thomson (Lord Kelvin), George Gabriel Stokes, and James Clerk Maxwell, were skeptical of Darwinian evolution. These, by Turner's standards, were the real professionals, practicing a laboratory-based, mathematical science from university positions. Yet they wanted to hold on to the religious element in science, and they were not content pursuing knowledge detached from wider meaning. The reason is not complicated: they were more comfortable with the established order, and their science was not threatened by highly placed opponents.[18] Naturalists like Huxley and Tyndall were, as Turner recognizes, more in need of protection from outside interference. They also used the evolution issue to enhance their status as public figures, treating science and the scientific spirit as the basis for a missionary campaign of reeducation.

The professional practice of biology was, in fact, beginning to develop in Britain during the years of the Darwinian controversies. Huxley, who had trained as a surgeon and whose scientific competence was strongest in morphology rather than natural history, was a prominent defender of this biological professionalism. The most vexed issue faced by his sort of biology was the defense of vivisection against a strong humanitarian movement animated by tender affection for pets. The career biologist as he was emerging in the 1880s was less often found crossing seas or tromping through fields and forests than on a boat or seashore collecting specimens or in a laboratory with a dissecting knife and a microscope. By this time, evolution had gained general acceptance among British biologists, while the mechanism of natural selection was mostly discredited.[19] Darwin's theory depended too much on adventitious interactions of organisms with the physical as well as the biological environment, events quite distinct from morphology and taking place far from the now-sacred space of the laboratory. Biologists in 1900 continued to prefer a mechanism of evolution that was internal to the organism, something more directed than the higgledy-piggledy of natural selection. This "higgledy-piggledy" comes from a critique of Darwin's theory by the quintessential gentleman of science from the generation before Darwin's, John Herschel, and conveyed his dissent from a theory that omitted providential design. God's role was less urgently at issue for biologists at the end of the century, but the purposefulness of life remained important for them. Their skepticism about Darwin's mechanism shows the looseness of the relation between the naturalism of the *Origin of Species* and the emergence of a biological discipline.

The religious element in British science did not go away in the decades after 1859. One of the more prominent movements in science from the 1880s to the Great War and beyond was the investigation of psychical phenomena. The Society for Psychical Research, founded in 1882, drew members from many fields of science, but especially from physics. While scholars have often treated it as a rearguard effort of refugees from naturalism triumphant, an escape into irrationalism, it is better understood as an affirmation of science tinged by disquiet. The more militant naturalists, including Galton and Pearson as well as Huxley and Tyndall, denounced the psychics, but traditionalists in religion were no more enthusiastic. To invoke psychical research in support of religion was to bypass theology and biblical history in favor of naturalistic, or at least experimental, evidence. Thus, the investigation of spiritualism involved no impassable rift, but a bridge between particular forms of science and Christianity. It did not split the soul of the scientist between rationalist research and mystical spiritualism. Instead, it affirmed the positive role of science in establishing a basis for a moderate religion supported by reliable experience. This, broadly, was the position of Oliver Lodge, one of the most successful spokesmen for science at the beginning of the twentieth century. Psychical research was more an assertion of the universal reach of science than a retreat from its exclusive authority. As a movement, it casts doubt on any conception of post-Darwinian science made safe for professionals by securing its independence from all religion.[20]

Scientists and Elites

If the late nineteenth century brought the rise of the professional scientist, it is striking how many of the leading spokesmen for science rejected even the label *scientist*. The word was coined by William Whewell about 1830, partly as an unsexed term that could encompass the likes of Mary Somerville, but mainly as a brake on galloping specialization. The proliferation within science of specialist occupational categories such as chemist, geologist, botanist, and astronomer, he explained, had created a need for one that could embrace all those engaged in the pursuit of natural knowledge. Whewell himself was known in those days for the "foible" of omniscience. His neologism was not taken up, and by 1890 it had come to seem narrowing, a professional deformation akin to *scientism* or the French *scientiste*. Many leaders of science disliked a name suggesting that their calling was no more than an occupation or that their character was distorted by a preoccupation with only one kind of activity. The naturalists preferred to be regarded as men of wide learning and to understand science as vital to a responsible, cultivated life. Huxley supposed that the word *scientist* must have been made up in America, as pleasing a neologism, he supposed, as *electrocution*, by which, in a self-consciously scientific move, New York State had begun putting criminals to death. This word was indeed mainly confined to American English before 1890, but Americans insisted on a more genteel characterization of the scientist, who would earn the name through the purity of his intentions and his avoidance of merely commercial research topics.[21] Whereas science should be ennobling, the *professional* engaged in fee for service, and on that account was in bad repute among would-be scientists of the fin de siècle in America and Britain alike. Practicing a profession implied the subordination of knowledge to earning one's bread, often by maintaining trade secrets. The Victorians idealized open science, an openness that should extend beyond communication among experts. Naturalists in the Darwinian era could advance their career goals while earning a decent living by lecturing and writing for a larger public, which indeed was encouraged, provided they were open and truthful and did not pander.[22]

Although many men of science in Huxley's generation earned their living from science, and some came up from very modest backgrounds, respectable science was by no means synonymous with careerism. Among elites, amateurs and professionals alike presented themselves as animated by a sense of duty to truth or to the nation, and the early modern supposition that an independent income made for disinterestedness had not yet disappeared in late Victorian Britain. In terms of social background, practitioners of science remained heterogeneous even at the level of its most prominent leaders, as recent scholarship clearly demonstrates.[23] The success and prominence of wealthy amateurs such as Francis Galton in late Victorian science may still appear puzzling from the standpoint of the professionalization thesis, but amateurs were not yet outliers in the scientific establishment.[24] Right into the twentieth century, some consummate scientific professionals idealized the career pattern followed by Darwin and his cousin Galton, whose independence permitted them to seek out truth wherever their studies led them and to keep clear of the petty distractions as well as the disciplinary narrowness of institutionalized science.[25]

This view of science was laid out with particular clarity by Karl Pearson, who made his scientific reputation in the quarter century beginning about 1890 in the fields of statistics, biometry, and eugenics. Pearson's scientific and scholarly interests were unbounded, yet he was a consummate professional in terms of his technical competence and his systematic immersion in each new topic he took up. Institutionally, he made his career as a professor at University College London, where he trained up the first generation of statistical scientists. In his book on the social construction of British statistics, Donald MacKenzie proposed a sociological reading of Pearson's career, identifying his program as an ideology of the professional middle classes. This formulation differs from Turner's by emphasizing the larger social role of science rather than the inward-looking politics of self-governing disciplines. It is thus better suited for understanding a man who envisioned science as nerve center of the social organism rather than as an institution requiring insulation so it could remain pure. Yet there was more of tradition in Pearson's program than even MacKenzie's thesis allows. Although Pearson valued science as a source of technical knowledge, he valued it still more as the basis of a new social ethic and a higher wisdom, something to elevate the common knowledge while enhancing and reconfiguring that of cultural leaders. *The Grammar of Science*, rather than of dead languages, was the proper basis of a modern education, he argued. Scientific understanding should, in the future, be rooted in the *method*, really the ethic, of science, and must incorporate but could not be replaced by specialized expertise.[26]

Bread and Laboratories

Despite all these problems with a "professionalization" thesis, Turner was I think correct to detect an intense concern for their own status in the naturalistic campaigns of the post-Darwinian men of science. They did indeed want access to careers and respect for their specialist knowledge, but they envisioned also a new turn of history that would make science the shared idiom of a public culture. By the time of Pearson, who held religiously to just this aspiration, it was becoming less plausible, and its decline has much to do with the waning of scientific naturalism as a public ideal. I propose here a twofold explanation for this decline, involving the adaptation of scientists to a more confined, largely technical, role in the world, along with a loss of faith in the educability of the public.

Some of John Heilbron's scholarship provides the basis for a counter-Turner thesis, by which professionalized science organized around career making and the gleaning of research funds can be seen to discourage practitioners from challenging the authority of established elites. Heilbron writes often of scientists on the make, like the advocates of a metric system based on the longitudinal circumference of the earth during the most dangerous moments of the French Revolution, who, he argues, wanted really to put new measuring instruments to use and to keep the institutions of science functioning. The late nineteenth century has special significance for him because science was then becoming seriously expensive. State funds, supplemented by rich patrons, made possible the increasing scale of physics and other sciences, and ideologically threatening researchers were unlikely to be esteemed as highly deserving recipients. For scientists as interested actors, it seemed best to lie low and profess humility rather than mount public challenges to established religious doctrines or aristocratic values. Under more democratic circumstances, scientists might for the same reason take measures to avoid offending against public opinion. Paul Forman's thesis regarding acausal physics in Weimar Germany is about scientists bending the knee to cultural and political ideologies for the sake of bread and laboratories. Heilbron attributes the influence beginning in the late nineteenth century of what he calls "descriptionism," a broadly positivist mode that draws back from anything so brash as a truth claim, to an effort to avoid giving offense to elites, and especially to the dominant religion. By this account, aggressive naturalism is precisely what a campaign of professionalization and discipline building should most urgently avoid.[27]

Turner's thesis overlaps in some ways with Heilbron's. Both pertain to scientists looking after their own collective interests, and tailoring their pronouncements to this end. If the outcomes they describe are sharply opposed, this may owe to the divergent cases they examine rather than a fundamental incompatibility of goals. Heilbron writes of conditions under which the identity of science and of the scientist appears unproblematic. Dukes and divines did not contest the mathematical theories of entropy or blackbody radiation, whereas they did claim competence regarding the functioning and adaptation of organisms to their environments. A broader campaign was perhaps required to secure for the naturalists what they regarded as their legitimate territory. Yet it is scarcely plausible to reduce such monumental efforts to a defensive maneuver, and even Turner does not construe the story so simply. Neither, in fact, did Heilbron's protagonists retreat to such a narrowly professional world. The flip side of the descriptionist independence from metaphysical commitments was its wide applicability, and hence the unboundedness of science. The humility of descriptionism included a good measure of cunning and ambition. Ernst Mach presented science as distilled from, and tightly connected to, practical life, while Pearson was strident in his all-encompassing naturalism. No topic was too humble or too exalted to fall within the purview of scientific method, which Pearson characterized very simply as observing, classifying, and giving formulas that conveniently summarize a wealth of experiences. Mastery of scientific method, by his reckoning, gave *scientists* (a label he did not reject) the competence to pronounce on health and medicine, social policy, cultural history, and the practical work of almost every trade or profession. Statistics, the mathematical science of measurement and analysis, was, for Pearson, practically coextensive with science, which is to say, universal. The scientism of Vienna Circle positivism in the twentieth century was continuous with the descriptionist aspirations of the fin de siècle.[28]

The invention of the *scientist* as a professional identity did not abruptly cut science off from its civic role, and the aspiration to universalize scientific method has endured, especially in science pedagogy. It would be absurd to claim that scientific naturalism breathed its last or that the public voice of science grew silent, in Britain or anywhere, at the end of the nineteenth century. The movement of eugenics after 1900 carried on in a new key the biological discussions of human progress and decay to which Darwin's name had long since become attached and that were of great concern to Huxley, Spencer, and Galton. Historians have been and remain too eager to consign this engaged form of science to a preprofessional past, and to regard its more contemporary manifestations as distortion, abuse, or misunderstanding of proper science.

Yet the extraordinary increase of scientists, of science teaching, and of research budgets over the century that began about 1880 was real enough, and was achieved through new alignments of science. As an attitude, naturalism among scientists became, during the twentieth century, more secure than ever. As the basis for an accommodation between professional science and its publics, naturalism was gradually drained of meaning in the face of a new keyword of science, *technicality*. One important sense of technical science was its usefulness in relation to technology. State funding agencies sometimes allowed scientists a relatively free hand in choosing topics and methods of research, but this generally presupposed the long-run economic value even of "pure," "basic," or "fundamental" science. The practical benefits of science extended to the training of engineers and industrial scientists as well as research products, but typically meant new or improved technologies, including a more skillful deployment of them, in manufacturing, agriculture, medicine, transport, communication, or military affairs. These goals necessarily brought science into close interaction with industry, engineering, and government agencies. The public was presumed to be interested in the products of this kind of science, but had little reason to explore its theoretical or material content. This, after all, was instrumental knowledge and not the stuff of intellectual growth or moral improvement. The immense increase in material support of science took place in the context of much closer relations between research science and industrialized production beginning about 1870 or 1880, the era of the Second Industrial Revolution.

Technologically oriented science thus overlapped extensively with the other sense of the *technical*, referring to knowledge that should hold little attraction except to those with a professional stake in the work. The Darwinian debates presumed throughout that the mechanisms as well as the natural history of creation or of progress were properly matters of general concern. The same applied to naturalism in general, not least because it bore directly on the character and justification of religious belief, which in turn was still taken to underlie the social order. On such grounds, Victorian naturalists worked to diffuse such attitudes and such knowledge as the foundation for a new culture of reason and progress. Of course the leaders of science did not claim that citizens should engage with it so intensely as did the true "man of science," but the public campaigns of scientific naturalists extended to almost every field. Scientific writers and lecturers expected, naturally, a suitable degree of deference from their audiences, especially from those of lower station, yet they could not be indifferent to public acceptance. The gradual ascent of technical ideals in science was, in one important sense, the death of naturalism.

To explain this waning of naturalism requires a clearer sense of what it was made of in the first place. Far from adumbrating the autonomy and self-determination of science, the rise of naturalism was linked to a search for new and larger audiences. By reaching out to a nonelite public, the spokesmen for science asserted their own important role, as expert intellectuals of a new type, in shaping the future. They did not present nature as detached, neutral, and value free, but as a key element of culture in an era of industrial and political transition. They came into conflict with religious authorities because they construed science to undermine many of the dogmas and some of the values of the official religion and the established political order. The most prominent Victorian naturalists were not democrats, but they appealed to the public on the expectation that science and its representatives should benefit

FIGURE 10.2. Karl Pearson, pioneer of the mathematical field of statistics that contributed so much to the technical development of the social sciences, idealized the scientific career and achievements of Francis Galton (pictured here as a man of eighty-seven), not least for the independence and freedom of movement that Galton's inherited wealth made possible. From Karl Pearson, *The Life, Letters, and Labours of Francis Galton*, vol. 3a (Cambridge: Cambridge University Press, 1930), facing p. 353.

from the advance of democracy.[29] In the next generation, Pearson argued that "we must aristocratise government at the same time as we democratise it," and that a general familiarity with scientific method would enable the people to distinguish genuine knowledge and competence from mere pretense. "The ultimate appeal to the many is hopeless, unless the many have foresight enough to place power in the hands of the fittest."[30]

Pearson's distinctive vision of the mission of science in this increasingly professional society offers an enlarged sense of what seemed possible in a formative era that now appears as the twilight of the public culture of scientific naturalism. His work abetted that decline, in fact, since his extraordinary success in establishing statistics as a mathematical field marked him as a heroic figure in the triumph of a more arcane method of scientific reasoning. The technicality of the new statistics played a key role in separating parts of biology and much of psychology and the social sciences from the domain of public discussion.[31] The growth of science, with its proliferating specialties, and the institutionalization of the scientific career as a progression through undergraduate coursework and graduate-level mentoring and research, provided fertile soil for the cultivation of technical tools. Scientists found that for most purposes they no longer required a public audience. It was enough if, beyond their disciplinary colleagues, they were able to demonstrate the value of their research to funding agencies and sometimes to officials or managers in schools, factories, hospitals, military organizations, or state bureaucracies.

The fin de siècle, finally, was a turning point in the discourse of science and values. Huxley, in his late essay on *Evolution and Ethics*, in 1893, expressed skepticism about any inference from what happens in nature as to how humans ought to act. Henri Poincaré's declaration early in the new century of the utter logical incompatibility between indicative and imperative statements was much admired by the logical positivists, while a similar methodological pronouncement by Max Weber has come to be widely cited in the social sciences.[32] The argument was of course not altogether original, and its triumph was not sudden. The success of the eugenics movement provides massive testimony to the limits of its persuasiveness during the early decades of the twentieth century. Pearson, one of the most influential eugenic spokesmen, reluctantly abandoned the enchantment of nature but insisted all through his life that scientific method provided a basis for the impersonal morality of socialism.[33] Like the split between science and its larger publics, the fact-value distinction in science developed gradually. It was an important element in the triumph of technicality as a scientific ideal, and was welcomed especially by social scientists as testimony to their independence from politics. It marked, if not the end, at least a new phase of scientific naturalism. Its service to the public could now be identified overwhelmingly with technological productiveness rather than with the projects of enlightenment that had animated British naturalists from the beginning of the nineteenth century, as with the Society for the Diffusion of Useful Knowledge, and that gained particular prominence in the Darwinian era.

Since 1965, when George Stocking applied the phrase to science, "whiggish" history has meant judging the ideas of past science strictly for what they contributed to present knowledge. Such whiggism long ago fell from favor within the mainstream scholarship of the history of science. More persistent has been the tendency to regard the modern social organization of science, with its universities, academic departments, and funding agencies, as appropriately modern and therefore natural.[34] The Victorians, we tend to suppose, yearned to build institutions like those that matured in America in the twentieth century. Ours is a system in which the "scientist" is defined rather clearly (though still with more ambiguity than commentators often recognize), and in which professional structures are supposed to insulate the scientist from the many forces at large that could threaten objectivity. This sociology of professionalization idealizes a finely structured science composed of neat disciplines surrounded by walls and barbed wire, and a police force within. In the face of changes in the organization of science since 1978, when Turner could confidently hold up professional autonomy as its telos, this conception is coming to appear more and more as backward-looking ideal rather than as living reality. For the naturalists who fought for science against clerical distortions in the late nineteenth century, the triumph of professionalization would have appeared as a disappointing defeat. The shift to professional science and scholarship, one of the most fundamental socio-intellectual transitions of modern times, cannot be assumed away as natural or inevitable and therefore uninteresting to history. Professional science provided more job security for scientists, but turned the Victorian naturalists' dream of public life suffused with scientific rationality into an absurd utopia or even a dystopia. At some moments, including the present day, it has even been difficult for professional science to defend its walls against those barbarians who were abandoned outside the gates. Who can believe now in the civilizing mission of science that for naturalists like Huxley and Pearson was so vital?

Notes

1. Frank Miller Turner, *Between Science and Religion: The Reaction to Scientific Naturalism in Late Victorian England* (New Haven, CT: Yale University Press, 1974), chap. 2, esp. p. 13.

2. Michael Latham, *Modernization as Ideology: American Social Science and "Nation Building" in the Kennedy Era* (Chapel Hill: University of North Carolina Press, 2000).

3. These sociological currents were influential then in the historical discipline generally. For two years, from 1978 to 1980, when I was a graduate student, the Davis Seminar at Princeton was organized around the topic of professionalization. Sociologist Bernard Barber came in from Columbia to introduce the topic. One of the participants there was Thomas Haskell, whose *The Emergence of Professional Social Science* (Urbana: University of Illinois Press, 1977) remains a landmark historical interpretation of American social science. More relevant for the topic of Victorian science is Harold J. Perkin, *The Rise of Professional Society: England since 1880* (London: Routledge, 1989), which, however, is long on professional ideals and very short on what work these rising professionals actually did.

4. Frank M. Turner, "The Victorian Conflict between Science and Religion: A Professional Dimension," *Isis* (1978), reprinted in Turner, *Contesting Cultural Authority: Essays in Victorian*

Intellectual Life (Cambridge: Cambridge University Press, 1993), 171–200, quotes on 176, 179. Similar in its view of professional science, though much grander in scope, is Joseph Ben-David, *The Scientist's Role in Society* (Englewood Cliffs, NJ: Prentice-Hall, 1971). Jack Meadows, *The Victorian Scientist: The Growth of a Profession* (London: British Library, 2004), offers something like a collective portrait but gives little attention to the public role of science. On the historiography of professionalization, see Paul Lucier's excellent paper "The Professional and the Scientist in Nineteenth-Century America," *Isis* 100 (2009): 699–732.

5. Adrian Desmond and James Moore, *Darwin* (London: Michael Joseph, 1991); also Desmond, *The Politics of Evolution: Morphology, Medicine, and Reform in Radical London* (Chicago: University of Chicago Press, 1989).

6. Adrian Desmond, *Thomas Huxley: From Devil's Disciple to Evolution's High Priest* (London: Penguin, 1998).

7. For example, James Moore, "Geologists and Interpreters of Genesis in the Nineteenth Century," in *God and Nature: Historical Essays on the Encounter between Christianity and Science*, ed. David C. Lindberg and Ronald L. Numbers (Berkeley: University of California Press, 1986), 322–50.

8. James Moore, "Deconstructing Darwinism: The Politics of Evolution in the 1860s," *Journal of the History of Biology* 24 (1991): 353–408.

9. Bernard Lightman, "Scientists as Materialists in the Periodical Press: Tyndall's Belfast Address," in *Science Serialized: Representations of the Sciences in Nineteenth-Century Periodicals*, ed. Geoffrey Cantor and Sally Shuttleworth, 199–237 (Cambridge, MA: MIT Press, 2004); James E. Strick, *Sparks of Life: Darwinism and the Victorian Debates over Spontaneous Generation* (Cambridge, MA: Harvard University Press, 2000); and Paul White, *Thomas Huxley: Making the "Man of Science"* (Cambridge: Cambridge University Press, 2003).

10. We might think here of Susan Faye Cannon, *Science in Culture: The Early Victorian Period* (Folkestone, UK: Dawson Publishing, 1978).

11. Roy M. MacLeod, "Whigs and Savants: Reflections on the Reform Movement in the Royal Society," in *Metropolis and Province: Science in British Culture, 1780—1850*, ed. Ian Inkster and Jack Morrell (Philadelphia: University of Pennsylvania Press, 1983), 55–90; Martin J. S. Rudwick, *The Great Devonian Controversy: The Shaping of Scientific Knowledge among Gentlemanly Specialists* (Chicago: University of Chicago Press, 1985); Jack Morrell and Arnold Thackray, *Gentlemen of Science: Early Years of the British Association for the Advancement of Science* (Oxford: Oxford University Press, 1981); James Secord, *Controversy in Victorian Geology* (Princeton, NJ: Princeton University Press, 1986); and Marie Boas Hall, *All Scientists Now: The Royal Society in the Nineteenth Century* (Cambridge: Cambridge University Press, 1984).

12. Rudwick, *Great Devonian Controversy*; Anne Secord, "Corresponding Interests: Artisans and Gentlemen in Nineteenth-Century Lancashire," *History of Science* 32 (1994): 269–315; and Anne Secord, "Coming to Attention: A Commonwealth of Observers during the Napoleonic Wars," in *Histories of Scientific Observation*, ed. Lorraine Daston and Elizabeth Lunbeck, 421–44 (Chicago: University of Chicago Press, 2011).

13. This is the subject of a wide literature on British science in the early nineteenth century, including Roger Cooter, *The Cultural Meaning of Popular Science: Phrenology and the Organisation of Consent in Nineteenth-Century Britain* (Cambridge: Cambridge University Press, 1984); Desmond, *Politics of Evolution*; James Secord, "Extraordinary Experiment: Electricity and the Creation of Life in Victorian England," in *The Uses of Experiment: Studies in the Natural Sciences*, ed. David Gooding, Trevor Pinch, and Simon Schaffer, 337–83 (Cambridge: Cambridge Univer-

sity Press, 1989); Simon Schaffer, "The Nebular Hypothesis and the Science of Progress," in *History, Humanity, and Evolution*, ed. James R. Moore, 131–64 (Cambridge: Cambridge University Press, 1989); and James Secord, *Victorian Sensation: The Extraordinary Publication, Reception, and Secret Authorship of "Vestiges of the Natural History of Creation"* (Chicago: University of Chicago Press, 2000).

14. Janet Browne, *Charles Darwin: Voyaging* (New York: Knopf, 1995), 468–72. These issues were discussed already by Charles Gillispie in *Genesis and Geology* (Cambridge, MA: Harvard University Press, 1951).

15. See Turner's chapter on scientific naturalism in *Between Science and Religion*.

16. Theodore M. Porter, *The Rise of Statistical Thinking, 1820–1900* (Princeton, NJ: Princeton University Press, 1986); and Theodore M. Porter, "Henry Thomas Buckle," in *The Dictionary of Nineteenth-Century British Scientists*, ed. Bernard Lightman, 4 vols. (Bristol, UK: Thoemmes Continuum, 2004), 4:333–35.

17. J. D. Y. Peel, *Herbert Spencer: The Evolution of a Sociologist* (New York: Basic Books, 1971); and Mark Francis, *Herbert Spencer and the Invention of Modern Life* (Ithaca, NY: Cornell University Press, 2007).

18. Crosbie Smith and M. Norton Wise, *Energy and Empire: A Biographical Study of Lord Kelvin* (Cambridge: Cambridge University Press, 1989); and Crosbie Smith, *The Science of Energy: A Cultural History of Energy Physics in Victorian Britain* (Chicago: University of Chicago Press, 1998).

19. Peter J. Bowler, *The Non-Darwinian Revolution: Reinterpreting a Historical Myth* (Baltimore: Johns Hopkins University Press, 1988).

20. The standard work on science and spiritualism remains Janet Oppenheim, *The Other World: Spiritualism and Psychical Research in England, 1850–1914* (Cambridge: Cambridge University Press, 1985), but here I follow the insights of Courtenay Raia, "From Ether Theory to Ether Theology: Oliver Lodge and the Physics of Immortality," *Journal of the History of the Behavioral Sciences* 43 (2007): 18–43. See also Peter J. Bowler, *Reconciling Science and Religion: The Debate in Early-Twentieth-Century Britain* (Chicago: University of Chicago Press, 2001).

21. Sydney Ross, "Scientist: The Story of a Word," *Annals of Science* 18 (1962): 65–85; Lucier, "Professional and Scientist," 727–28; and Ruth Barton, "'Men of Science': Language, Identity and Professionalization in the Mid-Victorian Scientific Community," *History of Science* 41 (2003): 73–119.

22. See Jim Endersby, *Imperial Nature: Joseph Hooker and the Practices of Victorian Science* (Chicago: University of Chicago Press, 2008), 21–25. Lucier, "Professional and Scientist," finds a similar disdain for "professional science" in America.

23. Barton, "'Men of Science'"; Steven Shapin, *The Social History of Truth: Gentility, Credibility, and Scientific Knowledge in Seventeenth-Century England* (Chicago: University of Chicago Press, 1994); Richard Bellon, "Joseph Dalton Hooker's Ideals for a Professional Man of Science," *Journal of the History of Biology* 34 (2001): 51–82; Katharine Anderson, *Predicting the Weather: Victorians and the Science of Meteorology* (Chicago: University of Chicago Press, 2005); and Endersby, *Imperial Nature*.

24. John C. Waller, in "Gentlemanly Men of Science: Sir Francis Galton and the Professionalization of British Life Sciences," *Journal of the History of Biology* 34 (2001): 83–114, and "Becoming a Darwinian: The Micro-Politics of Sir Francis Galton's Scientific Career," *Annals of Science* 61 (2004): 141–63, explains Galton's far-flung scientific efforts as the pursuit of professional career advancement in the form of social prestige rather than income.

25. On aristocratic sites of science: Simon Schaffer, "Physics Laboratories and the Victorian Country House," in *Making Space for Science: Territorial Themes in the Shaping of Knowledge*, ed. Crosbie Smith and Jon Agar, 149–80 (New York: St. Martin's Press, 1998); also Deborah Coen, *Vienna in the Age of Uncertainty: Science, Liberalism, and Private Life* (Chicago: University of Chicago Press, 2007).

26. Donald MacKenzie, *Statistics in Britain, 1865–1930: The Social Construction of Scientific Knowledge* (Edinburgh: Edinburgh University Press, 1981); and Theodore M. Porter, "Statistical Utopianism in an Age of Aristocratic Efficiency," *Osiris* 17 (2002): 210–27. *The Grammar of Science* is the title of Pearson's important book, first published in 1892.

27. See John L. Heilbron, "The Measure of Enlightenment," in *The Quantifying Spirit in the Eighteenth Century*, ed. Tore Frängsmyr, J. L. Heilbron, and Robin E. Rider, 207–42 (Berkeley: University of California Press, 1990). The interpretation has been contested, however: see Charles C. Gillispie, *Science and Polity in France: The Revolutionary and Napoleonic Years* (Princeton, NJ: Princeton University Press, 2004), 238. See also John L. Heilbron, "Fin-de-siècle Physics," in *Science, Technology, and Society in the Time of Alfred Nobel*, ed. Carl-Gustav Bernhard, Elisabeth Crawford, and Per Sèrböm, 51–73 (Oxford: Pergamon Press, 1982); Paul Forman, J. L. Heilbron, and Spencer Weart, "Physics ca. 1900: Personnel, Funding, and Productivity of the Academic Establishment," *Historical Studies in the Physical Sciences* 5 (1975): 1–185; and a much-controverted work by Paul Forman, "Weimar Culture, Causality, and Quantum Theory, 1918–1927: Adaptation by German Physicists and Mathematicians to a Hostile Intellectual Environment," *Historical Studies in the Physical Sciences* 3 (1971): 1–115.

28. Theodore M. Porter, "The Death of the Object: Fin-de-Siècle Philosophy of Science," in *Modernist Impulses in the Human Sciences*, ed. Dorothy Ross, 128–51 (Baltimore: Johns Hopkins University Press, 1994), and on the technicality of science in Theodore M. Porter, "How Science Became Technical," *Isis* 100 (2009): 292–309.

29. It seems, however, that their voices were not the most effective in communicating to the public about science; see Bernard Lightman, *Victorian Popularizers of Science: Designing Nature for New Audiences* (Chicago: University of Chicago Press, 2007).

30. Karl Pearson, "The Moral Basis of Socialism" (1887), in Pearson, *The Ethic of Freethought and Other Addresses and Essays* (London: T. Fisher Unwin, 1888), 322; and Theodore M. Porter, *Karl Pearson: The Scientific Life in a Statistical Age* (Princeton, NJ: Princeton University Press, 2004).

31. That is, the reasoning and argumentation of social sciences. Some of their products, especially in the form of numbers such as census and polling results or basic economic quantities, became emblematic. See, for example, Sarah Igo, *The Averaged American: Surveys, Citizens, and the Making of a Mass Public* (Cambridge, MA: Harvard University Press, 2007); Thomas A. Stapleford, *The Cost of Living in America: A Political History of Economic Statistics, 1880–2000* (Cambridge: Cambridge University Press, 2009); and Theodore M. Porter, "Statistics and the Career of Public Reason: Engagement and Detachment in a Quantified World," in *Statistics and the Public Sphere: Numbers and the People in Modern Britain, 1800–2000*, ed. Tom Crook and Glen O'Hara, 32–47 (New York: Routledge, 2011).

32. Robert N. Proctor, *Value-Free Science: Purity and Power in Modern Knowledge* (Cambridge, MA: Harvard University Press, 1991).

33. This was fundamental in the first chapter of all three editions of his *Grammar of Science* (e.g., the third edition, London: Adam and Charles Black, 1911).

34. George W. Stocking, "On the Limits of Presentism and 'Historicism' in the Historiog-

raphy of the Behavioral Sciences" (1965), in Stocking, *Race, Culture, and Evolution: Essays in the History of Anthropology*, 1–12 (New York: Free Press, 1968); and Theodore M. Porter and Dorothy Ross, "Introduction: Writing the History of Social Science," in *The Cambridge History of Science*, vol. 7, *Modern Social Sciences*, ed. Porter and Ross, 1–10 (Cambridge: Cambridge University Press, 2003).

11

The Successors to the X Club?
Late Victorian Naturalists and *Nature*, 1869–1900

MELINDA BALDWIN

In the early months of 1869, a thirty-three-year-old British astronomer named Norman Lockyer began asking his friends and colleagues to write articles he could publish in the first issue of a new weekly scientific periodical.[1] The new publication was not, Lockyer emphasized, a specialized scientific journal. Although Lockyer was soliciting contributions from Britain's most famous scientists and intended to print abstracts of technical papers and reports from foreign scientific societies, the journal was not affiliated with any scientific society and the audience for the new weekly was not solely other men of science. Rather, Lockyer hoped that his publication would be read by educated laymen of all trades and was publishing the weekly with the commercial London publishing house Macmillan and Company. Most of the people Lockyer consulted about his undertaking had at best modest expectations for the new publication. The market for Victorian science periodicals was highly competitive, and Lockyer already had one unsuccessful publication on his editorial résumé, a general-interest journal called the *Reader*. Lockyer's acquaintance Joseph Dalton Hooker, an eminent botanist and the director of Kew Gardens, pessimistically responded to the project by telling Alexander Macmillan, "By all means make public my good will to the Lockyer periodical . . . [but] the failure of scientific periodicals patronized by men of mark [has] been dismal. I do not see how a really scientific man can find time to conduct a periodical scientifically, or brains to go over the mass of trash."[2]

Today, Lockyer's periodical, *Nature*, is arguably the world's most prestigious scientific journal, and most would call the publication an unparalleled success—although not the kind of success its editor had initially envisioned. Early in its life, *Nature* underwent a significant change in content. The journal never acquired much of a following among broader reading audiences and quickly abandoned its plan to devote a large portion of its contents to popular science pieces. The first issue of *Nature* was published in November 1869; by 1875, the primary audience for *Nature* had shifted from laymen to men of science. It appears that Lockyer had difficulty balancing the two parts of his initial vision and that his contributors' preferences facilitated *Nature*'s transformation into a publication very different from the one he had founded.[3] But despite the shift in *Nature*'s content and audience, the journal continued publishing and became one of the central organs of scientific communication in Britain. Why did *Nature* survive when so many others—including Lockyer's previous magazine, the *Reader*—did not?[4]

Some historians have linked *Nature*'s success to the support of the X Club, whose scientific stars were on the rise during the 1870s and 1880s.[5] Frank M. Turner grouped Lockyer together with the X Club members as part of "the young guard of science" that sought to bolster the status of men of science by establishing scientific naturalism as the proper basis for scientific inquiry. Turner argued that *Nature* was the embodiment of the X Club's efforts to "professionalize" science and exclude the club's opponents from scientific discourse.[6] If Lockyer was indeed part of such a "young guard," it might be expected that he would favor the members of the X when they clashed with other men of science. But in reality, the combative Lockyer had difficulty remaining on good terms with the X Club. Rather than support X Club members over their opponents, Lockyer and his subeditors at *Nature* allowed both sides equal chance to savage one another—a strategy, as Ruth Barton has shown, that led to serious rifts with Hooker, John Tyndall, and Herbert Spencer.[7]

The key to *Nature*'s success was not the X Club, but the following generation of British men of science, a group of scientific practitioners who regarded the X Club generation as their mentors. When we examine patterns of contributions to *Nature*, it becomes evident that members of the X Club and their contemporaries viewed *Nature* as a place to publish popular pieces or participate in debates, but not as a desirable forum in which to announce their most important scientific work or to give substantial commentary on scientific theories. In contrast, the younger generation, men born in the 1840s and 1850s, saw *Nature* both as an ideal forum for scientific discussions and, increasingly, as a useful way to spread news of their original work. It was this younger generation who adopted *Nature* as a central organ of scientific communication, and their contributions to, and enthusiasm for, the journal that cemented *Nature*'s status as Britain's most important scientific publication.

Nature and a New Generation

By 1875, *Nature* had not only alienated its original audience by shifting toward more specialized articles—Lockyer's editorial practices had also alienated several prominent X Club members who had initially been enthusiastic contributors. And yet, losing the X's support was not a disastrous blow for Lockyer. Despite its success at obtaining influential positions in scientific societies, the X Club still had many important scientific opponents and these opponents continued to contribute to *Nature*. Furthermore, the X's stay atop Britain's scientific hierarchy would not last forever. By the mid-1880s, the members of the X were well into their sixties and many were in poor health. Spencer, Tyndall, Hirst, and John Lubbock could no longer travel to the club's dinners with regularity, and in 1886, Huxley told Lubbock that he feared the club would not endure much longer. In 1889, a disagreement between Huxley and Spencer over land nationalization policy blew up into a serious public argument; the two barely spoke for the next four years. Thomas Hirst's death in 1892 proved the final blow for the declining X, and the group ceased to meet after May of that year.[8] As the members of the X aged and grew ill, a new generation of British men of science, born after 1840, began to assume positions of leadership in the British scientific community. This new generation included some of *Nature's* most influential and prolific contributors in the late nineteenth century.

Given their importance to the journal, it is worth mentioning the backgrounds of a few key members of the younger generation: the zoologist E. Ray Lankester (1847–1929), the naturalist George John Romanes (1848–94), the botanist William Turner Thiselton-Dyer (1843–1928), and the industrial chemist Raphael Meldola (1849–1915). Some important common threads become apparent when we examine their biographies. All of these men had university degrees. Romanes attended Cambridge; Thiselton-Dyer and Lankester attended Oxford (and Lankester later spent an unhappy period as a professor there); Meldola took his degree at the Royal College of Chemistry. All of these men chose to pursue their careers in London and three of them did so in London colleges. Lankester eventually left his hated position at Oxford for one at University College London, and Romanes also established his scientific career at UCL. Meldola became a professor at Finsbury Technical College following a successful stint at the industrial firm of Brooke, Simpson, and Spiller. The exception was Thiselton-Dyer, who held a few brief university positions before making his career at Kew Botanical Gardens. None of these men had an aristocratic background, although Romanes's family was quite wealthy and Thiselton-Dyer and Lankester had family connections to science or medicine.

Compare this to a sample of older *Nature* contributors: Herbert Spencer (1820–1903), Joseph Hooker (1817–1911), Thomas Huxley (1825–95), Alfred Russel Wallace (1823–1913), and Charles Darwin (1809–82).[9] Huxley, Hooker, and Spencer were all X Club members; Darwin and Wallace were not. These older naturalists were the trusted mentors of the younger generation and there are many similarities between the two groups. Aside from Darwin, who came from a wealthy family, no one from the older group was well-off or aristocratic. The London connection can also be seen in this older group. Hooker, Huxley, and Spencer would eventually settle in London, and Darwin lived there for a time as a young man. But there are important differences. Of these five, only Hooker and Darwin had university degrees (Darwin from Cambridge, Hooker from Glasgow), and universities—especially Oxford and Cambridge—were far less central to the older generation's scientific careers. Wallace was a railroad surveyor who indulged his interest in natural history largely in his spare time, and Spencer also began his career in railway management before deciding to make his living as a writer and theorist in London's literary circles. Hooker pursued his career at Kew Gardens. With the exception of Darwin, whose family money enabled him to assume the role of gentleman naturalist, the older naturalists had to exercise a great deal of creativity and self-promotional skill in order to find ways to pursue science as a paying vocation, including finding ways to be paid for their scientific essays. Their younger counterparts seem to have faced fewer difficulties in building their scientific careers and were less reliant on income from their writing.

Turning to *Nature*, we see another contrast between these two groups: the younger group seems to have been more comfortable with, and attached to, *Nature* as a vehicle of scientific communication than the X Club and its contemporaries. The key difference was where these two generations chose to publish original scientific essays. Men of Huxley's, Hooker's, and Spencer's generation were certainly regular contributors to *Nature*. However, their contributions were usually in the form of lecture summaries, book reviews, and occasionally letters on controversies in the journal; this generation did not use *Nature* to spread news of advances in their scientific work, or even to print substantial critiques of others' scientific work. Instead, men of the older generation preferred to direct their discussions with fellow men of science to literary monthlies or other general-interest periodicals and saved their own scientific work for society journals, monographs, or a meeting of a scientific society. In contrast, the younger generation began to embrace the practice of sending a short notice to *Nature* as a way of making their findings public and adopted the journal as the primary forum for debating scientific theories in Great Britain.

George J. Romanes and Late Victorian Evolutionary Debate in *Nature*

One useful way to gain insight into these generational differences in *Nature* is to follow the work of George J. Romanes. *Nature* played an important role in Romanes's scientific career: it was what called Darwin's attention to the young naturalist in 1873. After reading Romanes's letter to the editor on "Permanent Variation of Colour in Fish," in which Romanes declared his allegiance to the Darwinian theory of natural selection, Darwin wrote a note of congratulations to Romanes expressing interest in his work and future career.[10] Darwin soon became Romanes's scientific mentor and close friend.[11] After Darwin's death in 1882, Romanes took it upon himself to defend natural selection both from critics like the Duke of Argyll and from former Darwinian allies such as Wallace, who had come to question whether natural selection applied to human evolution.

Romanes was gifted with a seemingly infinite capacity for correspondence. Between 1881 and his death in 1894, he was almost always involved in some type of discussion or debate in *Nature*. He engaged in print arguments with a variety of opponents about Darwin's theories, the epistemological implications of natural selection, and Romanes's own ideas about the evolutionary mechanism. Romanes's spirited tenacity drew equally impassioned responses from those who disagreed with him and made him a focal point of late Victorian evolutionary controversy in *Nature*.

In the 1880s and 1890s, Romanes advocated two additions to the theory of natural selection: physiological selection and panmixia. In August 1886, Romanes published a three-part abstract entitled "Physiological Selection: An Additional Suggestion on the Origin of Species," a shortened version of a paper he had read before the Linnean Society on 6 May.[12] In it, Romanes argued that Darwin had recognized three evolutionary facts that the theory of natural selection could not account for. First, domesticated species (such as different breeds of dogs) bred much more freely with one another than species that had evolved in the wild, which tend to have more selective fertility. Second, the theory of natural selection did not explain how crossbreeding between parents with different characteristics affected the development of species. Finally, the theory of natural selection could not account for the fact that many of the features that distinguished species from one another were useless from a survival standpoint.

Romanes concluded that there must be another evolutionary mechanism operating alongside natural selection. He proposed that this mechanism was something he called "physiological selection." Physiological selection, said Romanes, occurred when a new variety of animal was infertile with its parent form but fertile with other members of its own variety. This limited fertility, said Romanes, would cause a new variety to endure and become a species in its own right.

> When accidental variations of a non-useful kind occur in any of the other systems or parts of organisms, they are, as a rule, immediately extinguished by intercrossing. But whenever they happen to arise in the reproductive system in the way here suggested, they must inevitably tend to be preserved as new natural varieties, or incipient species. At first the difference would only be in respect of the reproductive system; but eventually, on account of independent variation, other differences would supervene, and the new variety would take rank as a true species.[13]

As we shall see, this new theory did not win many immediate converts and provided the basis of an intense ongoing debate about interspecies sterility and whether natural selection was sufficient to explain the origin of species.

The second evolutionary issue Romanes discussed in *Nature* was a theory called "panmixia," a doctrine first proposed by the German naturalist August Weismann.[14] The theory of panmixia stated that when an organ no longer conferred an evolutionary advantage to an animal (for example, if the horns of a species of sheep ceased to help the animal survive and reproduce), that organ would no longer be the subject of natural selection. Weismann argued that the cessation of selection could result in an organ significantly decreasing in size, or even vanishing altogether. Romanes was a supporter of this theory, but many other British naturalists were skeptical. E. Ray Lankester argued that the cessation of selection would mean that a now-useless organ (for example, a horn or a tail) was equally inclined to grow and to diminish, and that panmixia could not account for a decrease in the size of an organ unless there was an evolutionary advantage to having smaller horns or a shorter tail.

The debate over panmixia was closely related to the third (and arguably the most significant) point of evolutionary controversy in *Nature*: the inheritance of acquired characters. Like Weismann, Romanes maintained that cessation of selection could account for the dwindling in the size of a now-useless organ, but Romanes differed from Weismann in believing that the principle of panmixia alone could not fully explain why a useless organ might vanish altogether. In his first letter to *Nature* on the subject, Romanes wrote,

> While Prof. Weismann believes the cessation of selection to be capable of in-
> ducing degeneration down to the almost complete disappearance of a rudi-
> mentary organ, I have argued that, *unless assisted by some other principle*, it can
> at most only reduce the degenerating organ to considerably above one-half its
> original size—or probably not through so much as one-quarter. [Emphasis
> in original.][15]

Romanes believed that this "other principle" was the inheritance of acquired
characters. If a parent animal did not need its useless organ, suggested Romanes, the effects of this disuse would pass to its offspring, who would be
born with an even smaller version of the organ.[16]

Romanes's ideas on physiological selection, panmixia, and the inheritance
of acquired characters drew strong opposition from evolutionary theorists of
both generations. Wallace was one of the first to criticize Romanes's paper
on physiological selection. In an article in the *Fortnightly Review*, Wallace
denounced the idea that any principle aside from natural selection was necessary to account for the origin of species.[17] Lankester was deeply skeptical of the
principle of panmixia, which led to an eight-week exchange in *Nature* with Romanes in the spring of 1890. Meldola was more favorably disposed than Wallace
or Lankester toward Romanes's theory of physiological selection, but believed
that any physiological selection had to be dependent upon Darwinian natural
selection.[18] Wallace, Lankester, and Meldola also shared a belief that acquired
characters could not be passed down from parents to offspring, though it was
Meldola who engaged most extensively with Romanes on this point in *Nature*.[19]

Romanes's opponents often found him a somewhat annoying correspondent, both because of his seemingly insatiable appetite for debate and because of his argumentative tactics. In his exchanges with his fellow men of
science, Romanes often attempted to blunt their criticisms by claiming that
his opponents really agreed with him but had not realized it yet. During the
argument with Wallace, Romanes frequently insisted that the older naturalist
had acknowledged the importance of fertility and sterility in the evolution
of species and that Wallace's opposition to physiological selection was based
on a misunderstanding of Romanes's principles.[20] Romanes employed the
same strategy during an argument with Meldola in 1891 over whether two
apparently unrelated characters, neither of which was advantageous on its
own, might combine to provide an advantage and evolve concurrently. Romanes, drawing on the doctrine of use-inheritance, believed that such "co-
adaptation" could occur; Meldola believed that it was too unlikely for two
useless characters to occur in the same animal and combine to produce an
advantage. Romanes wrote that Meldola actually agreed with him: "As it appears to me, from his reply, that Prof. Meldola's views on the subject of 'co-

adaptation' are really the same as my own, I write once more in order to point out the identity."[21]

This strategy did not endear Romanes to his fellow naturalists. Responding to Romanes's argument that he had adopted the theory of physiological selection without realizing it, Wallace wrote that if Romanes continued to press the claim, "it will show that our respective standards of scientific reasoning and literary consistency are so entirely different as to render any further discussion of the subject on my part unnecessary and useless."[22] Wallace never forgave Romanes for claiming that he had adopted the theory of physiological selection. In an 1893 letter to W. T. Thiselton-Dyer, Wallace declined to write a letter of sympathy to the terminally ill Romanes, explaining,

> He made a very gross misstatement & personal attack on me when he stated, both in English & American periodicals, that, in my <u>Darwinism</u>, I <u>adopted</u> his theory of "Physiological Selection" and claimed it as <u>my own</u>. . . . I told him then that unless he withdrew this accusation as publicly as he had made it I should decline all further correspondence with him, & sh'd avoid referring to him in any of my writings.[23]

Although Thiselton-Dyer considered Romanes a friend, he well understood Wallace's frustration.[24] Four years earlier, he himself had written to Wallace to complain about a recent conflict with Romanes in *Nature*.[25] "To tell you the truth I was rather cut up about my controversy with Romanes," he admitted. "I will never engage in a discussion with Romanes again. He does not, I am persuaded, grasp his own views, much less those of other people. He is elusive as an eel."[26]

Meldola appeared to be equally annoyed by Romanes's persistence. During the 1891 coadaptation debate, Meldola declared that while he had hoped their two-month discussion might come to an end, "I very much regret to find, however, that Dr. Romanes—whose amount of spare time appears to be most enviably inexhaustible—still finds it necessary to prolong the correspondence."[27] At one point, Lankester wrote a letter to *Nature* about Romanes that was so sharply worded, the controversy-loving Lockyer overruled his subeditors and declined to print it. In a note to Lockyer, Lankester admitted it was probably best the letter was not made public, but saved some choice epithets for Romanes:

> You are quite right not to print my letter about Romanes—as it is not argumentative but purely denunciatory. I am glad he has seen it—as he will now know what a humbugging piece of foolery I consider his attempt to say "Darwin-and-I" and "the Darwin-Romanes theory"—is. It is time that he knew that I consider him a wind-bag.

> I think it is perhaps my duty to say so, urbi et orbi—or do you hold that a
> man is to be allowed to puff himself and falsely pass himself off on the crowd
> as a second Darwin—without protest—[28]

Romanes, a close friend of Lockyer's, also saw Lockyer's hand behind the disappearance of Lankester's letter.[29] He told Lockyer he was unconcerned with the opposition to his theory.

> My own pet theory about Physiological Selection has met, as you will have
> seen, with a storm of opposition. But this does not affect me in the least; seeing
> it is obvious that as yet there are no data for an adverse judgment. It can only
> be made or marred by a long course of verification. Am I right in connecting
> your return with the non-appearance of Lankester's letter to <u>Nature</u>—proof
> of which was sent me by letter? I had written such a beautiful reply; but all the
> while thought it would be a mistake to disfigure <u>Nature</u> with so unseemly a
> correspondence.[30]

Men of science of all ages may have been united in their annoyance with Romanes. However, their approaches to contributions in *Nature* indicate that the younger generation, the men of science born after 1840, seems to have been more comfortable with and attached to *Nature* as a vehicle of scientific communication than the X Club and their contemporaries, who were born in the 1820s or before. This older generation contributed book reviews and letters to the editor to *Nature*, but saved news of their own work or substantial criticisms of others' work for other publications. Notably, although Wallace was one of *Nature*'s most prolific book reviewers, when Wallace wished to criticize Romanes's theory on physiological selection, he did not write to *Nature*. Instead, his initial reply appeared in the *Fortnightly Review*, a prestigious liberal journal that was renowned for its literary and political commentary (and that paid writers like Wallace much more generous sums than *Nature* did).[31] Romanes's response to Wallace's choice is also telling: in his first letter discussing Wallace's criticisms, Romanes strongly implied that the older naturalist had done the *Nature* readership a disservice by moving the discussion to another publication and expressed surprise that "criticisms on the theory of physiological selection are flowing through channels other than the pages of NATURE."[32] Only when Romanes attacked him by name in *Nature* did Wallace choose to respond in that journal. By contrast, Romanes's younger critics, such as Meldola (born 1849), Lankester (born 1847), and Francis Darwin (Charles Darwin's son, born 1848), wrote their own thoughts on physiological selection for *Nature* first.[33]

It might be argued that Romanes's younger opponents were responding to Romanes in *Nature* simply because that was known as the combative researcher's publication of choice. However, members of the younger generation preferred *Nature* to publications like the *Fortnightly Review* even for debates that did not involve Romanes at all. We can see this in the responses to the anti-Darwinian George Douglas Campbell, the eighth Duke of Argyll (1823–1900), a prominent Scottish peer and a chancellor of the University of St. Andrews. Argyll, a member of the older generation, preferred publications like the *Nineteenth Century* when he sought to criticize Darwin's theories.[34] When Huxley responded to Argyll's ideas, he also chose general periodicals.[35] But when Thiselton-Dyer disagreed with Argyll over the inheritance of acquired characters, the younger botanist wrote to *Nature*.[36] Similarly, when Thiselton-Dyer felt that the Irish physicist George Gabriel Stokes had acted improperly in running for Parliament while holding the presidency of the Royal Society, he chose to make his views known in *Nature* rather than in a literary or political monthly.[37]

A similar gap can be seen when we consider where the older and younger generations chose to publish their original scientific research. Romanes had been contributing to *Nature* since the earliest years of his career; as previously noted, it was a letter to *Nature* in 1873 detailing some observations on fish that had brought his work to Darwin's attention. Romanes clearly believed that the *Nature* abstract of his physiological selection paper was an essential part of putting his theory before the British scientific community and even seemed startled when criticisms of his theory appeared in other publications. However, older evolutionary theorists such as Huxley, Spencer, and Wallace did not use *Nature* for this purpose. They continued to publish their scientific work either as monographs or in more established scientific publications and did not use *Nature* to announce their forthcoming works.[38]

Why *Nature*? Publishing, Priority, and Scientific Prestige in Late Nineteenth-Century Britain

Why did the younger generation adopt *Nature* as a central organ of scientific communication when their mentors had not done so? One obvious explanation might be that as a younger journal, *Nature* was more accessible to lesser-known young men than the established scientific journals or prestigious literary periodicals like the *Fortnightly Review*, but this argument does not hold up to scrutiny. All of the members of the younger generation had successful publication records in scientific society journals and George J. Romanes wrote prolifically for the literary periodicals, suggesting that access to these publications was not the determining factor in his, or his colleagues', attachment to *Nature*.[39] Furthermore, even after they became fellows of the

Royal Society and had built well-established scientific careers, men like Lankester, Meldola, and Thiselton-Dyer continued their frequent contributions to *Nature*.

Instead, *Nature*'s success with these men of science appears to have had a great deal to do with its publication speed. Unlike the literary periodicals, there was almost no delay between the submission of a piece and its appearance in the journal. *Nature* often printed letters and communications the same week they were received. Sir John Maddox, editor of *Nature* from 1966 to 1973 and 1980 to 1995, once suggested that one of *Nature*'s greatest early assets was the speed of the Royal Mail. British men of science knew that a contribution sent to the journal would reach its destination the day after it was posted.[40] The speed of publication created a sense of immediacy among the contributors to *Nature*—Romanes could write to the periodical and read a range of responses to his ideas less than two weeks later. *Nature* was the closest print substitute for a meeting of a scientific society, and unlike a discussion at a gentleman's club such as the Athenæum or a debate at the British Association, a letter to *Nature* would be printed and available to readers who were outside the membership or could not attend a particular meeting.[41] The publication speed may also have contributed to the occasionally combative tone of the publication, as the weekly schedule gave men of science less opportunity to rethink and rewrite harsh words.[42] Furthermore, an 1873 letter to Lockyer from the American astronomer Henry Draper indicates that *Nature* reached audiences outside Britain far more quickly than the older scientific publications:

> I wish that the publications of the great Societies could be made to reach those who are interested more quickly. The Transactions of the Royal Society take an incredible time to make their appearance here and we have really to depend on the abstracts that are published in scientific magazines for fresh information. In this respect "Nature" is invaluable.[43]

By the end of the nineteenth century, *Nature* had become the publication of choice for the discussion of controversies in the British scientific community.[44]

A second reason *Nature*'s speed of publication would have been compelling to younger men of science is that getting one's work into print quickly had become an increasingly essential feature of establishing priority for a scientific finding or theory.[45] Using *Nature* to announce a new finding or a forthcoming paper, which became one of that periodical's primary functions in the twentieth century, was still a developing use of the publication in the nineteenth century. But contributors such as Lankester, Thiselton-Dyer, and especially Romanes were beginning to use announcements of their work in *Nature* to advance their careers and their scientific reputations. It should be noted that this younger generation did not see an article in *Nature* as a substitute for delivering a full paper to a scientific society or writing for one of the scientific societies' journals. An abstract or letter to the editor in *Nature* simply could not convey the same amount of information as a seventy-page article in the *Philosophical Transactions* or a talk at the British Association. But although *Nature* was not a replacement for these forums, some of the older institutions began to resent the way in which announcements in *Nature* seemed to steal their thunder. As early as 1880, some within the British Association viewed *Nature* as a competitor, complaining that the ease of writing in to the weekly journal had stripped the BA meetings of their traditional significance—few saved new or provocative ideas for the annual meeting, instead preferring to initiate discussions immediately by submitting a piece to *Nature*.[46]

Why wasn't *Nature*'s speed of publication more attractive to the older generation? There appear to have been a number of factors at work. One compelling explanation for the gap is the disparity in pay between monthly periodicals like the *Fortnightly Review* and *Nature*. Men like Wallace, Huxley, and Spencer were accustomed to earning their livings from their pens. At the height of their careers these men could earn substantial payments for writing lengthy essays on scientific topics or book reviews for monthly periodicals; a five-hundred-word letter to the editor in *Nature*, however, was far less lucrative.[47] Furthermore, the members of the older generation were at a vastly different stage in their careers. They had already established their reputations and likely did not feel the same need to secure priority for their work. It is also possible that the older generation continued to utilize more established forms of communication because they remembered how Lockyer had initially advertised his journal. They may have viewed *Nature* as a popularizing periodical and preferred to direct their scientific essays to publications they felt were more intellectually prestigious.

Finally, the older generation's rejection of *Nature* as a major venue for scientific debates appears to signal that they and their younger counterparts held very different ideas about the appropriate audience for scientific debates. While many members of the X Club generation had been prolific contributors to literary periodicals, contributions to *Nature* appear to have replaced the literary periodicals for many members of the younger generation. A brief glance at the author listings in the *Wellesley Index to Victorian Periodicals* (which catalogs the contents and authors of forty-five general-audience Victorian periodicals) shows us that between 1870 and 1900, Huxley contributed seventy articles to publications included in the *Wellesley Index*; Wallace wrote thirty-six; Spencer wrote sixty-eight.[48] In the younger generation, Lankester wrote only five; Meldola and Thiselton-Dyer wrote none. Romanes was a notable and significant exception to this generational trend; in this time period, he wrote twenty-nine articles for *Wellesley Index* publications.[49] But as Joel Schwartz observes, Romanes used journals like the *Nineteenth Century* to popularize evolutionary theory and to promote his own image as Darwin's heir to a lay audience. Romanes did not view the lay publications as a place to publish substantial criticisms of others' theories, as Wallace and Huxley before him had.[50]

The choice of *Nature* versus the literary periodicals as a host for scientific discussions was not a mere aesthetic preference—it represented a choice between two fundamentally different types of scientific debate. When members of the older generation chose publications like the *British Quarterly Review* or the *Nineteenth Century* for an essay questioning a colleague's scientific theory, they were placing their work in a publication read by educated men of all trades, alongside articles on politics, religion, literature, and philosophy. Wallace, Huxley, and the rest likely would have seen this as a point in the general periodicals' favor. Publishing in these types of magazines helped them subtly press the argument that science was an intellectual endeavor worthy of equal standing with the more "classical" subjects, and that men of science ought to be voices of broader cultural authority. Furthermore, the same debate was frequently carried out in several journals at once. But younger men of science had reaped the rewards of the older generation's attempts to establish science as a respectable endeavor; as noted earlier in this essay, they faced far fewer difficulties in constructing their scientific careers than their mentors had. More secure in their status, they saw less reason to debate scientific questions before the largely nonscientific audience of the literary periodicals and preferred to direct their writings to a publication with a more specialized audience. Romanes's complaint that Wallace had moved the physiological selection debate outside *Nature* seemed valid to Romanes, who assumed scientific issues should be debated for an audience of men of science, but it would likely have been baffling to Wallace, who was accustomed to scientific debates that spanned several publications.

Nature and Darwin's Legacy

One important exception to the generational trend was Darwin, who occasionally prepared abstracts of his longer scientific papers for *Nature* and saw a short piece in the weekly journal as a useful way to announce a forthcoming study and, more frequently, to disagree with another man of science or call attention to another naturalist's paper that he thought was of particular interest.[51] *Nature* quickly replaced the *Gardener's Chronicle* as Darwin's publication of choice.[52]

After Darwin's death in 1882, his image took on a new life in the pages of *Nature*. The posthumous treatment of Darwin in *Nature* is extremely striking and deserves at least a brief discussion, as it sheds further light on the scientific self-images of several key *Nature* contributors. The famous naturalist was one of the most revered figures in Victorian science (whatever his reputation may have been in nonscientific circles) and Thiselton-Dyer, Lankester, and especially Romanes were all explicit in their desire to emulate his great scientific career. In the two decades following Darwin's death, the correspondents in *Nature* spoke of Darwin with the utmost respect, even reverence. Romanes's physiological selection abstract mentioned Darwin repeatedly, and Romanes attempted to cast his theory as the solution to a problem Darwin himself had identified.

> For he [Darwin] says and he says most truly, "We have conclusive evidence that the sterility of species must be due to some principle quite independent of natural selection." I trust I have now said enough to show that, in all probability, this hitherto undetected principle is the principle of physiological selection.[53]

Many of Romanes's critics also invoked the name of Darwin in their discussion of his theory. Wallace's *Fortnightly Review* critique of physiological selection was titled "Romanes *versus* Darwin" and attempted to show that Romanes's ideas were antithetical to Darwin's work.[54] Meldola wrote that while Darwin had acknowledged that natural selection might not be the only agent at work in evolutionary change, Romanes's contention that physiological selection was equally important was contrary to Darwin's beliefs:

> Darwin to the last considered natural selection as the *chief agency* in the evolution of species, and no one saw more clearly than he did the difficulties which surrounded the formation of incipient species, owing to the obliteration of new characters by intercrossing with the parent form.[55]

Lankester wrote to *Nature* to argue that Romanes's physiological selection paper was an attack on Darwinism: "He [Darwin] considered his theory of natural selection to be a theory of the origin of species. Mr. Romanes says it is not. I say this is an attack on Mr. Darwin's theory, and about as simple and direct an attack as possible."[56] Darwin's son Francis, now a respected botanist, agreed with Lankester; in a letter to *Nature* he argued that his father had considered and discarded a theory very much like Romanes's.[57] In response to such criticism, Romanes insisted that his theory was an addition to, not a replacement for, natural selection, and that his opponents were the ones who were anti-Darwinian: "My contention from the first has been that upon this point I am in full agreement with Mr. Darwin, and differ only from those Darwinians who differ from their master in holding that *all* specific changes are likewise adaptive changes, and *vice versâ* [*sic*]."[58]

Interestingly, many of the naturalists involved in the late Victorian debates about evolutionary theory in *Nature* were advocating ideas that drew them away from Darwin's original writings. However, the discussions in *Nature* suggest that those involved did not believe they were breaking with Darwin's ideas (or, at least, had no desire to create that impression). Instead, men like Romanes, Lankester, Spencer, and Wallace all sought to portray themselves as faithful Darwinians. These naturalists wished their readers to think they were carrying on Darwin's program of evolutionary work. This observation reinforces the point that the "younger generation" whose contributions were so essential to *Nature* saw themselves as members of an established scientific tradition rather than revolutionaries whose ideas represented a break with the previous generation.

Conclusion

Despite the differences in their publication strategies, it is clear that men like Romanes, Lankester, Thiselton-Dyer, and Meldola regarded themselves as the scientific heirs of mentors such as Darwin, Huxley, and Hooker—the men generally seen as the central figures of scientific naturalism. In the context of this volume, the obvious question becomes, were these younger men scientific naturalists themselves? The members of the younger generation were certainly methodological naturalists; they believed that proper science should be based on naturalistic explanations and should have no theological overtones. But whether they fit into Frank Turner's classic definition of *scientific naturalism* is more ambiguous. Turner himself, for example, used Romanes as a case study for his book on *reactions* to scientific naturalism and described the spiritually conflicted Romanes as someone who "served the cause of scientific naturalism, and then returned to a very nearly Christian position."[59]

In many ways, the men of science born after 1840 seem to be a generation to whom the category of *scientific naturalist* no longer applied. Turner's influential definition of the term hinged largely on the conflict between science and religion, a struggle that (as other chapters in this volume have pointed out) was never as simple as clerics versus men of science, and was no longer nearly as prominent by the time Lankester, Romanes, and the rest had established their scientific careers. These researchers came of age in a scientific tradition that assumed that science ought to be done in the way that men like Huxley had outlined; they were devotees of Darwin, and their theories relied solely on naturalistic explanations for scientific phenomena. Their career paths and scientific methods nicely underline Matthew Stanley's point (made earlier in this collection) that the X Club was more successful than its opponents at seizing the training of the next generation of scientific workers. But men like Thiselton-Dyer, Romanes, Lankester, and the others were not embedded in the same battle for intellectual and cultural esteem that the X Club had faced; they did not share their mentors' alarm that religious influence might be a threat to scientific respectability.

In many ways, although several of the most prominent original scientific naturalists found themselves at odds with Lockyer and his journal, *Nature*'s continued existence serves as evidence that their efforts to establish a cultural voice for British men of science had borne fruit. Their successors formed a group coherent and strong enough to sustain a for-profit magazine devoted to their interests. And although *Nature* would not clear its debts until 1890, Macmillan was willing to continue its publication largely because *Nature* helped give respectability and influence to the publisher's profitable list of scientific books.[60] By 1900, the patronage and participation of British men of science was a commodity worth cultivating.

Acknowledgments

Some of the material in this chapter was previously published in Melinda Baldwin, "The Shifting Ground of *Nature*: Establishing an Organ of Scientific Communication in Britain, 1869–1900," *History of Science* 50 (2012): 125–54. The author would like to thank the National Science Foundation for the funding that enabled the archival research cited in this chapter, as well as Bernard Lightman, Gowan Dawson, and the other attendees at the "Revisiting Evolutionary Naturalism" workshop at York University for extremely helpful feedback.

Notes

1. On Lockyer, see Thomazine Mary Browne Lockyer, Winifred Lucas Lockyer, and Herbert Dingle, *Life and Work of Sir Norman Lockyer* (London: Macmillan, 1928); and A. J. Meadows, *Science and Controversy: A Biography of Sir Norman Lockyer* (London: Macmillan, 1972).

2. Joseph Hooker to Alexander Macmillan, 27 July 1869, Norman Lockyer Papers, Special Collections, University of Exeter Library, Exeter, UK (hereafter NLP), MS 236. All material from

the Norman Lockyer Papers appears courtesy of Special Collections, University of Exeter; every effort has been made to trace the rights holders.

3. For more on the early shift in *Nature*'s content, see Melinda Baldwin, "The Shifting Ground of *Nature*: Establishing an Organ of Scientific Communication in Britain, 1869–1900," *History of Science* 50 (2012): 131–46.

4. Notably, the X Club, of which Joseph Hooker was a member, had been involved with the *Reader*'s failure. The X Club purchased the *Reader* in 1864, only to sell it in 1865 because the club could not sustain the costs of publication. This was not the X Club's only experience with failed publications; Hooker's friend and fellow X Club member Thomas Henry Huxley had been part owner of a periodical called the *Natural History Review*, but the *NHR* folded within four years.

5. Between 1869 and 1887, at least one member of the X Club was always on the Council of the Royal Society, and Joseph Hooker, William Spottiswoode, and Thomas Huxley held consecutive presidencies of the Royal Society between 1873 and 1884. In addition, between 1869 and 1881, Hooker, Huxley, Tyndall, Spottiswoode, and Lubbock all served as president of the British Association for the Advancement of Science (BA), and many other X Club members served as BA trustees, council members, and section presidents. See Ruth Barton, "'An Influential Set of Chaps': The X-Club and Royal Society Politics, 1864–85," *British Journal for the History of Science* 23 (1990): 53–81; and Ruth Barton, "The X Club: Science, Religion, and Social Change in Victorian England" (PhD diss., University of Pennsylvania, 1976), 116–91.

6. Frank M. Turner, "The Victorian Conflict between Science and Religion: A Professional Dimension," *Isis* 69 (1978): 362; Frank M. Turner, *Contesting Cultural Authority: Essays in Victorian Intellectual Life* (Cambridge: Cambridge University Press, 1993), 180. The word "*professionalize*" is placed in quotation marks because I feel it is a problematic and not entirely appropriate term in the context of Victorian science. For challenges to the concept of professionalization, see Ruth Barton, "'Men of Science': Language, Identity and Professionalization in the Mid-Victorian Scientific Community," *History of Science* 41 (2003): 73–119; Jim Endersby, *Imperial Nature: Joseph Hooker and the Practices of Victorian Science* (Chicago: University of Chicago Press, 2008), 7, 21–27; and Paul Lucier, "The Professional and the Scientist in Nineteenth-Century America," *Isis* 100 (2009): 699–732.

7. Ruth Barton, "Scientific Authority and Scientific Controversy in *Nature*: North Britain against the X Club," in *Culture and Science in the Nineteenth-Century Media*, ed. Louise Henson, Geoffrey Cantor, Gowan Dawson, Richard Noakes, Sally Shuttleworth, and Jonathan R. Topham, 223–35 (Aldershot, UK: Ashgate, 2004).

8. On the decline of the X Club, see R. M. MacLeod, "The X-Club: A Social Network of Science in Late-Victorian England," *Notes and Records of the Royal Society of London* 24 (1970): 314–16.

9. There is an enormous amount of secondary literature on these five men. Helpful biographies include M. W. Taylor, *Men versus the State: Herbert Spencer and Late Victorian Individualism* (Oxford: Oxford University Press, 1992); Rick Rylance, *Victorian Psychology and British Culture, 1850–1880* (Oxford: Oxford University Press, 2000); Endersby, *Imperial Nature*; Adrian Desmond, *Huxley*, 2 vols. (London: Penguin, 1994); Paul White, *Thomas Huxley: Making the "Man of Science"* (Cambridge: Cambridge University Press, 2002); Peter Raby, *Alfred Russel Wallace: A Life* (Princeton, NJ: Princeton University Press, 2001); Ross A. Slotten, *The Heretic in Darwin's Court: The Life of Alfred Russel Wallace* (New York: Columbia University Press, 2004); E. Janet Browne, *Charles Darwin: Voyaging* (Princeton, NJ: Princeton University Press, 1996); and E. Janet Browne, *Charles Darwin: The Power of Place* (Princeton, NJ: Princeton University Press, 2003).

10. George J. Romanes, "Permanent Variation of Colour in Fish," *Nature* 8 (5 June 1873): 101.

11. For a complete account of Romanes's correspondence with Darwin, see Joel S. Schwartz, "George John Romanes's Defense of Darwinism: The Correspondence of Charles Darwin with His Chief Disciple," *Journal of the History of Biology* 28 (1995): 281–316.

12. George J. Romanes, "Physiological Selection: An Additional Suggestion on the Origin of Species," pt. 1, *Nature* 34 (5 August 1886): 314–16; George J. Romanes, "Physiological Selection: An Additional Suggestion on the Origin of Species," pt. 2, *Nature* 34 (12 August 1886): 336–40; and George J. Romanes, "Physiological Selection: An Additional Suggestion on the Origin of Species," pt. 3, *Nature* 34 (19 August 1886): 362–65. For the full paper, see George J. Romanes, "Physiological Selection: An Additional Suggestion on the Origin of Species," *Journal of the Linnean Society* 19 (1886): 337–411.

13. Romanes, "Physiological Selection," pt. 1, 316.

14. On Weismann, see Frederick Churchill, "August Weismann and a Break from Tradition," *Journal of the History of Biology* 1 (1968): 91–112; Frederick Churchill, "The Weismann-Spencer Controversy over the Inheritance of Acquired Characters," in *Human Implications of Scientific Advancement: Proceedings of the XVth International Congress for the History of Science*, ed. Eric Forbes, 451–68 (Edinburgh: Edinburgh University Press, 1977); and Frederick Churchill and Helmut Risler, eds., *August Weismann: Ausgewahlte Briefe und Dokumente, Selected Letters and Documents*, 2 vols. (Freiburg, Germany: Universitätsbibliothek, 1999).

15. George J. Romanes, "Panmixia," *Nature* 41 (13 March 1890): 438.

16. Romanes's ideas on the inheritance of disuse and the shrinking of useless organs follow quite closely on Darwin's views in chapter 13 of *On the Origin of Species*, where Darwin wrote, "On my view of descent with modification, the origin of rudimentary organs is simple. . . . I believe that disuse has been the main agency; that it has led in successive generations to the gradual reduction of various organs, until they have become rudimentary, as in the case of the eyes of animals inhabiting dark caverns, and of the wings of birds inhabiting oceanic islands, which have seldom been forced to take flight, and have ultimately lost the power of flying. Again, an organ useful under certain conditions, might become injurious under others, as with the wings of beetles living on small and exposed islands; and in this case natural selection would continue slowly to reduce the organ, until it was rendered harmless and rudimentary. . . . An organ, when rendered useless, may well be variable, for its variations cannot be checked by natural selection. . . . If each step of the process of reduction were to be inherited, not at the corresponding age, but at an extremely early period of life (as we have good reason to believe to be possible) the rudimentary part would tend to be wholly lost." Charles Darwin, *On the Origin of Species* (1859; repr., Cambridge, MA: Harvard University Press, 1964), 454–55.

17. A. R. Wallace, "Romanes *versus* Darwin: An Episode in the History of Evolution Theory," *Fortnightly Review* 60 (1886): 300–316.

18. Raphael Meldola, "Physiological Selection and the Origin of Species," *Nature* 34 (26 August 1886): 384.

19. George J. Romanes, "Co-adaptation," *Nature* 43 (26 March 1891): 489–90; Raphael Meldola, "Co-adaptation," *Nature* 43 (16 April 1891): 557–58; George J. Romanes, "Co-adaptation and Free Intercrossing," *Nature* 43 (23 April 1891): 582–83; Raphael Meldola, "Co-adaptation," *Nature* 44 (7 May 1891): 7; George J. Romanes, "Co-adaptation," *Nature* 44 (14 May 1891): 28; and Raphael Meldola, "Co-adaptation," *Nature* 44 (14 May 1891): 28–29.

20. George J. Romanes, "Mr. Wallace on Physiological Selection," *Nature* 43 (11 December 1890): 127–28.

21. Romanes, "Co-adaptation and Free Intercrossing," 582.

22. Alfred R. Wallace, "Dr. Romanes on Physiological Selection," *Nature* 43 (18 December 1890): 150.

23. Alfred Russel Wallace to W. T. Thiselton-Dyer, 26 September 1893, Alfred Russel Wallace Papers, British Library, London (hereafter ARWP:BL), MS 46436.300. The copyright of literary works by Alfred Russel Wallace that were unpublished at the time of his death and that are published in this book belongs to the A. R. Wallace Literary Estate. These works are licensed under Creative Commons Attribution-NonCommercial-ShareAlike 3.0 Unported. To view a copy of this, visit http://creativecommons.org/licenses/by-nc-sa/3.0/legalcode.

24. In his reply to Wallace's letter of 26 September 1893, Thiselton-Dyer wrote, "Romanes is an old acquaintance of mine of many years standing. Personally I like him very much; but for his writings I confess I have a great admiration. . . . I must confess I was in total ignorance of what you tell me. I don't see how under the circumstance you can do anything. I was never more surprised in my life, in fact, than when I read your letter. The whole thing is too incredibly preposterous. Romanes laments over me because he says I willfully misunderstood his theory. The fact is poor fellow that I do not think he understands it himself." W. T. Thiselton-Dyer to Alfred Russel Wallace, 27 September 1893, ARWP:BL, MS 46436.301.

25. See W. T. Thiselton-Dyer, "Mr. Romanes's Paradox," *Nature* 39 (1 November 1888): 7–9; George J. Romanes, "Mr. Dyer on Physiological Selection," *Nature* 39 (29 November 1888): 103–4; W. T. Thiselton-Dyer, "Mr. Romanes on the Origin of Species," *Nature* 39 (6 December 1888): 126–27; and George J. Romanes, "Natural Selection and the Origin of Species," *Nature* 39 (20 December 1888): 173–75.

26. W. T. Thiselton-Dyer to Alfred Russel Wallace, 29 October 1889, ARWP:BL, MS 46435.213.

27. Raphael Meldola, "Co-adaptation," *Nature* 44 (14 May 1891): 29.

28. E. Ray Lankester to J. Norman Lockyer, 25 September [1886], NLP, MS 110. Courtesy of Special Collections, University of Exeter. Unfortunately, it does not appear that Lankester's rejected letter to *Nature* has survived.

29. Following the death of his wife, Winifred, in 1879, Lockyer became quite close to George and Ethel Romanes, and carried on a warm personal correspondence with them. See George J. Romanes to J. Norman Lockyer, various letters, NLP, MS 110.

30. George J. Romanes to J. Norman Lockyer, 30 October 1886, NLP, MS 110. Courtesy of Special Collections, University of Exeter.

31. For more information about the *Fortnightly Review*, see "The Fortnightly Review," in *The Wellesley Index to Victorian Periodicals, 1865–1900*, ed. Walter E. Houghton (New York: Routledge, 1999). The gap in pay between *Nature* and general periodicals will be explored later.

32. George J. Romanes, "Physiological Selection and the Origin of Species," *Nature* 34 (9 September 1886): 439.

33. E.g., Meldola, "Physiological Selection and the Origin of Species," 384–85; E. Ray Lankester, "Darwinism," *Nature* 41 (7 November 1889): 9; and Francis Darwin, "Physiological Selection and the Origin of Species," *Nature* 34 (2 September 1886): 407.

34. On Darwin's theory of coral reef formation and debates surrounding the theory, see Alistair Sponsel, "Coral Reef Formation and the Sciences of Earth, Life, and Sea, 1770–1952" (PhD diss., Princeton University, 2009), chaps. 2–4.

35. T. H. Huxley, "Science and the Bishops," *Nineteenth Century* 22 (1887): 637.

36. W. T. Thiselton-Dyer, "The Duke of Argyll and the Neo-Darwinians," *Nature* 41 (16 January 1890): 247.

37. W. T. Thiselton-Dyer, "Politics and the Presidency of the Royal Society," *Nature* 37 (1 December 1887): 103–4. Thiselton-Dyer sought Huxley's advice about his letter; see W. T. Thiselton-Dyer to T. H. Huxley, 7 December 1887, Thomas Henry Huxley Papers, Records and Archives, Imperial College London, London, 27.214.

38. This pattern also bears out in the physical sciences. See Baldwin, "Shifting Ground of *Nature*," 140–41.

39. On Romanes's writings in literary publications, see Joel S. Schwartz, "Out From Darwin's Shadow: George John Romanes's Efforts to Popularize Science in *Nineteenth Century* and Other Victorian Periodicals," *Victorian Periodicals Review* 35 (2002): 133–59. Romanes's work for these publications focused largely on popularization and self-promotion, and he did not use them as a means of making his research known to other men of science.

40. John Maddox, introduction to *"Nature," 1869–1879* (London: Palgrave Macmillan Archive Press, 2002), 3.

41. As James Secord has observed, scientific conversations and lectures were another important form of scientific communication in the early nineteenth century, and like monographs, lectures and conversations had become far less central by the end of the nineteenth century. James A. Secord, "How Scientific Conversation Became Shop Talk," in *Science in the Marketplace: Nineteenth-Century Sites and Experiences*, ed. Aileen Fyfe and Bernard Lightman, 23–59 (Chicago: University of Chicago Press, 2007).

42. I would like to thank Gowan Dawson for suggesting this point.

43. Henry Draper to Norman Lockyer, 8 November 1873, NLP, MS 110. Courtesy of Special Collections, University of Exeter.

44. Peter C. Kjærgaard, "'Within the Bounds of Science': Redirecting Controversies to *Nature*," in *Culture and Science in the Nineteenth-Century Media*, ed. Louise Henson, G. N. Cantor, Gowan Dawson, Richard Noakes, Sally Shuttleworth, and Jonathan Topham, 211–21 (Aldershot, UK: Ashgate, 2004).

45. On the relationship between priority and publication, see Alex Csiszar, "Broken Pieces of Fact: The Scientific Periodical and the Politics of Search in Nineteenth-Century France and Britain" (PhD diss., Harvard University, 2010), chap. 2.

46. William H. Brock, "Advancing Science: The British Association and the Professional Practice of Science," in *Parliament of Science*, ed. R. M. MacLeod and P. M. Collins (London: Science Reviews, 1981), 116.

47. I thank Michael Taylor and James Elwick for drawing my attention to this point. On the importance of publication in general periodicals to Huxley's career and finances, see White, *Thomas Huxley*, 69–75.

48. The *Wellesley Index to Victorian Periodicals* indexes the contents and authors in forty-five Victorian publications, including *British Quarterly Review*, *Contemporary Review*, *Cornhill Magazine*, *Edinburgh Review*, *Fortnightly Review*, *Macmillan's Magazine*, *Nineteenth Century*, and *Westminster Review*. The *Index* includes two popular science publications, *Macmillan's Magazine* and the *Cornhill Magazine*, but no scientific periodicals aimed at an audience of men of science.

49. "Index of Authors," in *Wellesley Index to Victorian Periodicals*.

50. Schwartz, "Out from Darwin's Shadow," 133–59.

51. See, for example, Charles Darwin, "Pangenesis," *Nature* 3 (27 April 1871): 502–3; Charles

Darwin, "Perception in the Lower Animals," *Nature* 7 (13 March 1873): 360; Charles Darwin, "On the Males and Complemental Males of Certain Cirripedes, and on Rudimentary Structures," *Nature* 8 (25 September 1873): 431–32; Charles Darwin, "Recent Researches on Termites and Honey-bees," *Nature* 9 (19 February 1874): 308–9; Charles Darwin, "Sexual Selection in Relation to Monkeys," *Nature* 15 (2 November 1876): 18–19; Charles Darwin, "Fertility of Hybrids from the Common and Chinese Goose," *Nature* 21 (1 January 1880): 207; Charles Darwin, "Inheritance," *Nature* 24 (21 July 1881): 257; and Charles Darwin, "On the Dispersal of Fresh-water Bivalves," *Nature* 25 (6 April 1882): 529–30. After they began working together, Romanes aided Darwin in the preparation of his abstracts for *Nature*. See Schwartz, "Romanes's Defense of Darwinism," 299. Darwin's fondness for using *Nature* to publicize his research and discuss current theories may be related to his distaste for traveling to meetings. See E. Janet Browne, "I Could Have Retched All Night: Charles Darwin and His Body," in *Science Incarnate: Historical Embodiments of Natural Knowledge*, ed. Christopher Lawrence and Steven Shapin, 240–87 (Chicago: University of Chicago Press, 1998).

52. I owe this observation to Michele Aldrich and I am grateful for her willingness to share her research on Darwin's publishing patterns with me.

53. Romanes, "Physiological Selection," pt. 2, 340.

54. A. R. Wallace, "Romanes *versus* Darwin: An Episode in the History of Evolution Theory," *Fortnightly Review* 60 (1886): 300–316.

55. Meldola, "Physiological Selection and the Origin of Species," 384.

56. Lankester, "Darwinism," 9.

57. F. Darwin, "Physiological Selection and the Origin of Species," 407.

58. George J. Romanes, "Physiological Selection," *Nature* 36 (11 August 1887): 341.

59. Frank Miller Turner, *Between Science and Religion: The Reaction to Scientific Naturalism in Late Victorian England* (New Haven, CT: Yale University Press, 1974), 133.

60. For a brief discussion of *Nature*'s role within Macmillan, see Frederick Macmillan to George Macmillan, 25 December 1886, Macmillan Papers, British Library, London, MS 54788.66.

12

From Agnosticism to Rationalism: Evolutionary Biologists, the Rationalist Press Association, and Early Twentieth-Century Scientific Naturalism

PETER J. BOWLER

In 1925 the noted biologist J. B. S. Haldane attended a dinner organized by the Rationalist Press Association (RPA) to commemorate the centenary of the birth of Thomas Henry Huxley, an event made notable by the fact that the chairman dropped dead in the course of the proceedings.[1] The RPA's dinners were black-tie affairs, suggesting that here was an organization in which the radical thinkers could socialize in a manner that proclaimed their membership in the country's cultural elite. Yet as Bernard Lightman has shown, only forty years earlier Huxley had refused to work with the father of the RPA's founder, the publisher Charles Watts, because of the latter's association with radical intellectual and moral opinions.[2] Watts had published work by the atheist Charles Bradlaugh, just the kind of person Huxley thought it necessary to dissociate his scientific naturalism from for fear that it would be tainted by the claim that materialist thought would undermine moral values.[3] Watts's son, Charles Albert Watts, freed agnosticism from the taint of radicalism, but also fell out with Huxley in 1883. Huxley's article on "Possibilities and Impossibilities" eventually appeared in the *RPA Annual* in 1892, leading some scholars to assume that Huxley was reconciled with Watts—although Lightman insists that the relationship was never good.[4] By the early twentieth century the RPA's fortunes had certainly improved, so that figures such as Haldane—scientists with prominent positions in London and Oxbridge—could feel comfortable expressing their radical views through an organization still defined by its association with Watts's publishing house. There is some evidence, though, that the hint of immorality had not gone away, although by the 1920s few seemed to care too strongly about this.

This chapter will chart this transition through the life and work of four eminent biologists, all closely associated with evolutionary biology and all of whom used the RPA as one means of publicizing their radical views. Their lives span the two generations following that of T. H. Huxley, literally so in the case of one, his grandson Julian Huxley. The intermediate generation is represented by the zoologist E. Ray Lankester (1847–1929), who was one of the elder Huxley's protégés and who became associated with the RPA only toward the end of his career. Arthur Keith (1866–1955) came into rationalism without encountering Huxley directly; he became famous for his work in paleoanthropology and was always an active popular writer and lecturer. Julian S. Huxley (1887–1975) was a founder of the modern Darwinian synthesis who became known for his writing on humanism. Last but not least is J. B. S. Haldane himself (1892–1964), who made important contributions to the genetical theory of natural selection but was also a radical writer throughout his career, ending up as a Marxist writing for the *Daily Worker*.[5]

All four biologists adopted a radical philosophy modeled on Huxley's scientific naturalism, with parallel views on the need to replace ecclesiastical authority with that of trained experts in running the affairs of a modern state. Their evolutionism, along with other commitments to a materialistic methodology in the life sciences, certainly played a role in shaping these wider opinions. Since all were anxious to communicate their views to the general public, it is hardly surprising that they should associate themselves with the RPA as it emerged as a leading focus for the promotion of radical opinion in the country.

C. A. Watts had founded the RPA in 1899 as a successor to the Rationalist Press Committee of 1893, itself a successor to an earlier Agnostic Press Fund. The Watts publishing house had issued works by Bradlaugh and was thus firmly associated with the most radical opinions, a factor that no doubt fueled Huxley's suspicions. Watts was concerned to circumvent laws restricting the publication of radical literature and the actions of booksellers who refused to sell it. His activities began to achieve success in the early years of the twentieth century, mainly through the publication of cheap reprints of classic texts of free thought from the Victorian era, including the work of Darwin and Ernst Haeckel. It was Watts who published Joseph McCabe's translation of Haeckel's *The Riddle of the Universe* in 1900, soon reissued as a paperback at the modest price of six pence. Watts also had a stable of writers including McCabe, W. S. Ross, and Samuel Laing, all promoting a position derived from scientific naturalism but increasingly known as rationalism. He also issued the *RPA Annual*, a magazine-format collection of short articles on rationalist themes. Watts's Thinker's Library continued to issue classic reprints and new contributions to rationalism into the middle decades of the century.

Building on Watts's skills as a publisher of cheap literature, the rationalist movement thus became a relatively successful vehicle for translating Huxley's naturalistic philosophy into a form of popular radicalism. An important transitional figure is Edward Clodd, who had produced a number of popular texts on evolutionism and other naturalistic themes.[6] Although originally close to Huxley, and considered by historians to be a scientific naturalist, Clodd subsequently ignored the latter's hostility to rationalism and threw in his lot with Watts and the RPA. He served as chairman of the RPA from 1906 to 1913, and Watts included some of Clodd's works in his cheap reprint series.

Whatever Huxley's suspicions, Watts and the RPA certainly made rationalism available to a wide circle of ordinary readers. That Victorian scientific naturalism had morphed seamlessly into early twentieth-century rationalism is suggested both by Clodd's defection and by the fact that Lankester, also a close friend of Huxley's, eventually began to work with Watts and the RPA. The younger biologists discussed in this chapter also promoted themes associated with the Victorian movement. Like T. H. Huxley, they were all evolutionists and enthusiasts for a materialistic approach to the study of biology. They were all suspicious of the idea of a personal God and immortality. They all supported a greater role for science in public affairs, and they all took an active role in promoting their beliefs to a wider audience.

Two points stand out, however, as indications of the changing circumstances surrounding the promotion of rationalism. The first is that Huxley's suspicion of Watts's populism had been overcome. Eminent scientists with influential positions in the great universities and teaching hospitals were now willing to work with Watts and the RPA to help promote their radical ideals. The main focus of this chapter is to trace how Lankester, Keith, the younger Huxley, and Haldane became a part of the rationalist movement and how they functioned within it. Coming from a wide range of backgrounds, they all made their way—or were born—into the intellectual elite and were willing to use that status to promote their radicalism. Not everyone accepted the changing social status of the RPA, of course. Gowan Dawson relates how, in 1904, Frederick Macmillan advised W. K. Clifford's widow not to allow the RPA to reissue his *Lectures and Essays* because of its radical associations.[7] But the continuing hostility of some conservative thinkers cannot offset the rise in the RPA's fortunes made possible by the support of the eminent scientists studied below. By the 1920s radicalism was accepted both as a strand, albeit still a controversial strand, of the country's intellectual culture and as a popular movement that was gaining wider support.

The second and subsidiary point to note is the changing relationship between intellectual and moral radicalism. Huxley was desperately anxious to head off any suggestion that scientific naturalists were inclined to moral laxity. But of the four biologists discussed here, only Keith seems to have led a life completely free of any scandal. The others all gained some notoriety—openly in Haldane's case, behind the scenes in the others'—for what was considered loose sexual conduct. This suggests that as rationalism became more firmly established, the fears originally articulated by Huxley became less crucial. This was no doubt partly because society's moral values had themselves changed over time, but it also suggests that eminent radical thinkers now had sufficient social status to ignore the hostility of the conservative wing of the establishment.

Rather than recounting four separate biographies, in this chapter I want to focus on a series of topics, comparing and contrasting the situations of the four biologists in each area, to see if the exercise throws light on the evolution of scientific naturalism into rationalism. I will argue that the similarities and differences between the four figures do tell us something useful about the fate of scientific naturalism in the twentieth century. We shall see how rationalist scientists and—by association—the RPA gained a degree of social respectability. Lankester and Keith came from relatively impoverished backgrounds like T. H. Huxley himself, but by the time we reach Julian, the Huxley family had moved firmly into the intellectual and social elite. Significantly, Huxley retained some sympathy with aspects of traditional religious feelings. In the case of Haldane we have someone from an ancient and landed family who moved steadily toward a more extreme radicalism, ending up as a Marxist. Here we see how rationalism was pulled apart by the tensions of the period, some arguing for the preservation of what could be salvaged from older traditions while others wanted a root-and-branch transformation of society. Whatever their positions, however, all four continued the Victorian tradition of promoting their naturalism as widely as possible, and they all took to popular writing like a duck takes to water.

Background and Lifestyle

E. Ray Lankester's early career was closely entwined with that of the mature T. H. Huxley. Although he was never his actual teacher, Huxley mentored Lankester and served as his role model. Edwin Ray was the son of Edwin Lankester, who—like Huxley—had gained his early training in medicine and who rose to become the coroner for Central Middlesex and an important figure in public health reform. He was also an expert microscopist and a founding member of the British Association for the Advancement of Science. Edwin was a Nonconformist in religion and (unlike his son) preserved his faith throughout his life. Nevertheless, he was on good terms with Huxley, and the young E. Ray Lankester grew up in a home environment where Huxley and other scientific luminaries were regular visitors. The family never had much money, and E. Ray was plagued by financial concerns throughout his career—he too had to make his own way as a professional scientist. He studied first at Cambridge but then moved on a scholarship to Oxford, where he did well but acquired a hatred of the university's association with the Church and landed gentry that was to reemerge later in his career when he became Linacre Professor of Comparative Anatomy there. After extensive travels in Germany and a period at the newly founded Stazione Zoologica in Naples, he became professor of zoology at University College London. Although he was very successful there, financial considerations led him to take the Oxford chair in 1890 followed by the directorship of the Natural History Museum in London in 1898. He was forced to retire in 1907 by a rigid application of the civil service rule requiring retirement at age sixty, although he was knighted in the same year.[8]

Lankester always looked up to Huxley and shared most of the latter's views on social and philosophical issues. Huxley certainly took Lankester under his wing and promoted his career—he was extremely annoyed when Lankester turned down a carefully prepared move to the Edinburgh chair of natural history in 1883. But even he regarded the younger man as a loose cannon, inclined to cause offense through intemperate reactions where a more diplomatic approach might have been more effective. Lankester's fulminations were directed principally at the straitjacket imposed on the emerging scientific profession by the old religious and social elite, a situation that was particularly obvious at Oxford. In this respect his career was closely parallel to that of Huxley, and as we shall see in the next section, he also came to share Huxley's naturalistic worldview.

In one area, though, Lankester's lifestyle was very different from that of the straitlaced Huxley. He never married, and acquired a dubious reputation for loose sexual morality. This reputation was quite justified, as it turns out—his private diary reveals that he regularly formed relationships with prostitutes on his summer visits to France.[9] Lankester's immorality was confirmed in his critics' eyes when he was arrested in Piccadilly Circus one night in 1895 for defending prostitutes against police harassment. There can be little doubt that—coupled with his intemperate behavior—this reputation was responsible for his being removed from his position at the Natural History Museum. The two previous directors, Richard Owen and W. H. Flower, had been allowed to serve long past their sixtieth birthdays. But whatever the views of the more old-fashioned members of the establishment, Lankester's rejection of the strict Nonconformist code of personal morality shared by his father and by T. H. Huxley was in tune with the more liberal attitudes of many members of the upper class. Whatever his sense of financial insecurity, he moved in the highest circles of the intellectual elite, and those circles increasingly allowed a more lax attitude in matters relating to sex. It should be noted that both of the younger members of our quartet participated in this move—the younger Huxley was a notorious womanizer and Haldane was involved in a much-publicized divorce case investigated by the "sex viri" of Cambridge University.[10] As the rationalist movement gained a more secure position in the social world, it evidently abandoned the code of personal morality that the elder Huxley had tried to preserve. This was no doubt made possible by wider changes in moral values that diminished the necessity for intellectual radicals to protect their position by maintaining an irreproachable personal life.

Both Haldane and the younger Huxley were raised from childhood as members of the country's social and intellectual elite.[11] Huxley was educated at Eton, where Haldane (a few years younger) eventually fagged for him. Both then moved on to Oxford and began careers as professional scientists. Huxley's first real job was at the Rice Institution in Houston, where he was already hailed as a celebrity because of his name (although he himself still felt intellectually insecure). He returned to Oxford after the Great War and eventually collaborated with Haldane on a successful biology textbook. Subsequently appointed to a chair at King's College London, he abandoned his formal career to work with H. G. Wells on the popular survey *The Science of Life* (1929–30). Keenly aware of his professional isolation (he was turned down at first for a fellowship of the Royal Society), he then became secretary of the Zoological Society. Tensions built up due to the amount of time he was spending on writing and other activities, and he was eventually forced to resign. He never held an academic post again, although he did serve as the director general of UNESCO.

Haldane's social background was even more secure. He came from an ancient and landed Scottish family: his father, John Scott Haldane, was professor of physiology at Oxford, and his uncle, R. B. Haldane, was a noted politician and intellectual who became Viscount Haldane of Cloan. After Eton he worked on physiology with his father but eventually became professor of genetics at University College London. Notoriously undiplomatic in personal interactions, he combined an immense productivity in science, including his work on genetics and natural selection, with an increasingly active political life, which moved steadily to the Left. In 1957 he quit Britain for India, partly in protest against the failed Anglo-French attempt to seize the Suez Canal.

Arthur Keith is in many ways the outlier of our group of scientists, both socially and professionally. He came from a poor Scottish farming family and scraped together the funding for a medical education at Aberdeen.[12] He then served as a medical officer with a mining company in Siam (Thailand), where he extended his interest in human anatomy to the study of gibbons and began to work seriously on the relationship between humans and apes. When he returned to Britain, he settled in London, where, after much difficulty, he established himself as an anatomy lecturer in the London Hospital. He also began to work and write on hominid fossils, gaining wide public recognition in this field (based in part on his involvement in the now notorious Piltdown discoveries). Although one might have expected someone from his insecure background to associate with figures such as the elder Huxley and Lankester, this was not the case. Keith only settled in London in the year of Huxley's death, and he later recalled that he was glad he learned his zoology from Alleyne Nicholson at Aberdeen, who did not teach using the "Huxleyan types"—single species employed as typical examples of a whole group. He also seems to have been intimidated by Lankester's forceful personality.[13] Keith's medical background led him into different professional circles from those occupied by Lankester, the Huxleys, and Haldane, although at the end of his life he did join the Darwinian circle in the sense that he took up residence at Down House after the British Association for the Advancement of Science acquired it.

Roads to Rationalism: Science

Whatever the differences in background between our four biologists, their trajectories toward rationalism exhibit remarkable similarity. One unifying factor is their scientific work, both in evolutionism and in areas such as physiology and biochemistry. All four were Darwinists by some definition of that term, following the transition from the loosely defined Darwinism of the 1870s and 1880s through to the emergence of what Julian Huxley called the "modern synthesis" of natural selection and genetics. This in itself does not guarantee hostility, at least toward liberal religious thought, and Julian Huxley's leanings in this direction allowed him to make common cause with those who sought to interpret evolution as a morally significant process. But all four were interested in the effort to create a scientific understanding of how the living body functions, and although Huxley was tempted by Henri Bergson's notion of a creative life force, all four were, in effect, on the mechanist side in the mechanist-vitalist debate.[14] The combination of materialism and Darwinism necessarily forced them toward the rationalist position.

Lankester followed the common trajectory of many late nineteenth-century morphologists into evolutionism, focusing more on the course of evolution than on its causes. He was of course influenced by T. H. Huxley, but also by Ernst Haeckel (he supervised the translation of the latter's *Natural History of Creation*). His research focused on evolutionary morphology, although he had done significant work in physiology as a young man and stayed in touch with developments in the biomedical field (at University College many of his students were from the medical faculty). He gradually abandoned an early willingness to tolerate some nonselectionist mechanisms, moving into what became known as the neo-Darwinian camp in the 1890s. But he seems not to have engaged with genetics and the early efforts to synthesize the new science of heredity with Darwinism.

Keith came into the field via human anatomy and a commitment to medical education and research. His evolutionism was very much driven by his interest in hominid fossils and the transition from apes to humans. He always called himself a Darwinist, but he seems to have had little interest in natural selection operating within populations and even suggested that variation and hence evolution was driven by the effects of hormones on the body. He focused on natural selection operating between groups, and developed a theory in which competition between races was the driving force of human evolution. His interest in physiology led him to adopt a materialist view of the operations of the human body, which in turn led him to reject any notion of a spiritual dimension to life.

Julian Huxley combined an interest in evolution with the study of individual development. His first major success was the revelation that hormones played a role in the process by which the organism matures. His work on the axolotl, an amphibian that becomes sexually mature while still having gills, led to newspaper headlines about the possibility of human rejuvenation. He was interested in animal behavior, a study that he eventually used as a springboard for his participation in the emerging synthesis, not just of genetics and natural selection, but of the technicalities of population genetics with a whole variety of field studies. *The Science of Life* (1929–30), which he wrote with H. G. Wells, provided the first account of the new Darwinism, and of many other developments in the life sciences, accessible by a wide readership. His 1942 book *Evolution: The Modern Synthesis* gave the rejuvenated Darwinian theory its unofficial title.

Haldane's first involvement with science was as an assistant to his father, who was particularly interested in respiration, with various practical applications, including the study of bad air in coal mines. He then moved on to genetics, where he soon began to appreciate the value of the new population genetics. He made important contributions to the emerging genetical theory of natural selection, his *The Causes of Evolution* (1932) providing a useful nonspecialist survey of the topic. Here again the combination of evolutionism, especially in its Darwinian form, with an awareness of the positive influence of the mechanist approach in the biomedical sciences, played an important role in supplying the foundations for a rationalist philosophy.

Roads to Rationalism: Doubts about Religion

If science led all four biologists toward a materialist position, their rejection of traditional religious values was also based on wider considerations. But here there were differences of emphasis, and the extent of their commitment to rationalism varied—Huxley in particular remaining uncomfortable with the extreme rejection of all religious belief favored by most members of the RPA.[15] Curiously, Huxley was the only one of the four not to be exposed to religious belief in childhood, acquiring a more sympathetic attitude through personal contacts with liberal Anglicans at Oxford. The others all abandoned any initial belief relatively early in their careers thanks to a variety of influences. All built up some relationship with the RPA, although this was a very late development in Lankester's career. Keith remained loyal throughout his life, with Watts publishing many of his later books, including his autobiography. Huxley, as just noted, gradually came to distrust the rabid nature of some of the RPA's publications, in this respect sharing his grandfather's suspicions. Haldane moved steadily further toward a radical ideology, culminating in his acceptance of Marxism, although he never severed his links with the RPA.

E. Ray Lankester may have been raised in a Dissenting household, but he came strongly under the elder Huxley's influence as soon as he entered university. When he took up his scholarship at Oxford, he was immediately aware of the university's position as a bastion of the social establishment and of the Church's role in maintaining the existing hierarchy. Not surprisingly, most of his early activities were associated with Huxley's campaign to get the professional classes inserted into that hierarchy and to ensure that science gained the recognition it deserved both in education and in society at large. During his early career he did not make a point of publishing attacks on the actual belief system of Christianity, although his commitment to Darwinism was made plain in his little book on *Degeneration*, published in 1880.[16] The one area in which his rationalist credentials did hit the headlines was in 1876, when he played a key role in unmasking the fraudulent spiritualist medium Henry Slade. Lankester remained a committed opponent of spiritualism into the movement's revival after the Great War.[17]

By this time he had emerged as a controversial exponent of free thought, most notably in his 1905 Romanes Lecture "Nature's Insurgent Son," which attracted much criticism in the press and was publicly repudiated by G. J. Romanes's widow.[18] Following his enforced retirement in 1907, Lankester began his popular "Science from an Easy Chair" articles in the *Daily Telegraph*, reprinted in a series of books. These frequently referred to issues that allowed him to make points favorable to the naturalistic perspective. Watts's publishing house subsequently issued a reprint of some of these articles and of *The Kingdom of Man*, Lankester's 1907 book containing his controversial Romanes Lecture of 1905. During the Great War, Lankester began sporadic contributions to the *RPA Annual*. In 1922 he contributed an article "Is There a Revival of Superstition?" and renewed his attack on spiritualism, which was now being promoted by Oliver Lodge and Arthur Conan Doyle.[19] He criticized J. Arthur Thomson for allowing Lodge to write a defense of spiritualism in his *The Outline of Science* (1922) and wrote a critical review of Conan Doyle's *History of Spiritualism*. Probably his most influential rationalist activity, in terms of reaching the public, was his involvement with the first part of H. G. Wells's *The Outline of History* (1920), detailing the evolutionary origins of humankind. Wells's 1895 story *The Time Machine* had drawn on Lankester's views on degeneration and the two men became friends. Wells used Lankester as a consultant when writing *The Outline of History*, which promoted a highly secularist vision of the rise of civilization. Watts subsequently published Wells's response to the critique of his book by Hilaire Belloc.[20] Lankester's involvement with Watts and the RPA was thus peripheral, but his contribution to the defense of the rationalist cause was not.

We have already seen that Arthur Keith became a rationalist without encountering Huxley's circle. As he built his reputation as a paleoanthropologist, he became associated in the public mind with the theory of evolution, but at first he did not go out of his way to emphasize the materialistic implications of his view of human origins. He later admitted that he had been reluctant to publicize views that he knew might upset some of his contemporaries, and his autobiography notes that when he did finally come out of the closet, some of his former friends rejected him. He certainly became active with the RPA during the 1920s, responding in the *RPA Annual* in 1922 to Hilaire Belloc's and G. K. Chesterton's attacks on Darwinism[21] and giving the RPA's Conway Memorial Lecture in 1925 on "The Religion of a Darwinist," which he admitted was actually an acknowledgment of his own lack of religion. Keith finally became publicly identified as a materialist when he gave his presidential address to the British Association for the Advancement of Science in 1927. Here he used both Darwinism and the materialistic tendencies of the modern biomedical sciences to demolish the idea of a spiritual component in the human mind. The address was broadcast by the BBC and generated a raft of newspaper headlines and critical responses by Oliver Lodge and many others. It was after this address, he later recalled, that he was cut dead by some of his friends.[22] Watts published the text of the address in its Forum series (curiously, the book was not a success), and subsequently issued a number of Keith's evolutionary writings.[23] Watts also published his autobiography in 1950.

In some respects Keith's aggressive materialism reflected the side of the RPA's activities that made Julian Huxley feel uncomfortable. We have noted that Huxley was the only one of the four scientists not to have been exposed to formal religion in childhood, yet when he went up to Oxford, he made friends with liberal Anglicans such as William Temple and B. H. Streeter and remained in contact with them for the rest of his career. Huxley was not religious in any orthodox sense—he did not believe in a personal God, or in an afterlife. But he did think that the universe had a moral purpose and he accepted that one might experience awe and wonder when confronted with its immensity, emotions very similar to those evoked by religion. His early exposure to Bergson helped to convince him that evolution was progressive, although not rigidly predetermined, and he retained this faith for the rest of his life. His books *Religion without Revelation* (1927) and *What Dare I Think?* (1931) were detailed expositions of this credo. Huxley certainly had links with Watts and the RPA—*What Dare I Think?* was based on his RPA Conway Memorial Lecture, and Watts published his *Science, Religion and Human Nature* in 1930. But his autobiography records his growing dissatisfaction with the RPA's strident rationalism, and in his later career he preferred to call himself a humanist.

If Huxley wanted a limited reconciliation with liberal religion, Haldane moved in exactly the opposite direction. His father had deep, if unconventional, religious beliefs and was a significant figure in the effort to forge a reconciliation between idealist science and liberal religion in the early decades of the twentieth century. The younger Haldane was turned off religion at Eton, partly as a result of the attentions of an overzealous matron who promoted Anglo-Catholicism. He immediately moved into the orbit of the RPA and was enthused by reading Haeckel's *Riddle of the Universe*.[24] An early notebook reveals him already speculating along the lines later developed in his *Daedalus* (1924), a visionary account of what humankind might achieve through control of the material and biological worlds.[25] Science, he argued, was essentially hostile to all conventional religions and to the fixed moral codes they tried to defend. At first he was not a fully fledged materialist, holding that on death the human spirit might merge with the infinite mind lying behind the material world. Nevertheless, he became a regular contributor to the *RPA Annual* and gave the Conway Memorial Lecture two years before Huxley. In the 1920s, he was still thinking along lines rather similar to those favored by Huxley, although far more radical in exposing the moral challenges that would be raised by increased scientific control of human life. In the 1930s, he began to move rapidly to the political Left and became increasingly involved with Marxism, although he did not join the Communist Party. He began to write regularly for the Communist newspaper, the *Daily Worker*. He did not, however, sever his links with the RPA, but his rationalism was now subsumed into dialectical materialism. His famous article on the origin of life was first published in the *RPA Annual* in 1929.[26]

The very different directions in which Huxley and Haldane moved reveal the tensions that enlivened the cultural ferment of the early twentieth century. In the early decades, there was a concerted effort by many thinkers to repudiate what was perceived as the extreme materialism promoted by Victorian thinkers such as T. H. Huxley. Keith offers perhaps the best example of a scientist who maintained this original form of rationalism into the new century. But Julian Huxley had some sympathy for the move to reconcile science and religion, and in the 1920s J. B. S. Haldane was sometimes confused with his father, the physiologist J. S. Haldane, because he still implied there might be a universal mind lying behind the material world. Huxley's diluted rationalism merged into his later humanism, allowing him to retain the idea that science could underpin a philosophy that would retain a sense of moral purpose in life via a belief in progress. Haldane, by contrast, followed the increasingly popular move to the political Left, which was driven by the rise of fascism and the economic problems of the old social order revealed by the Depression. This led him to Marxism and hence to a fully materialist view of the nature of life. The polarization of political and intellectual life in the 1930s effectively pulled rationalists in two opposing directions, while the old position still represented by Keith seemed increasingly out of touch with the modern world.

Rationalism and Politics

Haldane's transition to Marxism reminds us that scientific naturalism had not begun as a purely intellectual movement. The elder Huxley was active not only in promoting the interests of professional scientists as the source of expertise needed by a modern economy, but also in spreading information about science widely through the population in the hope that this would encourage wider recognition of its significance. Huxley was a meritocrat, his aim being to allow everyone, whatever their social origin, access to education and hence to professional influence. But he was not a liberal in all his political opinions, remaining suspicious of efforts to allow women equal access to education and endorsing the conventional view of his time that the white race was superior to all others. Following the opinions of our quartet of scientists will tell us something about how this package of social opinions fared in the transition to rationalism and its fragmentation under the social tensions of the 1930s. Ostensibly the RPA was not a political organization, but this merely allowed a plurality of opinions on such matters to flourish under its aegis. In the end, though, as politics came increasingly to dominate intellectual life, that plurality undermined the original sense of unity that its founders had hoped to maintain.

All four were essentially meritocrats, although Haldane's adoption of socialism moved him away from this position in his later life. In this respect the link with H. G. Wells is significant, since he was a staunch advocate both of wider education in the scientific way of thought and of giving power to the experts. Lankester certainly shared Huxley's view that clever people from poor backgrounds should be given the opportunity to improve themselves through education. Although a friend of Karl Marx—he was one of the few people at Marx's funeral—he was not a socialist. His main focus, like Huxley's, was to ensure that science should be recognized as a vital part of public life. He was probably more liberal than Huxley on the issue of allowing equality of opportunity to women, and seems to have expressed no strong opinions on the race question. In his later career he expressed suspicion of the plans for selective breeding of the human population advocated by the eugenics movement, and in this he both reflected Huxley's concerns and prefigured the attitude adopted by the three younger biologists.[27] All seem to have been convinced that in theory some individuals inherit a character superior to others, but all were certain that in the present situation any attempt to apply this principle in society was both impractical and immoral. Not enough was known about the relationship between heredity and environment, and the huge environmental differences between the social classes meant that any effort to identify the superior individuals was deeply flawed. Keith, like Lankester, expressed reservations about eugenics,[28] while both the younger Huxley and Haldane emerged as open critics of the movement.

One area where there were significant differences of opinion was the issue of race. Here Keith is the exception, mainly because the position he adopted on human origins committed him to the view that racial differentiation had occurred at a very early stage. As an anatomist, Keith did not take the latest advances in genetics on board, and he thus remained convinced that the races were, if not actually distinct species, certainly very well marked and ancient varieties. His support for Darwinism was also expressed through a theory in which struggle operated between races rather than individuals. He seems to have developed this view in the 1920s, although it was not expressed openly until the 1940s—Watts published his *New Theory of Human Evolution* in 1948. Julian Huxley also began his career accepting the common view that the white race was superior—he expressed quite derogatory opinions of black culture while in Houston.[29] But later on, his commitment both to the new Darwinism and egalitarianism allowed him to emerge as a leading opponent of Nazi race theory. Haldane's views on race are apparent from the fact that his move to India was in part an effort to identify with the downtrodden races of the empire.

Popular Writing

All four of our scientists were active in writing for the general public about science and its implications, in this respect continuing the tradition established by T. H. Huxley and many of the scientific naturalists.[30] In Lankester's case this was a late development—he began his "Science from an Easy Chair" series for the *Daily Telegraph* in order to supplement his pension following his enforced retirement. The series was a notable success and continued until the Great War. Lankester's subsequent work with H. G. Wells linked him with one of the great popular exponents of a secularist worldview. Keith also began popular writing under financial pressure, but in his case it was to make money when his career as an anatomy lecturer was still not secure. He began writing for periodicals such as the *Illustrated London News* and found that he could make a significant addition to his income in this way. Even after he became well established, he continued to accept commissions from editors to write popular pieces. When his rationalist opinions finally hit the headlines in 1927, he was thus well prepared to engage in a public debate with opponents such as Lodge.

Huxley too realized at a very early stage that he could make money and gain a wider reputation through popular writing. His work on growth hormones gave him access to the popular press, and from this point on he assiduously cultivated this kind of writing, culminating in his collaboration with Wells on *The Science of Life*. His engagement with popularization definitely damaged his career as a scientist, almost certainly playing a role in delaying the achievement of the FRS. After his expulsion from his position at the Zoological Society, he actually supported himself for some periods through his writing. Haldane was also an aggressive promoter of his writing skills, although in his case the motivation was more idealistic and he often gave away his royalties to good causes. In the 1920s he became anxious to extend public awareness of science and its implications and began to write for daily newspapers as well as for the more serious magazines. His long commitment to the *Daily Worker* was the culmination of this side of his career. Putting the work of our four scientists together, it can be said that they all made a significant impact in the area of the popularization of science, while at the same time exploiting their skills to increase public awareness of the opportunities and threats offered by scientific and technological advances. All four wrote either occasionally or regularly for Watts and the RPA, thus putting their writing skills at the service of the most public manifestation of the rationalist cause.

Conclusion

Whatever T. H. Huxley's suspicions of Watts and his rationalist publishing activities, the careers of the four biologists studied here show how professional scientists began to throw their weight behind the movement in the early twentieth century. All four were eminent figures with Oxbridge or powerful London connections, indicating that rationalism had become acceptable in the intellectual establishment—although there were still many critics, of course. In the 1920s the RPA enjoyed a brief period in the center of the intellectual stage. But while Lankester and Keith maintained a form of scientific naturalism recognizably similar to that advocated by T. H. Huxley and other Victorian figures, Haldane and the younger Huxley reveal the tensions that emerged in the radical community during the early twentieth century. There was a substantial reaction against naturalism, and the younger Huxley's evident discomfort at the extreme materialism expressed by some RPA members, including Keith, shows that he, at least, thought it was important to maintain links with the more liberal wing of this reaction. The elder Huxley too had remained on good terms with liberal religious thinkers, but Julian openly endorsed their vision of nature founded on ethical progress. Haldane by contrast moved steadily further toward a more radical position, eventually joining the substantial proportion of the scientific community that adopted Marxism as a counterweight to the threat of fascism.

The RPA remained active, but it no longer flourished as a major forum for the expression of radical views in the more turbulent years of the 1930s. The rationalist movement now seemed outdated, its message dispersed among the humanists (following Huxley) and the Marxists (including Haldane). Because the RPA's primary focus was nonpolitical, however, its membership increasingly overlapped with that of the British Humanist Association. It changed its name to the Rationalist Association in 2002 and now publishes the *New Humanist* magazine. The Watts publishing house continued after the war, and its Thinker's Library series remained popular, but it too was no longer seen as a cutting-edge promoter of radical ideals.

This study has also revealed other significant differences, in part reflecting the various professional interests of those concerned. While all four were deeply involved in the campaign for wider scientific education, they were for the most part convinced that the country should be led by professionals and experts. But Keith's views on the race question show how a particular scientific background (in his case in anatomy and paleoanthropology) could skew views on a particular social question. While Huxley and Haldane emerged as leaders in the fight against racism, Keith's theory of evolution driven by race conflict seems to reflect an ideology that could all too easily have been identified with the forces that most radicals were determined to resist.

There was always some diversity underlying the apparent unity of the rationalist movement, and in the end this diversity pulled the movement apart. In the case of Keith's views on race, this tension reflects disciplinary as well as ideological differences. But the increasing divergence between the intellectual and political views of Haldane and the younger Huxley suggests deeper tensions that emerged as the rationalists confronted the social traumas of the 1930s. It was no longer quite so obvious that the scientific elite had the power to transform society for the better, nor was it clear that a world freed from all vestige of irrationalism would necessarily be a better place. The most radical thinkers moved from rationalism to dialectical materialism because they saw Marxism as the only means of combating economic inequalities and fascism. But those who feared the unsettling effects of complete materialism on public life backed away from the more extreme expressions of materialism and followed the younger Huxley toward what became known as humanism.

Notes

1. This event, along with others concerning RPA dinners, is recounted in Ronald Clark, *J.B.S.: The Life and Work of J. B. S. Haldane* (London: Hodder and Stoughton, 1968), 62.

2. Bernard Lightman, "Ideology, Evolution, and Late-Victorian Agnostic Popularizers," in *History, Humanity and Evolution: Essays for John C. Greene*, ed. James R. Moore, 285–309 (Cambridge: Cambridge University Press, 1989).

3. Gowan Dawson, *Darwin, Literature and Victorian Respectability* (Cambridge: Cambridge University Press, 2007).

4. On the presumed reconciliation, see Michael Rectenwald, "Secularism and the Culture of Nineteenth-Century Scientific Naturalism," *British Journal for the History of Science* 46 (2013): 251–52; and Lightman, "Ideology, Evolution, and Late-Victorian Agnostic Popularizers," 303.

5. For background on the activities of the four scientists, see Peter J. Bowler, *Reconciling Science and Religion: The Debate in Twentieth-Century Britain* (Chicago: University of Chicago Press, 2001), and *Science for All: The Popularization of Science in Early Twentieth-Century Britain* (Chicago: University of Chicago Press, 2009).

6. On Clodd, see Bernard Lightman, *Victorian Popularizers of Science: Designing Nature for New Audiences* (Chicago: University of Chicago Press, 2007), 253–66. See also Frank Miller Turner, *Between Science and Religion: The Reaction to Scientific Naturalism in Late Victorian England* (New Haven, CT: Yale University Press, 1974), 9.

7. Dawson, *Darwin, Literature and Victorian Respectability*, 188.

8. For biographical information, see Joseph Lester (edited, with additions, by Peter J. Bowler), *E. Ray Lankester and the Making of Modern British Biology* (Stanford in the Vale: British Society for the History of Science, 1995).

9. Because of its sensitive nature, the contents of Lankester's diary (still in private hands) were not mentioned in the biography cited above and cannot be cited in detail here.

10. This was in 1925. As Haldane remarked, the "sex viri" were not the "sex weary" but the six men appointed under the statutes of the University of Cambridge who had the power to dismiss a staff member found guilty of gross or habitual immorality; see Clark, *J.B.S.*, 75.

11. For biographical information, see Julian Huxley, *Memories* (New York: Harper and Row, 1970); C. Kenneth Waters and Albert Van Helden, eds., *Julian Huxley: Biologist and Statesman of Science* (Houston: Rice University Press, 1992); and Clark, *J.B.S.*

12. Arthur Keith, *An Autobiography* (London: Watts, 1950).

13. Ibid., 73, 262.

14. Huxley's first book, *The Individual in the Animal Kingdom* (1912), was strongly influenced by Bergson. There was some input from the philosophy of monism into all four biologists' attitudes to this topic (although not necessarily Haeckel's version of the philosophy); see Peter J. Bowler, "Monism in Britain: Biologists and the Rationalist Press Association," in *Monism: Science, Philosophy, Religion, and the History of a Worldview*, ed. Todd Weir, 179–96 (New York: Palgrave Macmillan, 2012).

15. Huxley, *Memories*, 152. On his links with William Temple, see 150.

16. Lankester, *Degeneration: A Chapter in Darwinism* (London: Macmillan, 1880).

17. Lester, *E. Ray Lankester*, chap. 8 and 212–13.

18. Published as chap. 1 of Lankester, *The Kingdom of Man* (London: Archibald Constable, 1907). On the reaction, see Lester, *E. Ray Lankester*, 165.

19. E. Ray Lankester, "Will Orthodox Christianity Survive the World War?" *RPA Annual* (1917): 13–18; and "Is There a Revival of Superstition?," *RPA Annual* (1922): 3–10.

20. H. G. Wells, *Mr Belloc Objects to "The Outline of History"* (London: Watts, 1926).

21. Arthur Keith, "Why I Am a Darwinist," *RPA Annual* (1922): 11–14.

22. Keith, *Autobiography*, 519.

23. Arthur Keith, *Darwinism and What It Implies* (London: Watts, 1928) and *Darwinism and Its Critics* (London: Watts, 1935).

24. Clark, *J.B.S.*, 24.

25. Haldane's youthful notebook is in the J. B. S. Haldane papers, National Library of Scotland, MS20578. For more information on Huxley's and Haldane's views on religion, see Bowler, *Science for All*, 70–73.

26. Haldane, "The Origin of Life," *RPA Annual* (1929): 3–11, reprinted in Haldane, *The Inequality of Man and Other Essays* (London: Chatto and Windus, 1932), 148–60.

27. Lester, *E. Ray Lankester*, 176–77. On his links with Marx, see 185.

28. Keith, *Autobiography*, 552–53.

29. See Elazar Barkan, "The Dynamics of Huxley's Views on Race and Eugenics," in *Julian Huxley*, ed. Waters and Van Helden, 230–37. On eugenics, see Garland E. Allen, "Julian Huxley and the Eugenical View of Human Evolution," ibid., 193–222.

30. For details, see Bowler, *Science for All*, esp. chap. 11.

Acknowledgments

The editors would like to thank those who participated in the workshop on "Revisiting Evolutionary Naturalism: New Perspectives on Victorian Science and Culture," which took place at York University 6–7 May 2011, especially Robert Smith, Josipa Petrunic, Jonathan Smith, and Michael Taylor. We are also indebted to the workshop coordinator, Melinda Baldwin, and to PhD students Ali McMillan and Cam Murray, for the superb event that they organized on our behalf. We are grateful to those who sponsored the workshop, the Social Sciences and Humanities Research Council of Canada (SSHRC), the Situating Science SSHRC Knowledge Cluster Grant, the Office of the Vice-President Research and Innovation (York University), the Faculty of Graduate Studies (York University), the Humanities Department (York University), and the Faculty of Liberal Arts and Professional Studies (York University). Janet Friskney, the research officer for the Faculty of Liberal Arts and Professional Studies, gave invaluable aid to us as we drafted the application to SSHRC's Grant Aid to Research Workshops and Conferences in Canada. Gowan Dawson would also like to thank the Leverhulme Trust for the award of a Research Fellowship during which the final stages of the volume were completed.

Karen Darling and her staff at the University of Chicago Press have been, as usual, a joy to work with. Two anonymous readers provided extremely helpful suggestions for revisions that strengthened the volume.

This volume is dedicated to the late Frank M. Turner, whose work on scientific naturalism continues to challenge and inspire scholars in the field. The year 2014 will mark the fortieth anniversary of his book *Between Science and Religion: The Reaction to Scientific Naturalism in Late Victorian England.* He is sorely missed.

Bibliography of Major Works on Scientific Naturalism

Annan, Noel. 1984. *Leslie Stephen: The Godless Victorian.* Chicago: University of Chicago Press.

Barr, Alan P., ed. 1997. *Thomas Henry Huxley's Place in Science and Letters: Centenary Essays.* Athens and London: University of Georgia Press.

Barton, Ruth. 1983. "Evolution: The Whitworth Gun in Huxley's War for the Liberation of Science from Theology." In *The Wider Domain of Evolutionary Thought*, edited by David Oldroyd and Ian Langham, 261–87. Dordrecht, Netherlands: Reidel.

———. 1987. "John Tyndall, Pantheist: A Rereading of the Belfast Address." *Osiris*, 2nd ser., 3:111–34.

———. 1990. "'An Influential Set of Chaps': The X-Club and Royal Society Politics, 1864–85." *British Journal for the History of Science* 23:53–81.

———. 1998. "'Huxley, Lubbock and Half a Dozen Others': Professionals and Gentlemen in the Formation of the X Club, 1851–1864." *Isis* 89:410–44.

———. 2003. "'Men of Science': Language, Identity and Professionalization in the Mid-Victorian Scientific Community." *History of Science* 41:73–119.

———. 2004. "Scientific Authority and Scientific Controversy in *Nature*: North Britain against the X Club." In *Culture and Science in the Nineteenth-Century Media*, edited by Louise Henson, Geoffrey Cantor, Gowan Dawson, Richard Noakes, Sally Shuttleworth, and Jonathan R. Topham, 223–35. Aldershot, UK: Ashgate.

———. Forthcoming. *The X Club: Power and Authority in Victorian Science.*

Becker, Barbara. 2011. *Unravelling Starlight: William and Margaret Huggins and the Rise of the New Astronomy.* Cambridge: Cambridge University Press.

Beer, Gillian. 1996. *Open Fields: Science in Cultural Encounter.* Oxford: Clarendon Press.

Bicknell, John W., ed. 1996. *Selected Letters of Leslie Stephen.* 2 vols. London: Macmillan.

Blinderman, Charles. 1961. "John Tyndall and the Victorian New Philosophy." *Bucknell Review* 9:281–90.

———. 1966. "T. H. Huxley: A Re-evaluation of His Philosophy." *Rationalist Annual*, 50–62.

Bowler, Peter J. 1983. *The Eclipse of Darwinism: Anti-Darwinian Evolution Theories in the Decades around 1900.* Baltimore and London: Johns Hopkins University Press.

———. 2001. *Reconciling Science and Religion: The Debate in Early-Twentieth-Century Britain.* Chicago: University of Chicago Press.

Brock, William H., and Roy M. MacLeod, eds. 1980. *Natural Knowledge in Social Context: The Journals of Thomas Archer Hirst FRS.* London: Mansell.

Brock, William H., N. D. McMillan, and R. C. Mollan, eds. 1981. *John Tyndall: Essays on a Natural Philosopher.* Dublin: Royal Dublin Society.

Brown, Alan W. 1947. *The Metaphysical Society: Victorian Minds in Crisis, 1869–1880.* New York: Columbia University Press.

Budd, Susan. 1977. *Varieties of Unbelief: Atheists and Agnostics in English Society, 1850–1960.* London: Heinemann.

Burkhardt, Frederick H., et al., eds. 1983–. *The Correspondence of Charles Darwin.* 20 vols. Cambridge: Cambridge University Press.

Cantor, Geoffrey, Gowan Dawson, Graeme Gooday, Richard Noakes, Sally Shuttleworth, and Jonathan R. Topham. 2004. *Science in the Nineteenth-Century Periodical: Reading the Magazine of Nature.* Cambridge: Cambridge University Press.

Clark, J. F. M. 1997. "'The Ants Were Duly Visited': Making Sense of John Lubbock, Scientific Naturalism and the Senses of Social Insects." *British Journal for the History of Science* 30:151–76.

Clark, Ronald W. 1968. *The Huxleys.* London: William Heinemann.

Cockshut, A. O. J. 1966. *The Unbelievers: English Agnostic Thought, 1840–1890.* New York: New York University Press.

Collini, Stefan. 1991. *Public Moralists: Political Thought and Intellectual Life in Britain, 1850–1930.* Oxford: Clarendon Press.

Cosslett, Tess. 1982. *The "Scientific Movement" and Victorian Literature.* Brighton, UK: Harvester Press.

Cox, Jeffrey. 1982. *The English Churches in a Secular Society: Lambeth, 1870–1930.* Oxford: Oxford University Press.

———. 2006. "Modern European Historiography Forum. Provincializing Christendom: The Case of Great Britain." *Church History* 75:120–30.

Dale, Peter Allan. 1989. *In Pursuit of a Scientific Culture: Science, Arts, and Society in the Victorian Age.* Madison: University of Wisconsin Press.

Dawson, Gowan. 2007. *Darwin, Literature and Victorian Respectability.* Cambridge: Cambridge University Press.

Desmond, Adrian. 1982. *Archetypes and Ancestors: Palaeontology in Victorian London, 1850–1875.* Chicago: University of Chicago Press.

———. 1989. *The Politics of Evolution: Morphology, Medicine, and Reform in Radical London.* Chicago: University of Chicago Press.

———. 1997. *Huxley: From Devil's Disciple to Evolution's High Priest.* Reading, MA: Addison-Wesley.

———. 2001. "Redefining the X Axis: 'Professionals,' 'Amateurs' and the Making of Mid-Victorian Biology—A Progress Report." *Journal of the History of Biology* 34:3–50.

Desmond, Adrian, and James Moore. 1991. *Darwin.* London: Michael Joseph.

Di Gregorio, Mario A. 1984. *T. H. Huxley's Place in Natural Science.* New Haven, CT: Yale University Press.

Dixon, Thomas, Geoffrey Cantor, and Stephen Pumfrey, eds. 2010. *Science and Religion: New Historical Perspectives.* Cambridge: Cambridge University Press.

Durant, John R. 1979. "Scientific Naturalism and Social Reform in the Thought of Alfred Russel Wallace." *British Journal for the History of Science* 12:31–58.

Eisen, Sydney. 1964. "Huxley and the Positivists." *Victorian Studies* 7:337–58.

Elliott, Paul. 2009. *The Derby Philosophers: Science and Culture in British Urban Society, 1700–1850*. Manchester, UK: Manchester University Press.

Endersby, Jim. 2008. *Imperial Nature: Joseph Hooker and the Practices of Victorian Science*. Chicago: University of Chicago Press.

England, Richard. 1997. "Natural Selection before the *Origin*: Public Reactions of Some Naturalists to the Darwin-Wallace Papers (Thomas Boyd, Arthur Hussey, Henry Baker Tristram)." *Journal of the History of Biology* 30:267–90.

Eve, A. S., and C. H. Creasey. 1945. *Life and Work of John Tyndall*. London: Macmillan.

Everett, Edwin Mallard. 1939. *The Party of Humanity: The* Fortnightly Review *and Its Contributors, 1865–1874*. Chapel Hill: University of North Carolina Press.

Fichman, Martin. 2004. *An Elusive Victorian: The Evolution of Alfred Russel Wallace*. Chicago and London: University of Chicago Press.

Francis, Mark. 1986. "Herbert Spencer and the Mid-Victorian Scientists." *Metascience* 4:2–21.

———. 2007. *Herbert Spencer and the Invention of Modern Life*. Newcastle, UK: Acumen.

Friday, James R., Roy M. MacLeod, and Philippa Shepherd. 1974. *John Tyndall, Natural Philosopher, 1820–1893: Catalogue of Correspondence, Journals and Collected Papers*. London: Mansell.

Gay, Hannah, and John W. Gay. 1997. "Brothers in Science: Science and Fraternal Culture in Nineteenth-Century Britain." *History of Science* 35:425–53.

Gieryn, Thomas F. 1999. *Cultural Boundaries of Science: Credibility on the Line*. Chicago: University of Chicago Press.

Gruber, Jacob W. 1960. *A Conscience in Conflict: The Life of St. George Jackson Mivart*. New York: Columbia University Press.

Helfand, Michael S. 1977. "T. H. Huxley's 'Evolution and Ethics': The Politics of Evolution and the Evolution of Politics." *Victorian Studies* 20:159–77.

Helmstadter, R. J., and B. Lightman, eds. 1990. *Victorian Faith in Crisis: Essays on Continuity and Change in Nineteenth-Century Religious Belief*. Houndmills, UK: Macmillan.

Howard, Jill. 2004. "'Physics and Fashion': John Tyndall and His Audiences in Mid-Victorian Britain." *Studies in History and Philosophy of Science* 35:729–58.

Jacyna, Leon Stephen. 1980. "Scientific Naturalism in Victorian Britain: An Essay in the Social History of Ideas." PhD diss., University of Edinburgh.

Jarrell, Richard A. 1998. "Visionary or Bureaucrat? T. H. Huxley, the Science and Art Department and Science Teaching for the Working Class." *Annals of Science* 55:219–40.

Jensen, J. Vernon. 1970. "The X Club: Fraternity of Victorian Scientists." *British Journal for the History of Science* 5:63–72.

———. 1991. *Thomas Henry Huxley: Communicating for Science*. Newark: University of Delaware Press.

Jones, Greta, and Robert Peel, eds. 2004. *Herbert Spencer: The Intellectual Legacy*. London: Galton Institute.

Kim, Stephen S. 1996. *John Tyndall's Transcendental Materialism and the Conflict between Religion and Science in Victorian England*. New York: Edwin Mellen.

Kjærgaard, Peter C. 2004. "'Within the Bounds of Science': Redirecting Controversies to *Nature*." In *Culture and Science in the Nineteenth-Century Media*, edited by Louise Henson, Geoffrey Cantor, Gowan Dawson, Richard Noakes, Sally Shuttleworth, and Jonathan R. Topham, 211–21. Aldershot, UK: Ashgate.

Lane, Christopher. 2011. *The Age of Doubt: Tracing the Roots of Our Religious Uncertainty.* New Haven, CT, and London: Yale University Press.

La Vergata, Antonello. 1995. "Herbert Spencer: Biology, Sociology and Cosmic Evolution." In *Biology as Sociology, Sociology as Biology: Metaphors*, edited by S. Maasen, E. Mendelsohn, and P. Weingart, 193–229. Dordrecht, Netherlands: Kluwer.

Lester, Joseph. 1995. *E. Ray Lankester and the Making of Modern British Biology.* Edited, with additions, by Peter J. Bowler. London: British Society for the History of Science.

Levine, George. 2002. *Dying to Know: Scientific Epistemology and Narrative in Victorian England.* Chicago: University of Chicago Press.

Lightman, Bernard. 1987. *The Origins of Agnosticism: Victorian Unbelief and the Limits of Knowledge.* Baltimore: Johns Hopkins University Press.

———. 2007. *Victorian Popularizers of Science: Designing Nature for New Audiences.* Chicago: University of Chicago Press.

———. 2009. *Evolutionary Naturalism in Victorian Britain: The "Darwinians" and Their Critics.* Farnham, Surrey, UK: Ashgate.

Lightman, Bernard, ed. 1997. *Victorian Science in Context.* Chicago: University of Chicago Press.

Lucier, Paul. 2009. "The Professional and the Scientist in Nineteenth-Century America." *Isis* 100:699–732.

Maas, Harro. 2005. *William Stanley Jevons and the Making of Modern Economics.* Cambridge: Cambridge University Press.

MacKenzie, Donald. 1981. *Statistics in Britain, 1865–1930: The Social Construction of Scientific Knowledge.* Edinburgh: Edinburgh University Press.

MacLeod, Roy M. 2000. *The "Creed of Science" in Victorian England.* Aldershot, UK: Ashgate.

Maitland, Frederic William. 1906. *The Life and Letters of Leslie Stephen.* London: Duckworth.

McLaughlin-Jenkins, Erin. 2005. "Henry George and the Dragon: T. H. Huxley's Response to Progress and Poverty." In *Henry George's Legacy in Economic Thought*, edited by John Laurent, 31–52. Cheltenham, UK: Edward Elgar.

Meadows, A. J. 1972. *Science and Controversy: A Biography of Sir Norman Lockyer.* London: Macmillan.

Moore, James R. 1979. *The Post-Darwinian Controversies: A Study of the Protestant Struggles to Come to Terms with Darwin in Great Britain and America, 1870–1900.* Cambridge: Cambridge University Press.

———. 1986. "Crisis without Revolution: The Ideological Watershed in Victorian England." *Revue de Synthèse* 107:53–78.

———. 1986. "Geologists and Interpreters of Genesis in the Nineteenth Century." In *God and Nature: Historical Essays on the Encounter between Christianity and Science*, edited by David C. Lindberg and Ronald L. Numbers, 322–50. Berkeley: University of California Press.

———. 1991. "Deconstructing Darwinism: The Politics of Evolution in the 1860s." *Journal of the History of Biology* 24:353–408.

Moore, James R., ed. 1989. *History, Humanity and Evolution.* Cambridge: Cambridge University Press.

Morrell, Jack, and Arnold Thackray. 1981. *Gentlemen of Science: Early Years of the British Association for the Advancement of Science.* Oxford: Clarendon Press.

Morton, Peter. 2005. *"The Busiest Man in England": Grant Allen and the Writing Trade, 1875–1900.* New York and Basingstoke, UK: Palgrave Macmillan.

Numbers, Ronald L. 2003. "Science without God: Natural Laws and Christian Beliefs." In *When Science and Christianity Meet*, edited by R. L. Numbers and D. C. Lindberg, 265–85. Chicago: University of Chicago Press.

Owen, Alex. 2004. *The Place of Enchantment: British Occultism and the Culture of the Modern.* Chicago and London: University of Chicago Press.

Paradis, James. 1978. *T. H. Huxley: Man's Place in Nature.* Lincoln and London: University of Nebraska Press.

Paradis, James, and George C. Williams. 1989. *Evolution and Ethics: T. H. Huxley's "Evolution and Ethics," with New Essays on Its Victorian and Sociobiological Context.* Princeton, NJ: Princeton University Press.

Patton, Mark. 2007. *Science, Politics and Business in the Work of Sir John Lubbock: A Man of Universal Mind.* Aldershot, UK, and Burlington, VT: Ashgate.

Porter, Theodore M. 2004. *Karl Pearson: The Scientific Life in a Statistical Age.* Princeton, NJ: Princeton University Press.

Postelthwaite, Diana. 1971. *Making It Whole: A Victorian Circle and the Shape of Their World.* Columbus: Ohio State University Press.

Reidy, Michael S. 2010. "John Tyndall's Vertical Physics: From Rock Quarries to Icy Peaks." *Physics in Perspective* 12:122–45.

Richards, Evelleen. 1989. "Huxley and Women's Place in Science: The 'Woman Question' and the Control of Victorian Anthropology." In *History, Humanity and Evolution*, edited by James R. Moore, 253–84. Cambridge: Cambridge University Press.

———. 1989. "The 'Moral Anatomy' of Robert Knox: The Interplay between Biological and Social Thought in Victorian Scientific Naturalism." *Journal of the History of Biology* 22:373–436.

Richards, Joan. 1988. *Mathematical Visions: The Pursuit of Geometry in Victorian England.* Boston: Academic Press.

Richards, Robert. 1987. *Darwin and the Emergence of Evolutionary Theories of Mind and Behavior.* Chicago: Chicago University Press.

Royle, Edward. 1974. *Victorian Infidels: The Origins of the British Secularist Movement, 1791–1866.* Manchester, UK: Manchester University Press.

———. 1980. *Radicals, Secularists and Republicans: Popular Freethought in Britain, 1866–1915.* Manchester, UK: Manchester University Press.

Russell, Colin A. 1996. *Edward Frankland: Chemistry, Controversy and Conspiracy in Victorian England.* Cambridge: Cambridge University Press.

Sawyer, Paul. 1985. "Ruskin and Tyndall: The Poetry of Matter and the Poetry of Spirit." In *Victorian Science and Victorian Values: Literary Perspectives*, edited by James Paradis and Thomas Postlewait, 217–46. New Brunswick, NJ: Rutgers University Press.

Schlossberg, Herbert. 2009. *Conflict and Crisis in the Religious Life of Late Victorian England.* New Brunswick, NJ: Transaction Publishers.

Secord, James A. 2000. *Victorian Sensation: The Extraordinary Publication, Reception, and Secret Authorship of "Vestiges of the Natural History of Creation."* Chicago: University of Chicago Press.

———. 2004. "Knowledge in Transit." *Isis* 95:654–72.

Small, Helen. 2004. "Science, Liberalism, and the Ethics of Belief: The *Contemporary Review* in 1877." In *Science Serialized: Representations of the Sciences in Nineteenth-Century Periodicals*, edited by Geoffrey Cantor and Sally Shuttleworth, 239–57. Cambridge, MA: MIT Press.

Smith, Crosbie. 1998. *The Science of Energy: A Cultural History of Energy Physics in Victorian Britain.* Chicago: University of Chicago Press.

Smith, Crosbie, and M. Norton Wise. 1989. *Energy and Empire: A Biographical Study of Lord Kelvin.* Cambridge: Cambridge University Press.

Smith, Jonathan. 2006. *Charles Darwin and Victorian Visual Culture.* Cambridge: Cambridge University Press.

Stocking, George W. 1987. *Victorian Anthropology.* New York: Free Press.

Strick, James E. 2000. *Sparks of Life: Darwinism and the Victorian Debates over Spontaneous Generation.* Cambridge, MA: Harvard University Press.

Taylor, Michael W. 2007. *The Philosophy of Herbert Spencer.* London: Continuum.

Turner, Frank Miller. 1974. *Between Science and Religion: The Reaction to Scientific Naturalism in Late Victorian England.* New Haven, CT: Yale University Press.

———. 1993. *Contesting Cultural Authority: Essays in Victorian Intellectual Life.* Cambridge: Cambridge University Press.

White, Paul. 2003. *Thomas Huxley: Making the "Man of Science."* Cambridge: Cambridge University Press.

———. 2005. "Ministers of Culture: Arnold, Huxley and Liberal Anglican Reform of Learning." *History of Science* 43:115–38.

Whyte, Adam Gowans. 1949. *The Story of the R.P.A., 1899–1949.* London: Watts.

Wyhe, John van. 2004. *Phrenology and the Origins of Victorian Scientific Naturalism.* Aldershot, UK: Ashgate.

Young, Robert M. 1970. *Mind, Brain and Adaptation in the Nineteenth Century.* Oxford: Oxford University Press.

———. 1985. *Darwin's Metaphor: Nature's Place in Victorian Culture.* Cambridge: Cambridge University Press.

Contributors

MELINDA BALDWIN, Department of the History of Science, Harvard University, Cambridge, Massachusetts.

RUTH BARTON, Department of History, University of Auckland, Auckland, New Zealand.

PETER J. BOWLER, School of History and Anthropology, Queen's University Belfast, Belfast, Northern Ireland, United Kingdom.

GOWAN DAWSON, Victorian Studies Centre, University of Leicester, Leicester, United Kingdom.

JAMES ELWICK, Natural Science Division and Science and Technology Studies, York University, Toronto, Canada.

JIM ENDERSBY, History Department, University of Sussex, Brighton, United Kingdom.

GEORGE LEVINE, Department of English, Rutgers University, New Brunswick, New Jersey.

BERNARD LIGHTMAN, Department of Humanities, York University, Toronto, Canada.

THEODORE M. PORTER, Department of History, UCLA, Los Angeles, California.

MICHAEL S. REIDY, Department of History and Philosophy, Montana State University, Bozeman, Montana.

MATTHEW STANLEY, Gallatin School of Individualized Study, New York University, New York, New York.

PAUL WHITE, Darwin Correspondence Project, University of Cambridge, Cambridge, United Kingdom.

Index